CYTOCHEMICAL BIOASSAYS

BASIC AND CLINICAL ENDOCRINOLOGY

Editors

P. Reed Larsen
Brigham and Women's Hospital
Harvard Medical School
Boston, Massachusetts

David R. London
Queen Elizabeth Hospital
Edgbaston, Birmingham,
England

Peter Sönksen
St. Thomas's Hospital
Medical School
London, England

1. Radioassay Systems in Clinical Endocrinology, *edited by Guy E. Abraham*
2. Congenital Hypothyroidism, *edited by Jean H. Dussault and Peter Walker*
3. Cytochemical Bioassays: Techniques and Clinical Applications, *edited by J. Chayen and Lucille Bitensky*

Other Volumes in Preparation

CYTOCHEMICAL BIOASSAYS

Techniques and Clinical Applications

Edited by

J. CHAYEN
LUCILLE BITENSKY

Division of Cellular Biology
Kennedy Institute of Rheumatology
London, England

MARCEL DEKKER, INC.   New York and Basel
BUTTERWORTHS   London

Library of Congress Cataloging in Publication Data

Main entry under title:

Cytochemical bioassays.

(Basic and clinical endocrinology ; 3)
Includes bibliographical references and indexes.
1. Hormones–Analysis–Addresses, essays, lectures.
2. Cytochemical bioassays–Addresses, essays, lectures.
I. Chayen, J. (Joseph), [date]. II. Bitensky, L.
(Lucille), [date]. III. Series. [DNLM: 1. Biological
assay–Methods. 2. Cytological techniques. 3. Cyto-
diagnosis–Methods. 4. Hormones--Analysis. W1 BA813T
v.3 / QY 330 C997].
RB48.5.C95   1983   616.07'56   83-17661
ISBN 0-8247-7001-3

MARCEL DEKKER, INC.
270 Madison Avenue, New York, New York 10016

Current printing (last digit):
10 9 8 7 6 5 4 3 2 1

PRINTED IN THE UNITED STATES OF AMERICA

Preface

In recent years there has been a resurgence of interest in bioassays. Two factors have contributed to this interest. The first is the fact that, in a proportion of cases, immunoassay has yielded results that are at obvious variance with the clinical and physiological status of the patient. The second is the development of sensitive in vitro bioassays that are at least as sensitive as the equivalent radio-immunoassays, so permitting comparison between immunoactivity and bioactivity. The most sensitive of these in vitro bioassays are the cytochemical bioassays that are at least one thousand times as sensitive as the equivalent radio-immunoassays. They have the further advantage that the same apparatus, and the same expertise, can be used for the bioassay of any polypeptide hormone. The cytochemical bioassay system has also proved to be of special value in helping to elucidate the mode of action of hormones. This use of the system has disclosed the existence of immunoglobulins that can block the effect of hormones.

These bioassays, and the cytochemical bioassay system, are now being used widely both for clarifying clinical conditions, such as the role of blocking antibodies in thyroid pathology, and for more academic research. Consequently we agreed with the editors of this series that the time seemed ripe to review this rapidly expanding field. We therefore invited some of the leading investigators to describe the cytochemical bioassay relevant to their investigations and to discuss the clinical and research implications of their findings.

J. Chayen and Lucille Bitensky

Contributors

J. ALAGHBAND-ZADEH, Ph.D., F.R.C. Path., Department of Chemical Pathology, Charing Cross Hospital Medical School, London, England

DEREK R. BANGHAM, M.B.B.S., F.R.C.P., Head of Division of Hormones, National Institute for Biological Standards and Control, London, England

PETER H. BAYLIS, M.D., F.R.C.P., Consultant Physician and Senior Lecturer in Medicine, Department of Medicine, Royal Victoria Infirmary and University of Newcastle Upon Tyne, Newcastle Upon Tyne, England

G. M. BESSER, M.D., F.R.C. Path., F.R.C.P., Department of Endocrinology, St. Bartholomew's Hospital, London, England

LUCILLE BITENSKY, D.Sc., M.R.C.P., Head of Laboratory Medical Histochemistry, and Deputy Head of Division of Cellular Biology, Division of Cellular Biology, Kennedy Institute of Rheumatology, London, England

GIAN FRANCO BOTTAZZO, M.D., M.R.C. Path., Senior Lecturer in Clinical Immunology, Department of Immunology, Middlesex Hospital Medical School, London, England

J. CHAYEN, D.Sc., Head of Division of Cellular Biology, Division of Cellular Biology, Kennedy Institute of Rheumatology, London, England

KLAUS-DIETER DÖHLER, Ph.D., Professor of Experimental Endocrinology, Department of Clinical Endocrinology, Medizinische Hochschule Hannover, Hannover, Federal Republic of Germany

DEBORAH DONIACH, M.D., F.R.C.P., Emeritus Professor Clinical Immunology, Department of Immunology, Middlesex Hospital Medical School, Authur Stanley Institute, London, England

H. A. DREXHAGE, M.D.,* Senior Lecturer in Clinical Immunology, Department of Immunology, Arthur Stanley House, Middlesex Hospital Medical School, London, England

D. EMRICH, M.D., Professor of Medicine and Nuclear Medicine, Department of Nuclear Medicine, Universität Göttingen, Göttingen, Federal Republic of Germany

STEPHEN FENTON, B.Sc., Department of Medicine, Charing Cross Hospital Medical School, London, England

TAKUMA HASHIMOTO, M.D.,** Research Fellow of Alexander von Humboldt Stiftung, Department of Clinical Endocrinology, Medizinische Hochschule Hannover, Hannover, West Germany

RONALD W. HOILE, M.S., F.R.C.S., Lecturer in Surgery, Surgical Unit, St. Stephens Hospital, London, England

JULIA C. JONES, M.Sc., Department of Medicine, Charing Cross Hospital Medical School, London, England

G. NEIL KENT, B.Sc., Ph.D.,† Division of Inherited Metabolic Diseases, M.R.C. Clinical Research Centre, Harrow, Middlesex, England; Division of Cellular Biology, Kennedy Institute of Rheumatology, London, England; and Department of Endocrine Physiology and Pharmacology, National Institute for Medical Research, London, England

NIGEL LOVERIDGE, Ph.D., Deputy Head WHO Collaborating Centre for Cytochemical Bioassays, Division of Cellular Biology, Kennedy Institute of Rheumatology, London, England

CHRISTOPH LUCKE, M.D.,‡ Department of Clinical Endocrinology, Medizinische Hochscule Hannover, Hannover, Federal Republic of Germany

Present affiliations
 *Department of Pathology, Free University Hospital, Amsterdam, The Netherlands
**Assistant Professor of Medicine, The Central Clinical Laboratory, Kanazawa University Hospital, Kanazawa, Japan
†Department of Endocrinology and Diabetes, Sir Charles Gairdner Hospital, Queen Elizabeth II Medical Centre, Nedlands, Western Australia
‡Professor and Head Physician, Hagenhoff-Klinik, Langenhagen, Federal Republic of Germany

GRAHAM A. MacGREGOR, F.R.C.P., Senior Lecturer/Honorary Consultant Physician, Department of Medicine, Charing Cross Hospital Medical School, London, England

J. MAXWELL McKENZIE, M.D., Professor and Chairman, Department of Medicine, University of Miami School of Medicine, Miami, Florida

ALEXANDER von zur MÜHLEN, M.D., Professor of Internal Medicine, Chairman of the Department of Clinical Endocrinology, Medizinische Hochschule Hannover, Hannover, Federal Republic of Germany

THOMAS O. F. WAGNER, M.D., Department of Clinical Endocrinology, Center of Internal Medicine and Dermatology, Medizinische Hochschule Hannover, Hannover, Federal Republic of Germany

W. H. C. WALKER, F.R.C. Pathol., F.R.C.P.(C)., Professor of Pathology, McMaster University Medical Centre, Hamilton, Ontario, Canada

H. E. de WARDENER, M.D., F.R.C.P., Emeritus Professor of Medicine, Research Laboratories, Charing Cross Hospital Medical School, London, England

J. A. H. WASS, M.D., M.R.C.P., Senior Lecturer and Honorary Consultant Physician, Department of Endocrinology, St. Bartholomew's Hospital, London, England

JUDITH WEISZ, M.B., B.Chir., Chief, Division of Reproductive Biology, Department of Obstetrics and Gynecology, The Milton S. Hershey Medical Center, The Pennsylvania State University, Hershey, Pennsylvania

HANS K. WEITZEL, M.D.,* Department of Clinical Endocrinology, Medizinische Hochschule Hannover, Hannover, Federal Republic of Germany

MARGITA ZAKARIJA, M.D., Professor of Medicine, Department of Medicine, University of Miami School of Medicine, Miami, Florida

JOAN M. ZANELLI, Ph.D., Senior Scientific Staff, Hormones Division, National Institute for Biological Standards and Control, London, England

Present affiliation
*Professor and Chairman, Department of Obstetrics and Gynecology, Steglitz Clinic, Free University of Berlin, Berlin, Federal Republic of Germany

Contents

Contents xiii

CYTOCHEMICAL BIOASSAYS

1

General Introduction to Cytochemical Bioassays

J. Chayen and Lucille Bitensky / Kennedy Institute of Rheumatology,
London, England

ORIGINS

Quantitative cytochemistry, as it is used in the cytochemical bioassays, was
developed over many years as a form of truly cellular biochemistry, that is, the
measurement of metabolic activity or of active moieties in individual cells
within a histologically complex tissue. These developments have been discussed
elsewhere (Chayen and Bitensky, 1968; Chayen, 1978a; 1980). To achieve this
histological specificity, the sensitivity of measurement had to be increased so
that the activity of one cell could be measured, as contrasted with the mean activity of one million cells that is used in conventional biochemistry. This was
done by the use of scanning and integrating microdensitometry, which had been
developed earlier (Deeley, 1955) for measuring the amount of Feulgen stain (for
DNA) in individual nuclei. It is now clear that microdensitometry of individual
cells yields results that are quantitatively comparable to those obtained by more
conventional procedures, done on aliquots of 10^6 cells (Chayen, 1978b; Olsen
et al., 1981).

The methods of quantitative cytochemistry depend on chilling the tissue, and
sectioning it at low temperature, without producing any observable ice artifact.
Perhaps the best validation of these techniques (Chayen and Bitensky, 1968;
Chayen, 1978a) is the fact that sections, prepared by the techniques described
in Chapter 3, respond to the relevant polypeptide hormone with the same sensitivity as do the segments of the target organ. Methods were then devised for

retaining the integrity of the undenatured sections during the cytochemical reaction designed for disclosing the required enzymatic or other activity (as discussed in Chapter 3). The aim of quantitative cytochemistry is to precipitate the colored reaction product in the cell in which the chemical activity resides. The section is then inspected in the microdensitometer to determine the histology and to identify the target cells; the instrument can then measure the amount of reaction product specifically in these cells.

At the same time as the methods of quantitative cytochemistry were being developed, recourse was being made to the system of nonproliferative organ maintenance culture that had been developed by Trowell (1959). Thus, for example, samples of human synovial tissue were maintained in vitro, with no apparent change either in histology or in biochemical activity (Chayen and Bitensky, 1982), in order to test the effect of anti-inflammatory agents.

These methods, both of maintenance culture and of quantitative cytochemistry, found use in many diverse applications (Pattison et al., 1979). About 1970, the late Professor John Daly suggested that, because of their sensitivity and because they could measure changes solely in the target cells, they ought to be applicable to the development of very sensitive bioassays of polypeptide hormones. At that time, his interest in adrenocorticotropic hormone (ACTH) coincided with our rheumatological interest in this hormone. The first demonstration of the feasibility of this project was given in 1971 (Chayen et al., 1971). The first cytochemical bioassay, which was for ACTH, followed shortly (Chayen et al., 1972).

THE NEED FOR SENSITIVE BIOASSAYS OF POLYPEPTIDE HORMONES

For polypeptide hormones at least, there can be no doubt that "the hormone" is a biological concept and must be measured by the biological activity which it evinces. Consequently, for many years, such hormones were detected, defined, and measured by in vivo bioassay. However, these generally proved to be too insensitive for measuring normal circulating levels in humans or in animals; the best that they could achieve was to demonstrate excessively high circulating levels. [It may be remarked that, according to some authorities, such as Orth (1977), this is the most clinically useful purpose of assays.]

The advent of radioimmunoassay, and now of the other types of immunological assay, produced profound changes in the assaying of polypeptide (and other) hormones. The much improved sensitivity over the older in vivo bioassays, made it possible, in most cases, to measure the normal circulating levels of these hormones and to define conditions in which there was excessive, or too little, secretion of the hormone. The fact that these immunoassays could be automated gained them ready acceptance in routine clinical chemistry.

These immunoassays were able to become firmly established because of the great advances that had been, and were being, made in polypeptide chemistry. It was now possible, apparently, to isolate a "pure" peptide and say, for example, that "this" is the adrenocorticotropic hormone.

There is no intention in this volume to denigrate immunoassays. They have greatly advanced both the study of endocrinology and clinical chemistry; their place is inviolate. But it soon became clear that, while they gave an invaluable, rapid, first approximation to the endocrine status of patients, they occasionally gave results that conflicted with the clinical condition. Moreover, it seems that anomalous results were also found with some preparations that were candidates for designation as international standards because, as early as 1967, a special committee convened by the World Health Organization was concerned that the results of immunoassay should be capable of being checked by suitably sensitive bioassays (WHO Report, 1975). This concern was finally expressed by the WHO Expert Committee on Biological Standardization (WHO Report, 1975), which is responsible for international standards of polypeptide hormones. That committee recognized that "a limitation on the use of radioimmunoassays for evaluating hormonal bioactivity is that the methods measure a composite of antigenic activity, which is not necessarily related to the bioactivity of the hormone," and called for "the development of biological microassays, which should preferably have a sensitivity comparable with radioimmunoassays, with which they should be run in parallel."

FUNCTIONAL VERSUS ANALYTICAL ASSAYS

The need to be able to check the results of radioimmunoassay by bioassay is inherent in the different conceptual bases of these forms of assay. Radioimmunoassay is a good example of what have been called *analytical assays* (Chayen et al., 1976; Chayen, 1980). Fundamentally, these assays use methods of analytical or physical chemistry to determine the number of molecules of a more or less specific type or, more precisely, the number of antigenic determinants of a specific type, present in the sample. The basic assumption is that a particular molecule will produce the relevant hormonal activity, e.g., the 1-84 peptide of parathyroid hormone will have the biological activity of parathyroid hormone. The analytical, or physicochemical, procedures are considered to be adequate to determine the nature of the molecule; then it is assumed that such a molecule must assert its known biological properties. However, it is now apparent that this assumption is too facile. First, it seems clear that radioimmunoassay, in general, will not detect small changes in the intact molecule, such as oxidation of the methionine residues of parathyroid hormone (PTH) or of minor deletions from the N-terminus of PTH, both of which remove biological activity (Chapter 11). Second, radioimmunoassay can measure biologically inactive fragments of

the hormone, as discussed by Besser et al. (1971), in relation to ACTH. The
third objection to this assumption derives from the fact that all the longer polypeptide hormones can occur in various configurations, and it may be that not
all of these permit the molecule to exert its hormonal activity. An interesting
example of what may be configurational change leading to inactivation was reported by Orth (1977) for ACTH.

In contrast, the *functional assays* (Chayen et al., 1976; Chayen, 1980), such
as the in vivo bioassays, the isolated cell bioassays, and the cytochemical bioassays, make no assumptions about the biological activity of the material that
they assay: they assay the material by its functional activity. Clearly, they
must be well controlled to ensure that the activity being measured is caused
by a single biologically active agent that is congruent with the hormone. There
are many rules of bioassay for determining such congruity. In the general enthusiasm for immunoassay, and with the decline of in vivo bioassay, these rules
have been largely forgotten. They are clearly restated in Chapter 2 and are discussed in other chapters in relation to specific hormones.

THE ADVANTAGES OF CYTOCHEMICAL BIOASSAYS

For all the hormones studied up to now, the cytochemical bioassays have ful-
filled the requirements of the WHO Report (1975), namely, for "microbioassays" that are at least as sensitive as the equivalent radioimmunoassays and
that can be done, as required, in parallel with them. In fact, the cytochemical
bioassays are generally 1000 times as sensitive as the equivalent radioimmunoassay, so that not only can they be used to check the bioactivity of samples
measured by radioimmunoassay, but they can measure low-normal and subnormal circulating levels which, for some hormones (such as thyroid-stimulating
hormone, TSH, or PTH), are beyond the present sensitivities of immunoassay. It
should be noted that other microbioassays, that meet the requirements of the
WHO Report (1975), have been developed, notably the isolated-cell assays
for ACTH of Sayers et al. (1971) and for luteinizing hormone (LH) by Dufau
et al. (1974) and by Qazi et al. (1974). However, a special advantage of the
cytochemical bioassay system is that the same apparatus and the same general
techniques can be used for all polypeptide hormones.

Other advantages of the cytochemical bioassays, over immunoassays, are as
follows:

1. Because they measure a functional attribute they can be used to measure
biological activity even when it is either difficult or impossible to raise an antibody to the biologically active molecule. Thus, cytochemical bioassay has been
used to study hormone-like effects of immunoglobulins (as in Chapters 6, 7,
and 9) and to measure circulating levels of polypeptide hormones in small

animals where a specific antibody has not been readily available. Similarly, it was used to demonstrate the presence of the natriuretic hormone, or factor (Chapter 14), which has yet to be fully characterized chemically.

2. Because of its sensitivity, cytochemical bioassay normally requires very small samples of blood so that it can be used to assay hormones in heel-prick samples from neonates (Holdaway et al., 1973) or in samples from small animals. This ability to use small samples has been especially valuable when frequent serial sampling is required.

3. Probably the greatest advantage of the cytochemical bioassay system is its use in defining the mode of action of hormones. The basis of the cytochemical bioassay system is that, when a hormone binds to its receptor, it produces metabolic effects that lead to the response characteristic of the hormone. And the point of the cytochemical approach is that it allows the measurement of such metabolic changes solely in the target cells. Thus, the metabolic consequences attendant on exposure of the target cells to the hormone can be measured sequentially. Moreover, the flexibility of the cytochemical bioassay system makes it ideal also for studying the interaction between possible agonists (as discussed in Chapter 9); the effects of other components of the plasma (as discussed in Chapters 7, 9, and 11); and for investigating the possible presence of antibodies or other agents directed against the receptor (e.g., as in Chapter 9).

REFERENCES

Besser, G. M., Orth, D. N., Nicholson, W. E., Byyny, R. L., Abe, K., and Woodham, J. P. (1971). Dissociation of disappearance of bioactive and immunoreactive ACTH from plasma in man. *J. Clin. Endocrinol. Metab.* 32:595-603.

Chayen, J. (1978a). The cytochemical approach to hormone assay. *Int. Rev. Cytol.,* 53:333-396.

Chayen, J. (1978b). Microdensitometry. In *Biochemical Mechanisms of Liver Injury,* T. F. Slater (ed.). Academic, London, pp. 257-291.

Chayen, J. (1980). *The Cytochemical Bioassay of Polypeptide Hormones.* Monographs on Endocrinology, Vol. 17. Springer, Berlin.

Chayen, J., and Bitensky, L. (1968). Multiphase chemistry of cell injury. In *The Biological Basis of Medicine,* E. E. Bittar and N. Bittar (eds.). Vol. 1, Academic, New York, pp. 337-368.

Chayen, J., and Bitensky, L. (1982). Metabolism of rheumatoid and non-rheumatoid synovial lining cells. In *Articular Synovium,* P. Franchimont (ed.). Karger, Basel, pp. 59-74.

Chayen, J., Daly, J. R., Loveridge, N., and Bitensky, L. (1976). The cytochemical bioassay of hormones. *Recent Prog. Horm. Res.,* 32:33-79.

Chayen, J., Loveridge, N., and Daly, J. R. (1971). The measurable effect of low concentrations (pg/ml) of ACTH on reducing groups of adrenal cortex maintained in organ culture. *Clin. Sci.,* 41:2P.

Chayen, J., Loveridge, N., and Daly, J. R. (1972). A sensitive bioassay for adrenocorticotrophic hormone (ACTH) in human plasma. *Clin. Endocrinol. (Oxf),* 1:219-233.

Deeley, E. M. (1955). An integrating microdensitometer for biological cells. *J. Sci. Instrum.* 32:263-267.

Dufau, M. L., Mendelson, C. R., and Catt, K. J. (1974). A highly sensitive in vitro bioassay for luteinizing hormone and chorionic gonadotrophin: Testosterone production by dispersed Leydig cells. *J. Clin. Endocrinol. Metab.* 39:610-617.

Holdaway, I. M., Rees, L. H., and Landon, J. (1973). Circulating corticotrophin levels in severe hypopituitarism and in the neonate. *Lancet,* ii:1170-1172.

Olsen, I., Dean, M. F., Harris, G., and Muir, H. (1981). Direct transfer of a lysosomal enzyme from lymphoid cells to deficient fibroblasts. *Nature,* 291:244-247.

Orth, D. N. (1977). Assay of ACTH: Discussion. *Ann. N.Y. Acad. Sci.,* 297:260-262.

Pattison, J. R., Bitensky, L., and Chayen, J. (eds.) (1979). *Quantitative Cytochemistry and Its Applications.* Academic, London.

Qazi, M. H., Romani, P., and Diczfalusy, E. (1974). Discrepancies in plasma LH activities as measured by radioimmunoassay and an in vitro bioassay. *Acta Endocrinol. (Kbh),* 77:672-685.

Sayers, G., Swallow, R. L., and Giordano, N. D. (1971). An improved technique for the preparation of isolated rat adrenal cells: A sensitive, accurate and specific method for the assay of ACTH. *Endocrinology,* 88:1063-1068.

Trowell, O. A. (1959). Culture of mature organs in a synthetic medium. *Exp. Cell Res.,* 16:118-147.

WHO Expert Committee on Biological Standardization (1975). 26th Report. *WHO Tech. Rep. Ser. 565.*

2

What's in a Bioassay?

Derek R. Bangham / National Institute for Biological Standards and Control, London, England

GENERAL

Introduction

A chapter on biological assays must include consideration of what is meant by "an assay," for what purpose assays are done, the basic components of an assay method, and the nature of the difference between in vivo and in vitro hormone bioassays, in vitro ligand assays, and enzyme assays.

Biological assay systems are extraordinarily diverse, and it is essential to understand the common underlying principles which allow reliable and reproducible results to be obtained. This involves what is known as biological standardization, which consists of using a particular biometric discipline because of the variability of the assay systems involved. Such an analysis requires terms with clearly defined and understood meanings. Many of the terms used here are those recommended by the World Health Organization (WHO) and/or the International Federation of Clinical Chemistry (IFCC); for example, an *assay system* is all the components and procedures of an assay (see Appendix A).

Identification of the principles underlying standardization in biological assays received great stimulus in 1922-24 from the introduction of insulin extracts for the treatment of diabetics, when it was shown that accurate dosage of this protein extract was essential to sustain life. The biometric principles underlying bioassays have been reviewed (Gaddum, 1933; Jerne and Wood, 1949; Finney, 1978), and were later reconsidered (Ekins, 1970, 1976), for application in in vitro,

Table 1 Different Purposes of Assays

Purpose	Example
Quantify concentration of an analyte	Potency estimation of product for clinical administration
Observe changes in response to a stimulus	Assessment of pituitary hormone releasing factor
Detection of minimum quantity	Assessment of sensitivity of an assay system
Comparison of different substances on same system	Assessment of specificity of assay system for different substances
Comparison of different systems on same substance(s)	Calibration of international standard

ligand assay methods such as radioimmunoassays (see also Bangham and Coles, 1974). Although these principles are familiar in certain fields of biology, it is well to understand how they apply in new methods of assay, such as quantitative cytochemical procedures.

It is not the purpose of this chapter to restate comprehensively the well-thumbed precepts of biological standardization and still less to describe statistical treatment of results. Both subjects are treated extensively elsewhere (for example Finney, 1978). Practical and well-tried descriptions of classic bioassays of several hormones are given in certain pharmacopoeias (e.g., *British Pharmacopoeia*, 1980); the statistical appendix (BP Vol. II, Appendix XIV) provides an excellent, succinct description of statistical terms, tests, and procedures, with worked examples, for various designs of assays.

Assays: Different Purposes for Which Assays Are Done: Comparability, and Standardization

Assays are carried out for distinctly different purposes, and the validity and interpretation of the results depend on the purpose. While almost any experiment might be considered an assay, most assays are carried out (1) to estimate the amount or concentration (to quantify) of a substance (analyte) in a test specimen; (2) to compare the effects of two drugs or treatments or to assess the cross-reaction (or specificity) of an assay system with similar substances that might cross-react or interfere in it; or (3) to assess the response of two or more populations, individuals, or tissues to a common treatment (Table 1).

Although the same assay method may be used for any of these, the assumptions underlying the basis for the validity—and reproducibility—of the result may be different in each case. In the context of this chapter, however, an assay is considered primarily a procedure to quantify an analyte.

Many such assays carried out for clinical reasons are intended to measure the change of concentration of an analyte (e.g., in plasma) in response to a stimulus, for example, the estimation of growth hormone deficiency in a dwarf following an injection of insulin. Others are intended only to detect whether an analyte is present: a test for pregnancy depends on the detection of chorionic gonadotrophin in urine. Science requires that assays for all such purposes should give results which are reproducible and accurate—to confirm an observation, to allow comparison of results obtained at different times (e.g., to monitor a patient's progress), or to be able to specify the minimum concentration detectable by a given assay system. This requires that the assays are designed and carried out according to certain principles of standardization. Standardization does not mean that all assays are done by a particular and rigidly specified procedure; on the contrary, standardization provides guideline principles and procedures (technical and biometric) which, together with the requisite reference materials, improve accuracy and precision and hence reproducibility of results, from time to time, from place to place, and with different assay reagents. This type of standardization allows freedom for unlimited development of improvements and new procedures.

Another way in which assays can be categorized is according to whether the quantity of the component to which the analyte binds is present in the assay system in a limited amount, or in excess. The conventional radioimmunoassay is an example of the former; immunoradiometric assays and most bioassays are examples of the latter. Ekins (1976) has contrasted the differences in assay precision, validity, and sensitivity of these two categories, and they are referred to here in a later section.

To understand how standardization is applied in bioassay procedures, let us consider assays of various kinds, and the validity of comparisons.

"Validity": Biometric and Other Aspects

The validity (meaningfulness) of an assay result depends on certain assumptions about the nature and suitability of the reagents, design and conduct of the biological test system, and the correct numerical treatment of resultant data by statistical theory and procedures (Table 2). The three basic assumptions relating to the biometric validity of an assay system are (1) that known variables are reduced to a minimum where possible, and those that cannot be eliminated—as well as remaining unknown variables—are allocated treatments randomly; (2) that the response measured given by the assay system is indeed related to the dose (i.e., that the assay is suitably specific for the analyte and standard); and (3) that the analyte and standard behave identically in the test system.

Table 2 Aspects of "Validity" of Assays to Quantify an Analyte

Specificity of assay system shown to be appropriate for the purity of specimen
  and standard used
Linearity and parallelism of plot of log dose-response metameter for specimen
  and standard
Precision, stated as fiducial or confidence limits at stated level of probability,
  e.g., $P = 0.95$.
Significant level of homogeneity of results of repeated independent estimates,
  taking into account the precision of each of those estimates

The precision of an assay result could also be regarded as a component of
validity in the broader sense. The reliability, reproducibility, and precision of
an assay result depend as much on the design and dimensions of an assay as on
the reagents used and how it was carried out. The design of an assay involves an
understanding of the biology of the system used, the nature of the specimens to
be assayed, the precision required for the result, and the interpretation that
would be made of it. For example, high precision is desirable for the calibration
of a replacement international standard, whereas for many clinical purposes a
clinician needs to know only if a result is abnormally too high or too low, or
about normal.

BIOMETRIC REQUIREMENTS AND CONSIDERATIONS

Known and Unknown Variables, Their Treatment, and Randomization

The intrinsic variabilities of the biological components of assay systems are well
recognized. Animals and biological tissues vary widely and unpredictably in the
nature and extent of their response in all kinds of assay systems. Such variability
may be due to identifiable criteria, such as animal strain, age, sex, nutrition,
housing conditions, time of day, stress, or pretreatment. The precision of an assay may be improved by reducing known variables, for example, using animals
of a pure strain, of a narrow range of age or body weight, and treated with rigorously uniform and stress-free conditions before the assay.

  In comparisons involving biological materials it is axiomatic that not all the
variable factors in systems of comparison are known. The design and conduct of
an assay to quantify an analyte must therefore be planned so that the specimen
is compared with the standard in the same system at the same time and under
the same conditions. Because the variables are not all known, the only way to
treat them equally is to ensure that the treatment is allocated randomly to specimen and standard: for example, animals (or tissue sections in cytochemical assays) should be allocated using a suitable randomization procedure (e.g., tables

of random numbers) to their treatment with the various doses of test specimen
and standard. No matter how experienced or skilled the assay operator, unless
randomization is rigorously applied, unknown and unsuspected bias can influence an assay result and lead to unwarranted interpretations of it. As Finney
(1970) has diffidently remarked, it is surprising and inexcusable that, considering all the difficulties they overcome in doing a biological assay, biologists
so often needlessly jeopardize their results simply by inadequate randomization in the assay.

An analysis of variance of the raw data often helps to identify sources of
variability, such as differences between assays performed by different operators,
or with different batches of reagents, and also unsuspected correlations of
various factors.

On the other hand, true randomization makes the identification of variables
impossible.

Assay System Specificity for the Analyte

The second main assumption is that the measured response should relate directly or indirectly to the intended analyte. The assay system must have appropriate specificity, the ability to measure only (or at least mainly) the analyte, and not be influenced by other substances present in the system. The
specificity must be appropriately selective for the purity of the test specimen
and the standard (that is, the reference preparation) used. In general, providing
the analyte is the same in each, the greater the purity of both, the less specific
the assay system need be for reliable quantification, and vice versa. Thus, if
analyte and standard both consist of the same pure molecular species, say,
insulin, the one preparation may quite validly be compared with the other
by simple spectrophotometry or nitrogen determination. But analyte specificity
is essential when, for example, specimen or standard or other reagents contain
other substances which may react in the assay system. A naive but grave mis-
take is to assume that the use of a pure standard somehow conveys specificity
on an assay system.

The specificity of assay systems should be assessed, at least for substances
likely to be found in specimens or standard which are likely to cross-react in the
assay, using purified, well-characterized, and documented materials, preferably
attested reference preparations.

Similarity of Analyte in Test Specimen and Standard

The third basic assumption essential for the biometric validity of an assay result
is that the analyte in the specimen and in the standard are of identical molecular
species, or at least that they are similar to the extent that the less potent behaves
in the assay system as though it was a dilution of the other. If like is compared

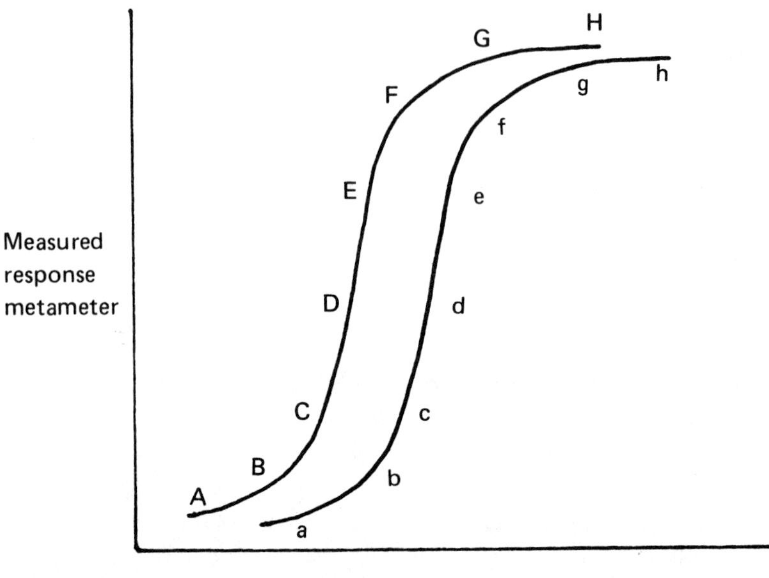

Figure 1A Graphical plots of sigmoid log dose-response curve in a comparison
of two preparations that behave similarly in the assay system. A represents
the shape of a typical log dose-response curve. The potency of one preparation
relative to the other is represented by the amount of one preparation which gives
the same response as a measured quantity of the other, i.e., the horizontal distance between the linear part of the curves BF and bf. Use of three dose levels
of each preparation gives the minimum information necessary to allow assessment of linearity and parallelism of the preparations.

with like, in the same system, under the same conditions, any difference in response in the test system then reflects only the difference in the doses. Finney
(1978) has called this an analytical dilution assay, to distinguish it from those
assays in which the materials compared are not identical and where responses
may be influenced by factors other than just the relative doses of a single active
component.

    In practice it is not possible to comply with this theoretical requirement in
many instances because of the molecular diversity, natural or artifactual, of
many hormones and other biological analytes (see under Standards). It is sober-
ing to realize how many commonly performed hormone assays to quantify an
analyte are "invalid analytical dilution" assays. When dissimilar materials are

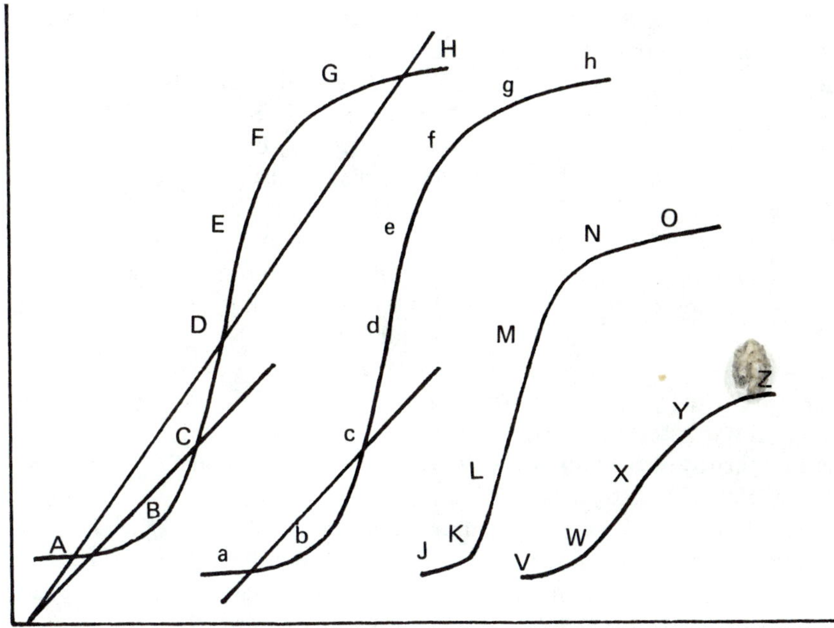

Figure 1B Plots of some log dose-response curves showing dissimilarity of preparations, and examples of unsuitable evidence for their identical behavior in the assay system. Curve AH is similar to curve ah; both are different from JO and VZ. Dissimilarity between preparations may be revealed only by experiments giving plots of the full response curve; otherwise, curve JO could be misleading, as part of it is similar to BF and bf, and VWX is similar to ABC. Examples of ill-chosen dose levels are: ABC and abc: give a low slope and thus lower precision, and test for parallelism may be less critical. ADH and adh: low slope and tests for validity may be less critical. BCD and adh: different parts of curve used which may actually be different, as in JO. Fewer than three doses of standard and three of specimen disallows test for linearity. Fewer than 2 doses of each disallows test for parallelism.

compared, reproducibility of the estimate cannot be expected: it will occur only if the assay system is exactly reproduced.

Despite this unpromising invalidity, such assays generally provide information which is sufficiently reproducible and reliable to be useful for the purpose for which they were done. But their theoretical shortcomings also account for many minor inconsistencies in the results and must always be kept in mind.

Evidence should be provided from the results of each assay that the analyte in specimen and in standard behaved similarly in the assay system. Such evidence is best displayed in the form of parallelism of the graphic plot of the log dose-response metameter curves (see Figure 1A). If the log dose-response plot does not give straight lines (linearity), a numerical factor may be applied to the response parameter (then called a *metameter*) so that parallelism can be assessed statistically and so derive a valid estimate of relative "potency."

Further critical evidence for identity is the similarity of the shapes of the full log dose-response curves of two preparations. This is not normally plotted except for certain specific purposes, such as the evaluation of a new standard or the characterization of a biologically active substance.

This requires that at least two dose levels of each specimen and standard are used in an assay (a 2 × 2 assay design), but at least a 3 × 3 dose design is needed to be able also to assess linearity. But where possible, a 4 × 4 design is better still in case the top or bottom dose turns out to be on the nonlinear part of the plot (see Figure 1B).

In many in vitro ligand assay systems used in routine clinical diagnosis, many specimens are compared at one dose level only, with the standard at several dose levels. The validity criterion of parallelism cannot be evaluated to show that each specimen behaves like the standard. This means that abnormal forms of analyte, e.g., from a tumor, which might be recognized by nonparallelism if two or more doses were assayed, may go unrecognized.

Precision

The precision of the result of an assay may also be considered as an aspect of validity. *Precision* means the consistency of agreement of repeated measurements, but a very precise result may nevertheless be very inaccurate: *bias* is the extent to which results are consistently inaccurate. In stating the results of a bioassay it is usual to calculate the precision from the internal evidence of the design and results in the form of fiducial limits at a given level of probability, generally 95%. (Precision can also be expressed as confidence limits, which are calculated in a slightly different way, but which mean much the same as fiducial limits.) This is the range of values between which the estimated potency could be expected to fall 19 times, if that assay, with the same precision, was done 20 times. Either confidence or fiducial limits should normally be calculated for the results of each assay, since a value for a relative potency by itself means little unless it is known how precise it is. This principle is critical, for example, in the interpretation of pass-fail limits for the potency specification of a pharmaceutical product.

A further test for the validity of assay estimates can be applied when results of a number of assays of a specimen are combined, to assess how well, having regard to the precision of each estimate, they agree with each other. This test, the χ^2 test for homogeneity, can be useful in revealing unsuspected trouble (for example, bias) in assay systems with good precise estimates which when combined turn out to be significantly heterogeneous by this test; when this occurs, and there is no reason to suspect there is something wrong with the assay system, the geometric mean of all the results is taken to be the best combined estimate value.

Replication and Precision

In planning the design of an assay there is often misunderstanding about where replication of measurements is most effective in improving its precision. Replication by itself does not necessarily increase accuracy: it may give an estimate of greater precision but of unknown bias. (Of course, it may be impossible to assess bias in an assay system unless there is some idea of what the true estimate should be.) As a rule, however, there is generally no point in repeating several times response measurements that already agree closely with each other (for example, making multiple response measurements on a large number of dose levels where closely similar values would be obtained), if bias to the estimate has already been caused by some other factor, such as inaccuracy in making dilutions of the standard. Repetition in an assay design should be made at those steps where greatest variability may occur. Thus, accuracy is much more likely to be improved by combination of the results of two or more smaller but completely independent assays than from a huge assay with massive replication within it. An independent assay is one which aims to include anew all steps that could be variables, starting with fresh solutions and dilutions of standard and specimen, fresh reagents, and a fresh assay system.

Other Assay Designs

Of course, other assay designs may be more appropriate for particular purposes or assay systems. For example, a twin cross-over design may yield much greater precision for a given amount of work. If the nature of the assay system permits, the same dose levels of specimen (or standard) may be applied subsequently to the same animals or organ or tissue preparation that had previously received doses of standard (or specimen); this greatly diminishes the effects of variability between the assay components. The cross-over design is suitable when the assay system (e.g., test animals, organ, or cells) can return to approximately its original condition before the second doses are given. It is not suitable if the first doses cause a significant change in the responsiveness or in the nature of the response, for example, due to tachyphylaxis. This design is particularly useful for those assay systems which involve components that are highly variable, scarce, costly, or difficult to prepare.

Description of the General Precision of an Assay Method

It is a common practice to describe the general precision of an assay method as a range of percentage of an estimate. Precision of an estimate is largely influenced by the dimensions and the design of the assay; an assay with four doses of test specimen and four of standard, and providing a large number of response measurements (e.g., from many animals), can be expected to give greater precision than one using two dose levels and fewer responses. Thus, the general precision of an assay method cannot be described in terms of the range of confidence limits unless the dimensions of the assay are defined. The term λ (lamda, the index of precision) is independent of the dimensions of the assay.

When large numbers of assays are carried out in several laboratories, it is possible to make general conclusions about the relative precision of methods. Thus in the collaborative assay of the European Pharmacopoeia standard for insulin, three bioassay methods were used.

General conclusions about the relative precision of well-defined bioassay methods are possible from the very large number of assay results, from many laboratories, in the international collaborative studies carried out to calibrate an international standard, e.g., human chorionic gonadotropin (HCG) (Bangham and Grab, 1964; Storring, Gaines-Das, and Bangham, 1980); for insulin (Bangham and Mussett, 1959; Bangham et al., 1978), or for oxytocin (Hartley et al., 1978).

The precision of in vitro protein-binding procedures is rarely derived from results of individual assays since the reagents and conditions of the assay system are held constant, and many specimens are assayed at a single dose level anyway. In this case the precision is based on the results of a typical series of assays with the system.

BIOLOGICAL ASSAY SYSTEMS

Components of Assay Systems: Analyte, Binding Agent, and Amplifier and Detector of the Response Signal

The basic components of all biomedical assay systems are essentially the same: (1) the analyte in the specimen and the standard; (2) the binding reagent (e.g., hormone receptor protein, plasma-binding protein, or antibody); and (3) a system which provides a signal related to the proportion of analyte bound, amplifies, and detects it. The great diversity of assay systems reflects the multifarious forms of the third component. Thus, in a conventional in vivo bioassay for insulin, the injected insulin (analyte) binds to tissue receptors and induces a sequence of interrelated metabolic events which cumulatively result in the signal of a hypoglycemic convulsion, or measured lowered blood glucose concentration.

Corticotropin binds to its receptors on adrenal cortical cells, which respond by producing cortisol (the signal), which is measured directly by a fluorimetric method which serves as the amplifier-detector system.

In cytochemical bioassays for a hormone, a component of the cell's response —for example, the activation of a particular intracellular enzyme—is "stained" quantitatively, and the intensity of the stain (the signal detector) is measured by a microdensitometer, which acts as the amplifier. In an assay for a blood-clotting factor, it is the natural enzyme cascade of the clotting system that amplifies, and the signal is the acceleration of clot formation.

Other amplification-detection systems used in in vitro ligand assay systems include red-cell agglutination and nephelometers and those designed to detect tracer labels, such as radioisotopes, fluophors, and enzymes, which serve to indicate the proportion of the analyte bound to the ligand protein and the proportion which is free.

Some assay systems are hybrids, in which the amplifier consists of a second, sensitive ancillary assay system to detect the response signal, e.g., radioimmunoassays to measure plasma growth hormone levels in an insulin stimulus test for growth hormone deficiency. But all signal detectors function to reflect directly or indirectly the amount of analyte bound to the ligand (binding) protein. And the fundamental event is the binding reaction of analyte to its binding protein.

Binding Reactions of a Hormone with Its Receptors of Antigen with Antibody, and of Enzyme with Substrate

A typical in vivo hormone bioassay depends on the binding reaction of the hormone with its target-cell receptor protein; the binding reaction with antibodies is typical of in vitro immunoassay. This binding is believed to be reversible, non-covalent, and effected by electrostatic and van der Waals forces acting at the sites where the hormone binds to the receptor; the strength of this binding depends on the closeness of fit of these complementary sites. The strength of the association of a hormone to its identified target-cell receptors differs with different hormones: some have an affinity constant of some 10^{-5}, others of 10^{-12} M/liter. This last is about the same order of affinity of antibodies in antisera selected for use in ligand assay systems. Monoclonal antibodies have lower affinity constants of about 10^{-6}.

What happens to a peptide hormone molecule once it is bound to its cell receptor is not yet fully understood (Schulster and Levitzki, 1980). It is believed that it associates for an extremely short time only; it has been suggested that the act of association induces a change of configuration of the receptor and possibly the hormone, which allows or facilitates the prompt dissociation and release of the hormone. Thus, peptide hormones in a solution are in dynamic equilibrium with accessible cell receptors.

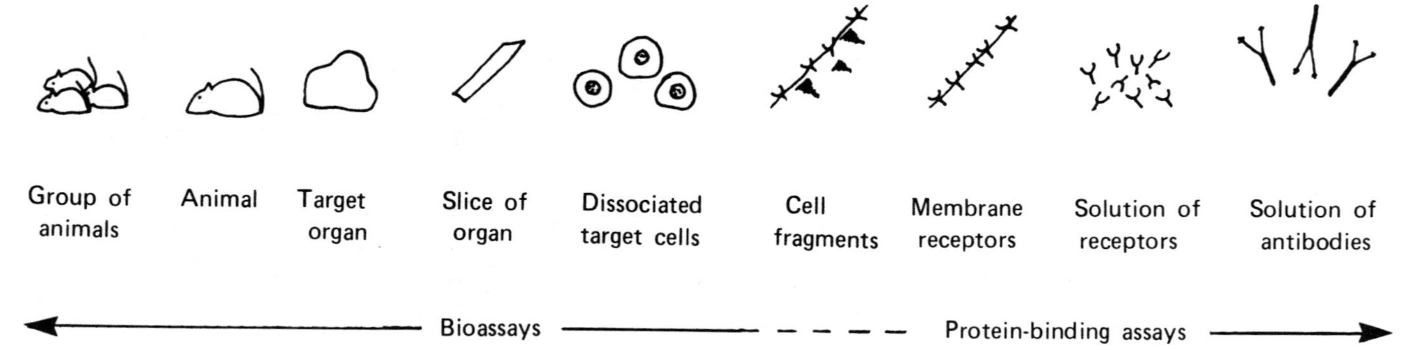

Figure 2   There is no agreed distinction between a bioassay and an in vitro binding assay.

Table 3   Comparison of Protein-Binding, Receptor, and Biological Assays

Example of typical hormone assay	Radioimmunoassay	Immuno-radiometric assays	Receptor assays	Bioassay
Type of assay	Limited binding reagent	Excess binding reagent	Limited or excess binding reagent	Excess binding reagent
Component measured as response parameter	Labeled analyte	Labeled binding reagent	Labeled analyte or labeled receptor	Response of biological matrix (cells or enzyme system activated by analyte)

On the other hand, there is some evidence that in certain hormone-cell interactions (e.g., insulin), a hormone bound to itc cell membrane receptor is taken into the cell by internalization, and the hormone is thus "consumed." In contrast, steroid hormones and tri-iodothyronine (T_3) and thyroxine (T_4) "actively diffuse" through the membrane of their target cells and reach their binding receptors in the cytoplasm. These hormones may then be transferred to the cell nucleus.

In assays involving the reaction of an enzyme with its substrate, the binding is rapid and transitory and the enzyme is unchanged, but the alteration of the substrate is effectively irreversible, the reaction is progressive, and the rate depends on the substrate concentration; other "cofactors" may be necessary for optimal action of any particular enzyme.

In Vivo and In Vitro Bioassays, Receptor Assays and Other In Vitro Protein-Binding Assays

There is no general agreement on what constitutes "a biological assay", or on the distinction between bioassays, receptor assays, and protein-binding assays: they appear to form a continuous spectrum (Figure 2), and it is interesting to analyze what basic differences there are between them (Table 3) (Bangham, 1982). Essentially all bioassays involving living cells are binding reagent excess assays; in many hormone assays perhaps at most only 1% of hormone receptor sites are involved even when the response is maximal. Perhaps one could distinguish bioassay systems broadly from protein-binding systems by the criterion that

Table 4 Some Differences Between Various Types of In Vivo and In Vitro Assay Systems: Which Assay Most Completely Characterizes a Hormone?

In vivo		
Long-term in vivo assays (days)		Reflect metabolism of hormone dose and its longer metabolic effects
Acute in vivo assays (minutes per/hour)	IV	Dose reaching target tissue may be influenced by the dilution and destruction in the plasma circulation
	SC	Dose may be influenced by its destruction at the site of injection and rate of absorption from it
In vivo-in vitro		
Direct injection into target organ artery (or retrograde injection in vein)		Most direct form of application of dose, but assay systems rarely reflect long-term slow response, or effects of metabolism of the dose
Direct application over surface of target organ		
In vitro		
Dose applied to tissue segments		Mechanism of access of hormone to cells not known; mechanism of activation of cells at center of tissue unknown
Dissociated intact cells		Tissue cells and hormones are liable to artifacts and damage due to reagents such as enzymes and procedure used in their preparation
Dissociated fragments of cells		Also susceptible to effects of liberated intracellular products, such as enzymes; only part of normal intracellular response involved
Solution of hormone receptors		Reflects only the ability of the hormone to bind to receptor, not necessarily the ability to activate intracellular events
Antiserum		Has specificity due to mixture of antibodies of various specificies; antibodies rarely both combine and activate receptor site on hormone target cell—exceptions are certain thyroid autoantibodies
Monoclonal antibody		Antibody from each clone has its own specificity, i.e., each binds to the same binding site on the antigen

they are involved with living cells, or at least with intracellular response mechanisms, but it is debatable how much intracellular biochemical reaction should be regarded as the critical minimum to constitute a "biological assay". Some further differences between the assays in the spectrum in Figure 2 are given in Table 4. Which type of assay most closely characterizes "the hormone" depends on the particular hormone.

The simple observation can be made that natural receptors in animals, their tissues and hormone receptors, tend to have a more reproducible specificity: whereas in contrast, each antiserum and each monoclonal antibody has its own individual spectrum of specificity.

Biological Assays and the "Definition" of Hormones, Hormone Analogs and Fragments

The discovery of a hormone generally starts with an observation that an extract of a tissue has a particular biological effect. The extract is then purified and a bioassay system devised which is convenient, sensitive, and specific. Until there is sustained general agreement on its exact chemical constitution, the "hormone" is understood as the natural substance or group of substances which shows the highest specific activity in that bioassay system. Thus the assay system, and the reference material used with it, together define the hormone. The early name of a hormone may reflect this activity (e.g., vasopressin was first assayed for its ability to raise blood pressure), although it may later be shown that it has other, perhaps more vital effects, such as its antidiuretic properties.

With the increased evidence for the intrinsic heterogeneity of several hormones (discussed also under Standards) it has become essential to agree on which biological assay system is used to define what group of the closely related molecules that together constitute the hormone. For example, in selecting materials to be used for international standards for HCG, follicle-stimulating hormone (FSH), and thyroid-stimulating hormone (TSH), candidate preparations were found to vary greatly in different bioassay systems (e.g., Storring, Zaidi, Mistry, Fröysa, et al., 1981). The precept followed by the WHO Expert Committee on Biological Standardization is that the material selected should consist of the natural and unaltered preparation which showed the highest potency in a classic in vivo biological assay system generally recognized by the scientific community as definitive for that hormone (WHO ECBS, 1982). Thus for HCG and LH, the in vivo assay employing the increase in weight of prostate or seminal vesicles in immature male rats is used, and for TSH, the in vivo McKenzie assay. While a wide variety of local modifications are used, pharmacopoeias (e.g., British Pharmacopoeia, 1980) contain good descriptions of several of the main pharmaceutical hormone products, e.g., corticotropin, cosyntropin (tetracosactide), insulin, glucagon, HCG, LH, FSH, oxytocin, vasopressin, desmopressin, gonadorelin, and calcitonins.

In addition, the international collaborative studies carried out to characterize an international standard offer unique opportunity for the comparison of various bioassay systems on a huge scale rarely otherwise performed. Reports of such studies (e.g., Bangham et al., 1978) contain useful information about characteristics such as the relative precision of various assay systems.

A contemporary and increasing problem is the substitution of bioassays by in vitro assays which are simple, rapid, inexpensive, and easily automated. Such systems do not directly measure the same functional property as that which causes a desired biological activity in bioassays in vivo, or in the patient treated. While these simpler assay systems are invaluable for some purposes, they cannot be relied on to measure the critical biological activity of a product, and thus cannot be relied on for characterization of a product for clinical treatment, or of a reference material.

Last, in this connection, many newly discovered naturally occurring substances, such as peptides, which are fragments of recognized peptide hormones, do not yet have identified biological effects, let alone recognized "classic bioassays." Evidence for the identity of preparations of such substances must rely heavily on physical, chemical, and immunological criteria, and on careful comparisons of the extracted substance with at least one other sample of the most authentic preparations available.

SOME PRACTICAL POINTS

Setting up an Assay

Use a stable standard from the beginning, if possible consisting of a purified preparation, preferably a well-characterized and well-documented recognized reference material. If an attested standard is not available, set up a laboratory standard—in small aliquots stored at as low a temperature as facilities allow. If it is stable, it will be possible to calibrate it against another standard later on.

Try to eliminate or minimize any recognizable variables: ensure all other known or possible variables are allocated randomly using tables of random numbers.

An important cause of variability among animals is stress: the extent to which animals are stressed before and during assays is probably not generally realized except by those doing research involving the pituitary-adrenal system.

In trying out a new assay, work out at an early stage a full log dose-response curve for the working standard and a typical specimen. Thereafter, until there is confidence and experience in an assay system, it is best to use a 4 X 4 (or at least a 3 X 3) design so that linearity and parallelism can be checked in each assay result.

The specificity of the assay system should be evaluated using well-characterized and well-documented, recognized purified reference materials where they are available.

Statistical Advice

Although modern computing facilities make calculations easy, it is wise to discuss the design, calculations, and interpretations with a statistician experienced in biometrics. Moreover, with powerful computers available, it is strongly recommended to use full statistical analyses instead of the simplified short-cut calculation procedures devised when computation was laborious.

The statistical appendix of the *British Pharmacopoeia* (1980), for example, contains descriptions of assay designs, worked examples of many types of bioassays, statistical tables, guidance on assessment of validity [tests for linearity, parallelism, the analysis of variance (to determine the design of variability of responses between different factors)], calculation of the confidence limits, the validity of combination of results, and a useful glossary.

Stability

Instability has various causes:

1. Chemical change due to attack by contaminating enzymes, such as peptidases, peroxidases from tissues, or contaminating live or fragmented bacteria; oxidation due to atmospheric oxygen; pH. Moisture facilitates and heat accelerates such changes.
2. Change of molecular form without chemical addition or subtraction, such as the β-aspartyl shift (Naughton et al., 1960), that may cause fundamental change of structure leading directly (e.g., with angiotensin II) or indirectly (due to increased chemical instability, e.g., with secretin) to alteration of structure and activity in various assay systems.
3. Adsorption onto surfaces (see below):

Chemical changes occur more slowly at low temperatures, so stability is better. Conversely, change occurs more rapidly at higher temperatures, which is the basis on which accelerated degradation tests are designed, for assessing the stability of standards (see below).

Storage in reliably stable conditions is required for many preparations of hormones and specimens, especially those of scarce and costly materials. It needs less cost and effort to take steps which can be relied on to ensure stability than to compromise with conditions where reliability is not known and risk loss, or have to assay on each occasion how much loss has occurred.

Thus, it is recommended that valuable solutions are kept in concentrated form, with suitable buffer, a bacteriostat such as 0.1% azide (unless they are freeze-dried), or 1% merthiolate, and a suitable carrier protein free from enzymes; and that this solution is then subdivided into several small containers and frozen to about $-40°C$ during 2-10 min; it may be advisable to freeze to a really low temperature, say, $-70°C$, to avoid supercooling effects, and then store at as low a temperature as feasible—in or above liquid nitrogen, in or over solid CO_2, or at about $-40°C$. Temperatures close to the eutectic freezing point of solutions should be avoided: thus, $-20°C$ is not advisable for saline solutions, whose eutectic temperature is $-23°C$. Also, the pH solutions of certain buffers changes at temperatures below freezing (Fishbein and Winkert, 1979): the pH of neutral sodium phosphate buffers becomes markedly acidic at about $-40°C$.

Adsorption of substances onto surfaces varies individually with each substance, each surface, and on the composition, ionic strength, and pH of the solution. Ideally, each substance should be tested individually with any surface with which it comes into contact, particularly in dilute solutions, to see if it adsorbs and to assess what carrier substance prevents its adsorption. As a rule, however, 0.1-1.0% protein prevents adsorption of many proteins and peptides on most surfaces. Recrystallized bovine albumin is useful, but every batch must be checked to ensure that it contains no trace of peptidase (Caygill, 1977). Human albumin prepared for IV injection has been heat treated and should be free of peptidase and pyrogens. Hydrolyzed gelatin (USP) is free from enzymes, but the molecular sizes and composition of the mixed peptides are ill characterized and not reproducible from batch to batch. An amino acid such as glycine may be a preferable carrier in some cases.

Traces of enzymes, e.g., peptidases and peroxidases in dilute solutions can lead to unexpected loss of peptides, especially in dilute solutions. Peptidases can be tested for with suitable synthetic chromogenic peptide substrates (Caygill, 1977).

Traces of Pyrogens

Lipopolysaccharide endotoxins from bacteria must be regarded as universal contaminants of laboratory solutions (including distilled water), materials, and containers unless proved to be absent. In mammals they generate a general systemic inflammatory reaction, with fever, and may interfere with certain assays, especially those involving in vivo metabolic responses. Lipopolysaccharide pyrogens are very stable molecules and tend to stick firmly to many materials, including chromatography columns. Pyrogens on glass may be destroyed by baking at $110°C$ for 20 min, but are best avoided by the use of scrupulously cleaned glassware. Glass-distilled water which has been redistilled is generally (but not necessarily) free of pyrogens.

A useful screening test for lipopolysaccharide endotoxin pyrogens from gram-negative bacteria is the in vitro test with *Limulus* amebocyte lysate reagents. However, standardization of the sensitivity of the reagents has not been achieved and it is advisable to obtain experienced help in doing any tests for pyrogens.

Pipetting

In making the high dilutions used in very sensitive in vitro assay systems it is advisable to avoid the preparation of solutions by a sequential dilution procedure. A small error (say, 5%) may be compounded to become a significant error after several successive dilution steps. Dilution accuracy is better if a stock solution is made first and small, accurately measured volumes of it diluted in only one or two dilution steps (Gaines-Das, 1980).

STANDARDS AS CALIBRATION MATERIALS

A *standard* is used here as a nonspecific term for a reference material or calibrator material, not a set of written specifications of quality, its other common meaning. A standard here refers to an identified preparation of material, containing a specified analyte, and intended for quantification in an assay system, or used in the assessment of some quality (such as specificity) of an assay system.

A traditional and pragmatic distinction is made between "pure chemical" substances, such as a steroid, whose whole exact structure can be determined by chemical and physical means alone, and substances such as proteins of complex molecular nature, and usually of biological origin, whose structure cannot be so determined. In the biomedical field these latter are called *biological substances.* As a standard for a pure chemical analyte, any good preparation of it could be used in an assay provided it complies with suitable specifications of identity and purity. But for the biologicals of complex structure, it is necessary to use a suitable and characterized reference preparation, which in the interests of international standardization must ultimately be calibrated in terms of a single one—the international standard. It is for this purpose that the World Health Organization provides international standards and reference preparations (see Appendix B).

The essential properties of a standard are that it should be—or contain—the right material, preferably in purified form; it should have high measured stability and any one sample of it should be identical with another sample; it should be well characterized by chemical, physical, and biological procedures and assessed in several laboratories, for its suitability to serve as a standard in the biological assay systems in which it is intended to be used. And it should have a value

Table 5 Forms of Heterogeneity

"Intrinsic"	TSH, FSH, LH, HCG; immunoglobulins; heparins; genetic, e.g., salmon calcitonins, rat insulins
Metabolic	Precursor and metabolized forms, e.g., peptides of human growth hormone, prolactin
Artifact	Denatured forms due to extraction and storage, e.g., deamidated forms, dimers; changed configuration but not composition (β-aspartyl shift); error peptide impurities in preparations of synthetic peptides

assigned to it for its unitage (in terms of units per mass of the standard, or units per ampule); or, if it is a chemical, a value of the mass or molar concentration. A detailed description of the setting up of international and laboratory standards has been published (WHO ECBS, 1978).

Sometimes the selection of the "right material" presents philosophical as well as practical problems. Several hormones, e.g., the glycoproteins TSH, FSH, LH, and HCG, exist naturally in heterogeneous forms (see Table 5) and artifact forms also occur in many extract preparations. Preparations of such hormones from even the most reputable laboratories consist of different mixtures of these forms, each of which may have substantially differing biological potencies in different assay systems when assayed against the same standard, in the same laboratory. Moreover, there is evidence suggesting that each group of similar component molecules separable by physicochemical means, may have its own slightly differing spectrum of biological activities. The basis for selection of a preparation for an international standard for a hormone is a preparation of the natural and unaltered material which shows the highest potency in a classic in vivo assay system which is generally recognized by the scientific community as defining that hormone (WHO ECBS, 1982). The material is, if possible, highly purified, unless high purity is associated with instability, as it is with TSH (for example, preparations of TSH of more than 6 IU/mg readily lose biological activity). But the material is not necessarily of pharmacopoeia quality and is never intended for administration to humans. "The standard" refers to all the material, which may consist of added carrier substances and buffer salts; it is generally not possible to specify an accurate mass concentration of a biological substance, in a biological standard, which is why only approximate figures are given.

For long-term stability and for ease of distribution, the candidate material for a standard is generally freeze-dried in several thousand ampules. The material is dissolved in a solution containing a suitable buffer and a carrier material, such as albumin, lactose, or mannitol. This solution is then distributed in identical volumes into 3-4000 hard glass ampules and freeze-dried slowly from a low temperature. After the material has been further desiccated, the ampules are filled with pure dry nitrogen and sealed by fusion of the glass. The material is essentially free from tissue and contaminating bacterial enzymes; it contains less than 1% moisture; it is in an atmosphere of a dry inert gas; and the ampules are thereafter stored at –20°C in the dark. The five factors that commonly cause instability—enzymes, moisture, oxygen, heat, and light—are thus reduced.

Following various check tests, the candidate standard is then evaluated in a planned international collaborative study in which some 10-30 academic, government, and manufacturers' laboratories take part. The study is designed to compare the candidate standard with the previous international standard, and with coded ampules of national or research standards, and specimens of appropriate sera or tissue extracts. The detail of assay systems is largely left to the participating laboratories, but all the raw result data are analyzed statistically independently at the organizing laboratory.

A full report, describing the material; its handling, characterization, and stability; the assay results and their analysis; and the interpretation of the data, is submitted to the Expert Committee on Biological Standardization of WHO, together with recommendations on the suitability of the ampuled preparation for use as a standard and a proposal for its unitage, all with the agreement of the participants in the assay. It is the WHO Committee that makes the decision on the establishment of the international standard and the international unit defined by it. This is significant, since sometimes the scientific data are such that arbitrary decisions have to be made, and the numerical value of the international unit assigned may have important medical, industrial, and legal implications. But while it is clear that such decisions do have to be made, scientifically the international standard is really only a working hypothesis, a preparation which, according to the best contemporary scientific expertise, is agreed to be used for estimates of the named analyte hormone. It is understood that new evidence may at any time reveal that a standard is unsatisfactory in some respect, and steps would then be taken to replace it. The replacement is often a more highly purified material which has become available: presently, new standards for human pituitary TSH, FSH, LH, growth hormone, and prolactin, and human, porcine, and bovine insulins, are under preparation.

The selection and obtaining of suitable material may also present practical difficulties. The materials are donated free to WHO; substantial quantities are needed, and several of the materials are scarce and very costly. Moreover, costly assays may be required to select a suitable preparation. For example, the large amounts of human pituitary hormones required mean that they can be supplied only from laboratories with access to large (e.g., national) collections of glands. On the other hand, human parathyroid hormone is obtainable only from rare

Bangham

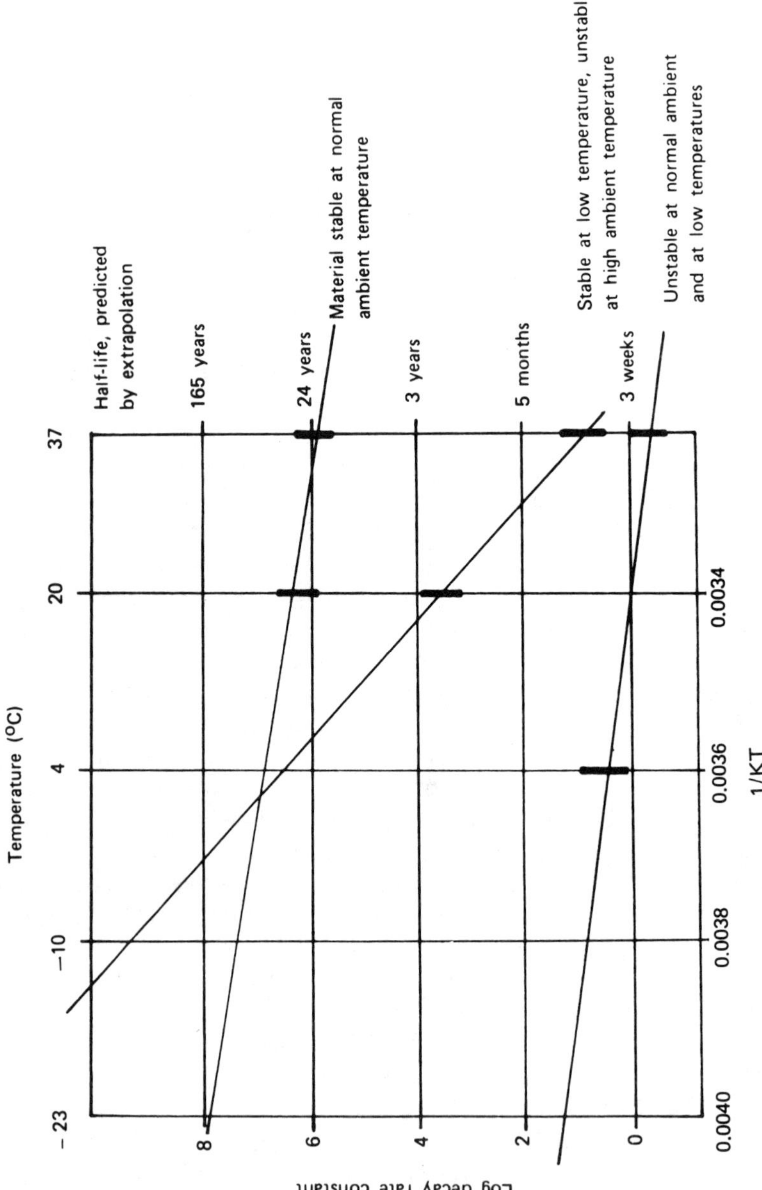

Figure 3 Assessment of stability by accelerated thermal degradation using the Arrhenius plot. Examples of three materials whose potency (95% confidence interval indicated by vertical bars) were estimated after storage at 4, 20, and 37°C for several months.

adenomas removed at operation and extracted by particularly exacting procedures. An International Reference Preparation of this hormone was established in 1981 for which the bulk starting material was only some 0.15 mg of purified peptide (Zanelli and Gaines-Das, 1983). It was essential to obtain evidence for identification to show that this was indeed biologically active, and this was done in various in vitro bioassays against a prior research standard and other local preparations. For such evidence—and for the detection of contaminating hormones in, e.g., pituitary preparations, various very sensitive in vitro bioassay procedures are invaluable. Examples of the painstaking collaborative work of analyzing candidate materials are the studies on LH (Storring, Zaidi, Mistry, Lindberg, et al., 1982) and on FSH (Storring, Zaidi, Mistry, Fröysa, et al., 1981).

Stability of Standards

The stability of international standards must be suitable and is evaluated by accelerated degradation studies. Instability represents chemical change of the substance, and change is accelerated at higher temperatures. Stability is measured by comparison of characteristics of the material from sealed ampules which have been stored at temperatures such as 20, 37, 45, and 56°C for several weeks, months, or years, with ampules kept at low temperature, say, -20°C (at which they are normally stored (e.g., Cotes, Gaines-Das, Burbach, et al., 1978). Using the Arrhenius equation (Jerne and Perry, 1956) or an improved iterative procedure based on it (Kirkwood, 1977), an estimate of the rate of loss at various temperatures may be made (see Figure 3). In certain cases the predicted loss has been confirmed by assays in "real time," i.e., after 10-15 years or so (Storring, Gaines-Das, Tiplady, et al., 1980). This procedure has proved invaluable for making predictions about materials in which instability is due to a first-order chemical reaction (e.g., in dried reference preparations), but it may not be suitable for solutions of mixtures of interacting substances.

International standards are generally very stable and may lose about 0.1% activity per year at -20°C. They are also stable enough even at ambient temperatures encountered when they are distributed to users by airmail. Practical advice on the subsequent solution and handling of standards is given in the next section.

Use of Standards and Units and Interpretation of Results

The concept of measuring biological activity in terms of units is well established for hormones such as insulin, for antitoxins, antibiotics, heparins, and many other biological substances. But the introduction of radioimmunoassays and

other protein-binding assays led to a widespread belief that these methods measured substance, not activity, and that this was fundamentally different and the
result of a "determination" could therefore legitimately be expressed in mass
concentration. However, the unrecognized nonspecificity of many assay systems,
and the heterogeneity of many of the specimens and standards compared, led to
huge inaccuracies, and this has caused great confusion, for example, in understanding the reasons for the discontinuities that may occur in the unitage assigned to a replacement standard (Bangham & Cotes, 1974; WHO ECBS, 1975;
Storring, Gaines-Das, and Bangham, 1980).

As explained in Biological Assay Systems, an agonist hormone combining
with its target-cell receptor initiates a signal expressed by the biological response
of the cell. For the estimation of analyte concentration the purpose this serves
is essentially similar to that provided by a signal which indicates the proportion
of hormone bound in an excess reagent ligand assay, although the signal is generated in a different way. In vitro ligand assays could be used interchangeably
with bioassays if the binding protein that recognized that particular structure of
an agonist molecule which initiated the intracellular biological response, when
bound to a target cell, had the identical specificity of the receptor complex.
Antibodies that may do this are the so-called long-acting thyroid-stimulating
autoimmune antibodies (LATS). And the fact that TSH has a time course of
action different from that of LATS emphasizes that the overall in vivo biological
action of an agonist reflects many other factors, such as its mode and rate of
catabolism, as well as its ability to activate a target cell.

I suggest that one main difference between "classic" bioassays and in vitro
ligand assays is in the nature of their respective specificities.

Bioassays have the advantages that the reagents (animals and their tissues) are
endlessly reproducible and widely available, and their specificity stays approximately constant. The cells' response signal is influenced by many other inter-
relating factors (such as the metabolism of the hormone), all of which contribute
to a particular specificity of the assay system. And the "response" signal is generally selected as a function which itself has a certain biological significance,
whether the response is desirable, as in the case of a therapeutic product, or un-
desirable, e.g., toxic. There are relatively few widely used bioassay systems, and
their specificities are well documented and understood. In contrast, the specificity of each immunoassay system differs and each requires detailed characterization which in practice it seldom if ever receives; the supply of each reagent is
limited. Monoclonal antibodies are theoretically unlimited in supply; but, again,
each has its own spectrum of cross-reaction, and narrow specificity will probably
only be achieved by assay systems which combine two or more cloned antibodies of differing specificity. Hormone-cell receptors are theoretically reproducible, but in practice preparations of them are less so; they can also bind

molecular forms which do not activate a live cell, and do not carry all the characteristics of the specificity of a biological assay.

Thus in reporting the result of an assay, a statement that it was obtained with a well-known bioassay method immediately indicates the type of specificity: but a statement of estimates "by immunoassay" conveys no concept of its specificity and validity.

Mass is a precise physical concept, and in science its measurement should not be ridiculed by grossly inaccurate methods of estimation.

The philosophy underlying the use of "units" and biological standards is that we do not know exactly what is being quantified nor how accurate or precise is the result: hence the concept of units defined by a reference material, instead of mass; the use of the word *estimate* instead of *measure* or *determine*; and the qualification of the precision of the estimate as confidence or fiducial limits.

Of course, many immunoassay systems have a narrower specificity than bioassay systems; this has great advantages for certain purposes, such as the quantification of precursors and metabolite molecules, and molecules of no known bioassayable activity.

Be this as it may, because many bioassay systems and essentially all immunoreactive systems have differing specificity and because many specimens and analyte standards contain closely similar natural or artifact molecules, certain important consequences follow. These include: invalidities of assay results due to nonparallelism, or significant heterogeneity of results of repeated assays; and distinctly differing estimates when different assay systems with differing specificities are used, that is to say, which detect different properties of the analyte in the specimen and in the standard.

For many assays the differences are minor and the discrepancies do not cause problems in the use and interpretation of the results for many purposes. More consistent and serious discrepancies can arise in the calibration of secondary standards (national and laboratory standards) against the international standard. International standards are assayed with a wide range of bioassay methods which may give a spread of estimates, but because we do not know which of them is the most accurate, the unitage assigned is generally based on the combination of *all* the values. In contrast, national and laboratory standards are usually calibrated with only the assay system(s) with which they will be (most) used (e.g., a pharmacopoeial method); this method may have its own significant bias, whose difference from the bias of systems used in other laboratories could cause serious discrepancies. Indeed, the replacement of an international standard could make it necessary to check or recalibrate a secondary standard.

But perhaps the most important implication is that a statement on the result of an assay should always also include the assay system, the standard, and its unitage, and indicate the confidence limits of that assay or of the assay system.

ACKNOWLEDGMENTS

I would like to thank all those who have argued with me on these topics over the last 25 years, especially those who were on the staff of the Division of Biological Standards at the National Institute for Medical Research, and the National Institute for Biological Standards and Control, and also others in other places and contexts.

I would like to thank Fiona Forrester for her excellent assistance in assembling and typing this typescript.

APPENDIX A: GLOSSARY*

Analyte: Substance in test specimen to be determined

Analyte Specificity: The specificity of an assay system for estimation of a given analyte in a given type of test specimen, e.g., a specified biological fluid.

Assay System: All the components and procedures of an assay.

Assayist: Person responsible for the design and performance of an assay.

Bias: The numerical difference between the average of a series of estimates and the true or accepted value. This is equivalent to the definition of inaccuracy given by the International Federation of Clinical Chemistry.

Calibration Curve: The relationship displayed graphically between the amount (dose metameter) of reference material (horizontal axis) and the response of the detector (response metameter; vertical axis). (The term *standard curve* has been used, but should be discouraged wherever there could be ambiguity.)

The dose may be expressed as amount of specified substance or mass of the reference chemical, or as amount ("units") defined by a specified material.

Dose metameter is a transformation of the numerical value of the dose variable, usually log dose. The response metameter is a transformation of the numerical value of the response variable, usually adopted either to linearize the log dose relation or to stabilize the variance.

The response variable can be, for example, the (uncorrected) activity signal in the free or bound fraction.

Cross-Reaction: Ability of substances other than the analyte to bind to the binding reagent, and ability of substances other than the binding reagent to bind to the analyte.

Such substances, if present in a test sample, may compete with the analyte for the binding site, thus leading to an erroneous potency estimate.

---

*Taken from 31st ECBS Report, *WHO Tech. Rep. Ser. No. 658*, 1981.

These substances may be natural precursor forms of the analyte or binding protein, degradation products (from in vivo or in vitro degradation), or other substances that carry on their surfaces a molecular configuration similar to the binding (immunoreactive) site(s) of the analyte or binding protein.

Detection Limit: The smallest amount or concentration of analyte which, with stated confidence (commonly 2 standard deviations, or expressed as confidence or fiducial limits), can be distinguished from zero. This value depends upon the precision of the measurements of zero-dose solution and of the specimen.

In some detector systems the detection limit is determined by the signal/noise ratio in the measurement device.

Ligand: Substance that is reversibly and noncovalently bound by a binding agent. A general term used for analyte, cross-reactant, or calibrant that binds to the binding reagents.

Mass Concentration: Mass of a component (e.g., solute) divided by volume of system (e.g., solution).

Parallelism: Extent to which dose-response curves of two substances are identical, except for displacement along the dose axis of one relative to another. If the curves are curvilinear, this condition is described as *generalized parallelism.* Parallelism is one test of identity of two preparations, e.g., of analyte and calibrator. It is a prerequisite for the calculation of a single valid value of relative potency of one substance compared with another in an assay. Strictly, it applies only when the dose-response curves are linear.

Performance Characteristics (of an assay): Properties of an assay system relating to reliability and practicability. Characteristics of the reliability of an assay include precision, bias, sensitivity, specificity, validity, and ruggedness.

Characteristics of the practicability of an assay include speed, technical simplicity, cost, resources required, and service availability.

Precision or Imprecision (of a measurement): Precision is the closeness of agreement between the results obtained by applying a given experimental procedure several times under prescribed conditions.

As precision has no numerical value, the term *imprecision,* to which a numerical value can be assigned, may be preferable in some contexts.

Imprecision consists of disagreement between, or variability among, replicate measurements (of the same material), and is usually expressed as the standard deviation or variance or coefficient of variation. Such variability can occur within a single assay series, between assay series, between batches of reagents, between operators or laboratories, and between assay systems. Confidence limits or fiducial limits quantify the uncertainty about the values of a parameter after estimation from an experiment or sample.

Whenever a figure for variability is given, it should be stated whether it applies within a single assay, between assays, between batches of reagents, between operators or laboratories, or between assay systems.

Reproducibility: Closeness of agreement between the results of measurements of the same quantity, where the individual measurements are made: by different methods, with different measuring instruments, by different observers, in different laboratories, after intervals of time quite long compared with the duration of a single measurement, under different normal conditions of use of the instruments employed.

Can be expressed quantitatively as standard deviation or coefficient of variation and may be referred to as between-assay series variation. Reproducibility may be dependent on dose and thus may vary at different analyte concentrations.

Robustness (of data-processing procedures):   Insensitivity of estimates and statements of precision to small variations in assumptions used in statistical analysis, such as choice of metameters and form of dose-response relation.

Ruggedness: Characteristic of an assay system that makes the results obtained unaffected by changes in assay reagents and procedures.

Ruggedness can be assessed experimentally by observation of the influence on assay results (either from assay tube to assay tube or from assay batch to assay batch) or changes in the amount or quality of the assay reagents, or in the assay procedure.

In practice, nonruggedness is manifested by poor precision, poor interassay variability, and poor interlaboratory agreement.

Specificity: The specificity (structural) of a protein-binding reagent is the degree to which it is not influenced by cross-reacting substances.

The specificity of an assay system is the degree to which the results are not influenced by cross-reacting substances, or other noncompeting substances, such as plasma proteins and small ions present in the system, or other factors, such as the pH or the temperature of incubation, that affect the reaction.

Specificity may be evaluated: (1) by the introduction of potential cross-reacting substances and other substances likely to affect the reaction; and (2) by examination of the similarity of the performance of the assay system with test specimens and with reference materials, under minor differences of assay conditions, e.g., at different concentrations of analyte or parallelism.

Stability: Lack of alteration under defined conditions. Thermal stability of a reagent, for example, may be quantified as rate of chemical change at stated temperature(s). Often measured by accelerated degradation studies involving comparisons of samples of a material maintained at different temperatures for one or more periods of time.

Appendix B Table of Biological Standards and Reference Materials (National Institute for Biological Standards and Control)

Standard	Ampule Code No.	Defined Activity	Approximate Composition of Ampule Contents	Other Information
Angiotensin I (Asp Isoleu[5])	71/328	9 μg/amp nominal	9 μg synthetic angiotensin I 2 mg mannitol	Bangham et al. (1975)
Angiotensin II (Asp Ileu[5])	70/302	24 μg/amp nominal	24 μg synthetic angiotensin II 2 mg mannitol	Bangham et al. (1975)
1st IS for arginine vasopressin	77/501	8.2 IU/amp	20 μg synthetic arginine vasopressin 5 mg human albumin citric acid	WHO Unpublished Working Document WHO/BS/78.1231
1st IRP of calcitonin, human, for bioassay	70/234	1.0 IU/amp	8.5 μg synthetic calcitonin of sequence found in tumors 5 mg mannitol	Gaines-Das and Zanelli (1980)
1st IRP of calcitonin, porcine, for bioassay	70/306	1.0 IU/amp	10 μg purified extract 5 mg mannitol	T_3 and T_4 trace amounts WHO/BS/74.1077
1st IRP of calcitonin, salmon, for bioassay	72/158	80 IU/amp	20 μg synthetic salmon calcitonin 2 mg mannitol	WHO Unpublished Working Document WHO/BS/74.1077
1st IRP of chorionic gonadotropin for immunoassay	75/537	650 IU/amp	70 μg chorionic gonadotropin 5 mg human albumin	Canfield and Ross (1976) Storring, Gaines-Das, and Bangham (1980)
1st IRP of α-subunit of chorionic gonadotropin	75/569	70 IU/amp	70 μg chorionic gonadotropin, α-subunit 5 mg human albumin	Canfield and Ross (1976) Storring, Gaines-Das, and Bangham (1980)
1st IRP of β-subunit of chorionic gonadotropin	75/551	70 IU/amp	70 μg chorionic gonadotropin, β-subunit 5 mg human albumin	Canfield and Ross (1976) Storring, Gaines-Das, and Bangham (1980)

Appendix B   (Continued)

Standard	Ampule Code No.	Defined Activity	Approximate Composition of Ampule Contents	Other Information
2nd IS for chorionic gonadotropin for bioassay	61/6	5300 IU/amp	2 mg chorionic gonadotropin 5 mg lactose	Bangham and Grab (1964)
Corticotropin, human	74/555		11.6 μg corticotropin 5 mg human albumin 2.5 mg mannitol	Bioassay vs 3rd IS: SC adrenal ascorbate depletion assay = 3.19 IU/amp
3rd IS for corticotropin (ACTH), porcine, for bioassay. International Working Standard	59/16 Various	5.0 IU/amp 5.0 IU/amp	50 μg pituitary extract 5 mg lactose	Bangham et al. (1962) Storring, Gaines-Das, Tiplady, et al. (1980) Vasopressin <25 mU/amp by rat blood pressure assay in terms of IS
1st IS for desmopressin	78/573	27 IU/amp	27 μg purified desmopressin 5 mg human albumin citric acid	WHO Unpublished Working Document WHO/BS/80.1226
2nd IRP of erythropoietin, human, urinary, for bioassay	67/343	10 IU/amp	2 mg urinary extract containing erythropoietin 3 mg sodium chloride	Cotes et al. (1972)
2nd IRP of FSH-LH, human pituitary, for bioassay	78/549	FSH: 10 IU/amp LH: 25 IU/amp	0.5 mg FSH-LH 1.25 mg lactose	Bangham et al. (1973) WHO ECBS (1981)
IS FSH-LH human, urinary, for bioassay	70/45	FSH: 54 IU/amp LH: 46 IU/amp	1 mg human postmenopausal urine extract 5 mg lactose	Storring et al. (1976)
Gastrin, human	68/439	12 U/amp	12.6 μg synthetic gastrin, as hexamonium salt 5 mg lactose phosphate buffer	

Gastrin II, porcine	66/138	10 U/amp	10 μg gastrin II 5 mg sucrose phosphate buffer	
1st IRP of glucagon, porcine, for bioassay	69/194	1.49 IU/amp	1.5 mg glucagon 5 mg lactose 0.24 mg sodium chloride	Bangham et al. (1974) Calam and Storring (1975)
1st IS for glucagon, porcine, for immunoassay	69/194	1.49 IU/amp	1.5 mg glucagon 5 mg lactose 0.24 mg sodium chloride	Bangham et al. (1974) Calam and Storring (1975)
1st IRP for gonadorelin for bioassay	77/596	31 IU/amp	31 nmol synthetic gonadorelin 2.5 mg lactose 0.5 mg human albumin	Ovine/porcine sequence WHO ECBS (1981)
1st IRP of growth hormone, human, for immunoassay	66/217	0.35 IU/amp	175 μg growth hormone 5 mg sucrose phosphate buffer	Prolactin activity: Pigeon, 1 IU/amp; Rabbig, 7 IU/ amp; decidual response (mice) 1 IU/amp
Insulin C-peptide for immunoassay	76/561	Approx. 2.5 nmol/amp	10 μg synthetic human insulin C-peptide analog 50 μg human albumin; 0.05 nmol phosphate buffer	Synthetic (64-formyllysine) human proinsulin 31-65 Caygill et al. (1980)
1st IRP of insulin, human, for immunoassay	66/304	3.0 IU/amp	130 μg insulin 5 mg sucrose	WHO Unpublished Working Document WHO/BS/ 74.1084
4th IS for insulin, bovine and porcine, for bioassay	58/6	24.0 IU/mg	110-125 mg crystals 48% porcine insulin 52% bovine insulin	Bangham and Mussett (1959) Storring et al. (1975)
1st IRP of LH, human pituitary, for immunoassay	68/40	77 IU/amp	11.6 μg LH 5 mg lactose 1 mg human albumin 1 mg sodium chloride	Storring et al. (1978)

Appendix B (Continued)

Standard	Ampule Code No.	Defined Activity	Approximate Composition of Ampule Contents	Other Information
1st IS for lysine vasopressin	77/512	7.7 IU/amp	30 μg synthetic lysine-vasopressin acetate 5 mg human albumin citric acid	WHO Unpublished Working Document WHO/BS/ 78.1230
4th IS for oxytocin	76/575	12.5 IU/amp	24 μg synthetic oxytocin acetate 5 mg human albumin citric acid	WHO Unpublished Working Document WHO/BS/ 78.1227
Parathyroid hormone, human, for immunoassay	75/549	0.025 U/amp	250 ng extract of human adenomata 250 μg human albumin 1.25 mg lactose	Estimated to contain approx. 25 ng parathyroid hormone Zanelli and Gaines-Das (1980)
1st IRP of parathyroid hormone, human, for immunoassay	79/500	0.100 IU/amp	100 ng purified extract of human adenomata 250 μg human albumin 1.25 mg lactose	Zanelli and Gaines-Das (1983)
1st IRP of parathyroid hormone, bovine, for bioassay	67/342	200 IU/amp	0.6 mg gland extract 5 mg lactose	WHO Unpublished Working Document WHO/BS/ 74.1078
1st IRP of parathyroid hormone, bovine, for immunoassay	71/324	2.0 IU/amp	1 μg purified extract 200 μg human albumin 1 mg lactose	WHO Unpublished Working Document WHO/BS/ 74.1078
IRP of placental lactogen, human, for immunoassay	73/545	0.000850 IU/amp	850 μg placental lactogen 5 mg mannitol	Cotes and Gaines-Das (1978)
1st IRP of prolactin, human, for immunoassay	75/504	650 mIU/amp	20 μg extract 1 mg human albumin 5 mg lactose	Gaines-Das and Cotes (1979)

2nd IS for prolactin, ovine, for bioassay	57/8	22 IU/mg	10 mg extract	Bangham et al. (1963)
1st IRP of renin, human, for bioassay	68/356	0.1 IU/amp	0.27 mg renal extract 5 mg lactose phosphate buffer	Bangham et al. (1975)
2nd IS for serum gonadotropin, equine, for bioassay	62/1	1600 IU/amp	0.8 mg extract 5 mg lactose	Bangham and Woodward (1966)
1st IRP of cosyntropin (tetraco-sactride), for bioassay	80/590	490 IU/amp	490 μg cosyntropin 20 mg mannitol	
1st IRP of TSH, human, for immunoassay	68/38	150 mIU/amp	46.2 μg TSH extract 5 mg lactose 1 mg human albumin	Cotes, Gaines-Das, Kirkwood (1978)
1st IS for thyrotropin, bovine, for bioassay	53/11	74 mIU/mg	1 part pituitary extract 19 parts lactose	Mussett and Perry (1955)

REFERENCES

Bangham, D. R. (1982). Biological standards in clinical endocrinology: Some soluble and insoluble problems. In *The Eighth Tenovus Workshop* D. W. Wilson, S. J. Gaskell, and K. W. Kemp (eds.). Alpha Omega Publishing, Cardiff, Wales, pp. 43-57.

Bangham, D. R., Berryman, I., Burger, H., Cotes, P. M., Furnival, B. E., Hunter, W. M., Midgley, A. R., Mussett, M. V., Reichert, L. E., Rosemberg, E, Ryan, R. J., and Wide, L. (1973). An international collaborative study of 69/104: A reference preparation of human pituitary FSH/LH. *J. Clin. Endocrinol. Metab.*, 36:647-660.

Bangham, D. R., and Cotes, P. M. (1974). Standardization and standards. *Br. Med. Bull.* 30:12-17.

Bangham, D. R., de Jonge, H., and van Noordwijk, J. (1978). The collaborative assay of the European Pharmacopoeia biological reference preparation for Insulin. *J. Biol. Stand.* 6:301-314.

Bangham, D. R., and Grab, B. (1964). The second international standard for chorionic gonadotrophin. *Bull. WHO,* 31:111-125.

Bangham, D. R., and Mussett, M. V. (1959). The fourth international standard for insulin. *Bull. WHO,* 20:1209-1220.

Bangham, D. R., Mussett, M. V., and Stack-Dunne, M. P. (1962). The third international standard for corticotrophin. *Bull. WHO* 27:395-408.

Bangham, D. R., Mussett, M. V., and Stack-Dunne, M. P. (1963). The second international standard for prolactin. *Bull. WHO,* 29:721.

Bangham, D. R., Robertson, I., Robertson, J. I. S., Robinson, C. J., and Tree, M. (1975). An international collaborative study of renin assay: Establishment of the international reference preparation of human renin. *Clin. Sci. Mol. Med.,* 48:135S.

Bangham, D. R., Salokangas, A. A., Annable, L., and Storring, P. L. (1974). The first international standard for glucagon. *Acta Endocrinol (Kbh),* 77:705-714.

Bangham, D. R., and Woodward, P. M. (1966). The second international standard for serum gonadotrophin. *Bull. WHO,* 35:761.

Calam, D.H., and Storring, P. L. (1975). The heterogeneity and degradation of glucagon studied by polyacrylamide gel electrophoresis. *J. Biol. Stand.,* 3: 263-265.

Canfield, R. E., and Ross, G. T. (1976). A new reference preparation of human chorionic gonadotrophin and its subunits. *Bull. WHO,* 54:463-470.

Caygill, C. P. J. (1977). Detection of peptidase activity in albumin preparations. *Clin. Chim. Acta,* 78:507-509.

Caygill, C. P. J., Gaines Das, R. E., and Bangham, D. R. (1980). Use of a common standard for comparison of insulin C-peptide. *Diabetologia,* 18:197-204.

Cotes, P. M., Annable, L., and Mussett, M. V. (1972). The second international reference preparation of erythropoietin, human, urinary for bioassay. *Bull. WHO*, 47-99-122.

Cotes, P. M., and Gaines, Das, R. E. (1978). An international collaborative study of the assay of human placental lactogen: Establishment of WHO IRP of human placental lactogen. *Br. J. Obstet. Gynaecol.* 85:451-459.

Cotes, P. M., Gaines-Das, R. E., Burbach, J. P. H., and Bartlett, W. A. (1978). The radioimmunoassay of human prolactin: An interim report of findings from an international collaborative study of the proposed international reference preparation of human prolactin for immunoassay. In *Progress in Prolactin Physiology and Pathology,* C. Robyn and M. Harter (eds.). Elsevier/ North Holland, Amsterdam.

Cotes, P. M., Gaines-Das, R. E., Kirkwood, T. B. L., Bennie, J. G., and Hunter, W. M. (1978). The stability of standards for radioimmunoassay of human TSH: Research standard A and the IRP initially 68/38. *Acta Endocrinol. (Kbh),* 88:291-297.

Ekins, R. P. (1970). Theoretical aspects of saturation analysis. In *In Vitro Procedures with Radioisotopes in Medicine.* International Atomic Energy Agency, Vienna, pp. 325-353.

Ekins, R. P. (1976). In *Hormone Assays and Their Clinical Application.* J. A. Loraine, and E. T. Bell (eds.). Churchill Livingstone, London, pp. 1-72.

Finney, D. J. (1970). Covariance analysis in bioassay and related procedures. In *Statistics in Endocrinology,* J. W. McArthur and T. Colton (eds.). MIT Press, Cambridge, Massachusetts.

Finney, D. J. (1978). *Statistical Method in Biological Assay,* 3rd ed. Charles Griffin, London.

Fishbein, W. N., and Winkert, J. W. (1979). Parameters of freezing damage to enzymes. In *Proteins at Low Temperatures,* M. J. Comstock (ed.). Advances in Chemistry Series Vol. 180, pp. 55-82.

Gaddum, J. H. (1933). Methods of biological assay depending on a quantal response. In *Reports on Biological Standards,* MRC Spec. Rep. Ser. No. 183.

Gaines-Das, R. (1980). Dilution as a source of error: Implications for preparation and calibration of laboratory standards and for quality control of radioimmunoassay. *Clin. Chem.,* 26:1726-1729.

Gaines-Das, R. E., and Cotes, P. M. (1979). International reference preparation of human prolactin for immunoassay: Definition of the international unit, report of a collaborative study and comparison of estimates of human prolactin made in various laboratories. *J. Endocrinol.,* 80:157-168.

Gaines-Das, R. E., and Zanelli, J. M. (1980). The IRP of calcitonin, human, for bioassay: Assessment of material and definition of the international unit. *Acta Endocrinol. (Kbh),* 93:37-42.

Hartley, R. E., Gaines-Das, R. E., and Bangham, D. R. (1978). Unpublished WHO Working Document, WHO/BS/78.1227.

Jerne, N. K., and Perry, W. L. M. (1956). The stability of biological standards. *Bull. WHO,* 14:167-182.

Jerne, N. K., and Wood, E. C. (1949). The validity and meaning of the results of biological assays. *Biometrics,* 5:273.

Kirkwood, T. B. L. (1977). Predicting the stability of biological standards and products. *Biometrics,* 33:736-742.

Mussett, M. V., and Perry, W. L. M. (1955). The international standard for thyrotrophin. *Bull. WHO,* 13:917-929.

Naughton, M. A., Sanger, F., Hartley, B. S., and Shaw, D. C. (1960). The amino acid sequence around the reactive serine residue of some proteolytic enzymes. *Biochem. J.,* 77:149-163.

Schulster, D., and Levitzki, A. (1980). *Cellular Receptors for Hormones and Neurotransmitters.* John Wiley, Chichester, England.

Storring, P. L., Bangham, D. R., Cotes, P. M., Gaines-Das, R. E., and Jeffcoate, S. L. (1978). The international reference preparation of human pituitary luteinizing hormone for immunoassay. *Acta Endocrinol. (Kbh),* 88:250-259.

Storring, P. L., Dixon, H., and Bangham, D. R. (1976). The first international standard for human urinary FSH and human urinary LH (ICSH) for bioassay. *Acta Endocrinol. (Kbh),* 83:700-710.

Storring, P. L., Gaines-Das, R., and Bangham, D. R. (1980). International reference preparation of hCG for immunoassay: Potency estimates in various bioassay and protein-binding assay systems; and the international reference preparations of the α and β subunits of hCG for immunoassay. *J. Endocrinol.* 84: 295-310.

Storring, P. L., Gaines-Das, R. E., Tiplady, R. J., Stenning, B. E., and Mistry, Y. G. (1980). Stability of the third international standard for corticotrophin: Accelerated degradation study using different bioassays and isoelectric focusing. *J. Endocrinol.* 85:533-539.

Storring, P. L., Greaves, P. L., Mussett, M. V., and Bangham, D. R. (1975). Stability of the fourth international standard for insulin. *Diabetologia,* 11:581-584.

Storring, P. L., Zaidi, A. A., Mistry, Y. G., Fröysa, B., Stenning, B. E., and Diczfalusy, E. (1981). A comparison of preparations of highly purified human pituitary follicle-stimulating hormone (FSH): differences in the FSH potencies as determined by in vivo bioassay, in vitro bioassay and immunoassay. *J. Endocrinol.* 91:353-362.

Storring, P. L., Zaidi, A. A., Mistry, Y. G., Lindberg, M., Stenning, B. E., and Diczfalusy, E. (1982). Preparations of highly purified human pituitary luteinizing hormone (LH): Differences in the LH potencies as determined by in vivo bioassays, in vitro bioassay and immunoassay. *Acta. Endocrinol. (Kbh),* 101:339-347.

World Health Organization Expert Committee on Biological Standardization, 26th report (1975). *WHO Tech. Rep. Series No. 565*, p. 21.

World Health Organization Expert Committee on Biological Standardization, 29th report, Annex 4 (1978). *WHO Tech. Rep. Series No. 626*.

World Heath Organization Expert Committee on Biological Standardization, 31st report (1981). *WHO Tech. Rep. Series No. 658*, p. 24.

World Health Organization Expert Committee on Biological Standardization, 32nd report (1982). *WHO Tech. Rep. Series No. 673*.

Zanelli, J. M., and Gaines-Das, R. E. (1980). International collaborative study of NIBSC Research Standard A for human PTH for immunoassay. *J. Endocrinol.*, 86:291-304.

Zanelli, J. M. and Gaines-Das, R. E. (1983). The first International Reference Preparation of human parathyroid hormone for immunoassay: Characterization and calibration by international collaborative study. *J. Clin. Endocrinol. Metab.* In press.

3

The Techniques of Cytochemical Bioassays

Nigel Loveridge / Kennedy Institute of Rheumatology, London, England

INTRODUCTION

The word *hormone* was first used by Starling (1905) to describe the nature of
the substance "secretin" initially reported by Bayliss and Starling (1902a,b).
Starling (1905) defined a hormone as "a substance normally produced in the
cells of some part of the body and carried by the blood stream to distant parts
which it affects for the good of the body as a whole." In so doing the hormone
must cause some metabolic alteration in the target organ, or more particularly
the target cell, upon which it acts so that the final physiological response to
the hormone can be achieved. Therefore it would seem reasonable to assume
that increases in the order of sensitivity of the measurement of the response
to the hormone would be achieved by changing from an in vivo assay (i.e.,
measuring the response in a whole animal) to an assay using the response of the
target organ and finally to an assay where the response was measured in the
target cell.

An advantage in dealing with an assay based on tissue from a single animal
is that any interanimal variation in response is avoided. Furthermore, if the target organ or cell is first isolated from the animal, then the hormone will not be
diluted, in the circulation, prior to reaching the target organ or cell. When considering the difference in sensitivity of response between a target organ and a
target cell it should be noted that not all the cells within a given target organ
may respond to the hormone with the same degree of sensitivity. For instance,

the major effect of gastrin is on the parietal cell of the gastric fundus, which is only one of the three main cell types which constitute the mucosal layer of the stomach.

Isolation of the target cells from the surrounding tissue can be achieved in two distinct ways. First, the target cells can be physically isolated from the surrounding nonresponsive tissue and separated out to give a relatively pure (usually around 95%) population of cells that respond to the hormone. However, problems have occurred with such techniques in that separation of the cells usually requires exposure of the tissue to some form of digestive enzymatic activity, usually trypsin or collagenase. Hormone receptors have been shown to be susceptible to such treatment [e.g., the insulin receptor in fat cells (Cuatrecasas, (1971)], and the degree to which the digestion affects the sensitivity of the response to the hormone is unknown. Furthermore, in certain cases, such as adrenocorticotropic hormone (ACTH), isolated cells, while responding well to standard hormone preparations, seem to be affected by the presence of plasma (Sayers, 1977). This necessitates the extraction of the hormone from plasma prior to assay, as described by Liotta and Krieger (1975).

It would seem, therefore, that if the responsive cell could be kept within the tissue matrix, increased sensitivity of response to the hormone, due to minimal disruption of the tissue environment, might be achieved. This approach is embodied in the techniques of quantitative cytochemistry where the biochemical changes in particular cells can be measured, not by prior physical isolation of that cell, but by optically isolating the cell from the surrounding nonresponsive cells.

Thus, the basic concept behind the use of quantitative cytochemistry for hormone bioassay is as follows. A hormone (e.g., thyroid-stimulating hormone, TSH) acting on its target cell (a thyroid follicle cell) must produce some alteration in the biochemistry of that cell in order that the final response (thyroid hormone secretion) can occur; the particular biochemical change is then measured only in the target cell without its prior isolation from the surrounding tissue. Three aspects of cellular biology are required for such assays, and these will be discussed in separate sections.

MAINTENANCE CULTURE

By the methods developed by Trowell (1959), segments of the target organ can be maintained in nonproliferative culture. Using a medium containing amino acids, salts, glucose, insulin, and phenol red as a pH indicator, Trowell (1959) showed that it was possible to maintain explants of a large number of organs of the rat. It should be stressed at this point that the culture procedure is nonproliferative, so allowing the use of fully differentiated cells kept within their tissue matrix.

The Function of Maintenance Culture

The use of organ culture for bioassay of hormones has several distinct advantages. First, it allows all the assays to be "within-animal assays," so eliminating the variations in response between animals. For instance, in a series of 36 segment assays of gastrin-like activity, done with tissue from different animals, the slope of the response to gastrin (that is, the change in activity to each logarithmic dose) varied from 1.1 to 7.3 (Loveridge, 1977). Therefore it would not have been possible to use the tissue from these different animals for a single between-animal assay.

Second, organ culture removes the tissue from the endogenous hormone influence of the animal. During life, the target organ is exposed to the hormone to be assayed. At death the activity of the processes involved in hormone action, including those being measured as the end point of the assays, will reflect the concentration of the hormone present in the circulation. Therefore it would not be possible to measure the effect of lower hormone concentrations than those in the circulation at the time of removal of the target organ from the animal. Consequently, the use of organ culture is one of the factors that contribute to the sensitivity of the cytochemical bioassay system by allowing the removal of the target organ from the hormonal influence of the animal and its subsequent maintenance for a period of time prior to exposure to the hormone.

Third, the maintenance culture procedure allows the tissue to recover from the trauma of excision. The removal of tissue from an animal and the subsequent handling of it will inevitably cause some degree of trauma. For instance, in early developmental studies on the assay of ACTH, it was found that segments of adrenal were still insensitive to ACTH after the first 3 hr of culture but after 5 hr the responsiveness had returned.

Procedure

Cleaning of Glassware

Although Decon 75 has been recommended for cleaning tissue culture glassware (Paul, 1975), in this laboratory the use of Decon has proved to be deleterious to the maintenance of tissue segments that are to be responsive to hormones at the end of the culture period. All tissue culture glassware is washed in Pyroneg (Diversey U.K. Ltd.), as recommended by Cumming (1970).

Selection of Animals

For most assays developed to date, the animal of choice has been the guinea pig. Although the sex of the animal is not critical, the age and body weight are significant. For instance, the thyroid glands of the Hartley strain of guinea pig with a body weight greater than 250 g are usually more cystic than those of younger

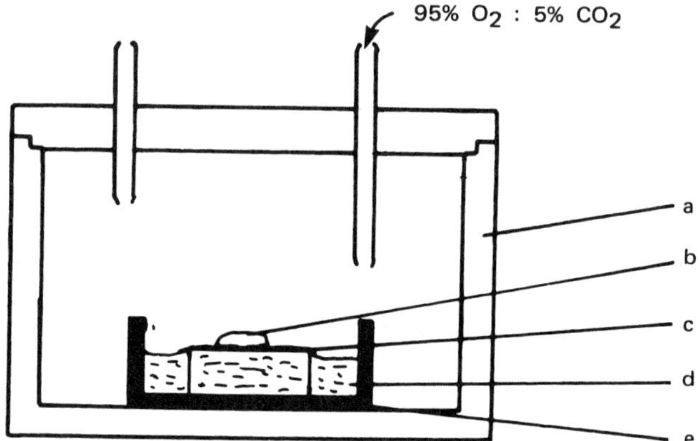

Figure 1   Organ culture. Segments of the target organ (b) are placed on a de-fatted lens tissue on top of a metal grid (c) placed in a vitreosil dish (e). Trow-ell's T8 culture medium (d) is added up to the level of the grid, and the vitreosil dish is placed in an outer chamber (a), which is filled with 95% O_2:5% CO_2, sealed with lanolin, and left at 37°C.

animals and are therefore not as good for bioassay purposes. On the other hand, the adrenal glands of older animals are larger so that animals of 400-500 g body weight are selected for the ACTH assay. In all cases it is recommended that the animals are brought in at least 1 week prior to use so as to allow the animal to settle before being used for an assay.

Preparation of Segments
The animals are killed by asphyxiation in nitrogen and the target organ removed as quickly as possible. The organ is then cleaned of fat and connective tissue. In the case of the fundus of the stomach, the mucosa is carefully cleared of debris by the use of a paintbrush. In the assays of parathyroid hormone (PTH) and arginine vasopressin (AVP) the kidney is decapsulated prior to being segmented. Segments of roughly equal size are then prepared. Care should be taken at all times to avoid excess handling of the tissue and that the target organ is kept moist with culture medium throughout the preparative procedures.

Organ Culture
Individual stainless steel mesh grids (mesh size 1.1 × 2 mm) are placed in vitreosil dishes, and culture medium (Trowell's T8) is added. The composition of the medium is given in the Appendix. Care must be taken to ensure that the level of the medium is not above that of the grid. The grid is then covered by a lens

tissue which has previously been defatted by immersion in three changes of diethyl ether followed by three rinses in distilled water. Segments of the target organ are then placed on the grid so that the segments are at the interface between the liquid and gas phases (Figure 1). The vitreosil dish is then placed in a chamber which is sealed with lanolin and is gassed with a mixture of 95% O_2: 5% CO_2 for up to 10 min. The chamber is then sealed and left at 37°C. The time of culture is dependent upon the particular assay system being used or developed. However, a period of 5 hr has usually been found to be both sufficient and convenient.

Modifications to the Culture Medium

It has been found that certain modifications to the culture medium are required in order to have a fully responsive segment of tissue at the end of the culture period. Thus, segments of adrenal require the addition of sodium ascorbate (10^{-3} M) in order that the initial loss of ascorbate from the segment during the first 3 hr of culture is replaced, so allowing full responsiveness to be achieved (Chayen et al., 1976). Similarly, in the bioassay for luteinizing hormone (LH), ascorbate is added to the culture medium (Rees et al., 1973). In the case of stomach, it has been shown that carbonic anhydrase activity in the parietal cells is higher in tissue cultured at pH 7.6 as opposed to that cultured at pH 7.0 (Loveridge et al., 1974; Chayen et al., 1976). Furthermore, it has been shown that the binding of gastrin to a gastrin receptor in rat parietal cells is maximal at pH 7.0 (Takeuchi et al., 1979). Therefore, for the segment assay of gastrin-like activity, the strips of fundus are maintained in medium that has had its bicarbonate concentration reduced to 13 mM (from 33 mM) by the addition of hydrochloric acid. With this concentration of bicarbonate and with a 95% O_2: 5% CO_2 gas mixture, the pH is maintained at pH 7.0. In the assays of TSH and PTH, no modifications of the culture medium is required.

Exposure to Hormone

It is at this point that the section and segment assay procedures differ. Section assays will be discussed later in this chapter. As the name implies, the segment assay system requires that segments of the target tissue are exposed to the hormone prior to being sectioned and reacted for a particular biochemical response. Therefore, at the end of the culture period each segment is exposed either to one of a number of graded concentrations of the standard hormone or to a dilution of the test sample. Both the standard hormone and dilutions of the test sample are made up in T8 medium with, if necessary, the modifications discussed previously. In the case of the PTH assay it has been found necessary to wash the segments in fresh medium prior to exposure to the hormone in order to remove any metabolites that may impair the sensitivity of the response (Chambers et al., 1978a).

The time of maximal response to each hormone (as well as for each cyto-chemical parameter being tested) has to be determined prior to the establishment of an assay system. This is achieved by exposing cultured segments of the target organ to a single concentration of the hormone for various times. This concentration is usually around one-tenth of the normal circulating level of the hormone. A further time course should then be done with a lower concentration of hormone. The segments are then chilled, sectioned, and reacted for the particular biochemical response being examined, and the time of maximal response is determined for subsequent use in the assay.

The time of maximal response may not necessarily be the same in different laboratories. This is presumably due to slightly different techniques, animal strains, or the standard preparation of hormone being used. For instance, in the assay of PTH described by Chambers et al. (1978a), the peak response was at 8 min. Essentially similar assays described by Goltsman et al. (1980) and Fenton et al. (1978) had peak response at times of 6 and 12 min, respectively. As already mentioned, the time of maximum response will also vary according to the particular biochemical parameter being used. Thus, again with the PTH assay, the peak responses to the same concentration of PTH shown by alkaline phosphatase activity, carbonic anhydrase activity, and glucose-6-phosphate dehydrogenase activity in different tubules were 4, 1, and 8 min, respectively (Chambers et al., 1978b).

One other factor to be taken into consideration is the pH of the medium used in presenting the hormone to the segments. Except for the assay of gastrin-like activity, all the assays previously described are done using T8 medium at pH 7.6. However, it should be noted that because the culture medium is buffered by bicarbonate and dissolved CO_2 (5%), exposure of the medium to the air will result in a loss of CO_2 to the atmosphere with a consequential rise in the pH of the medium. Therefore, over long periods of exposure to the hormone (as for the assay of TSI described in Chapter 6), the pH of the medium will rise. To retard this effect it is advisable to cover the vitreosil dish during the exposure to hormone so as to decrease the loss of carbon dioxide. Another approach that has met with some success is to gas the medium with a mixture of 95% air: 5% CO_2 prior to use (Allgrove et al., 1983) and to maintain the medium in that atmosphere. It should be noted that a 95% O_2:5% CO_2 mixture should not be used for this purpose as it has been shown, in some cases, to be deleterious to the response of the target organ to the hormone (Ealey, 1979). Alternative methods of stabilizing the pH of the medium involve the use of Hepes buffer (Fenton et al., 1978) or the equilibration of the medium with the atmosphere and subsequent reduction of the bicarbonate content with hydrochloric acid (as will be discussed later).

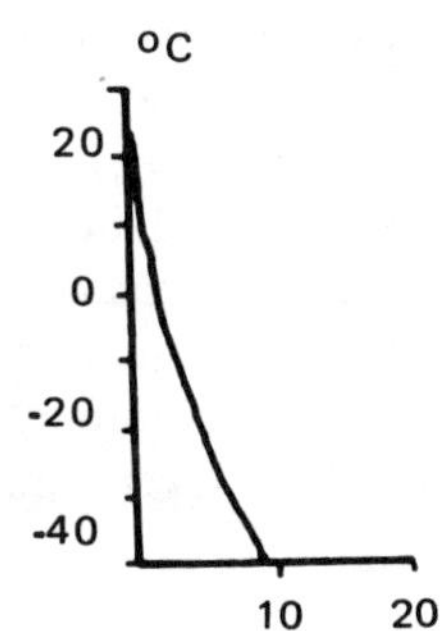

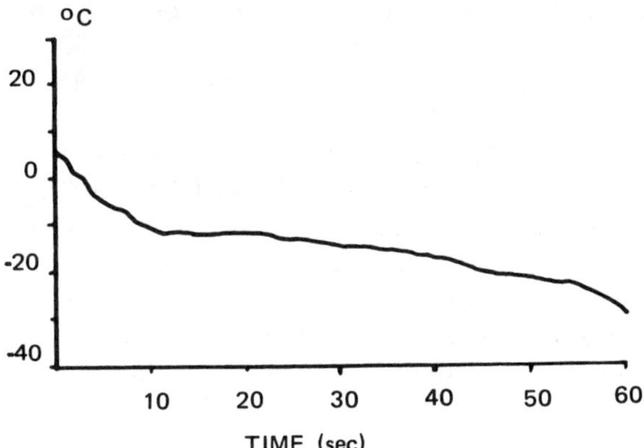

Figure 2 Chilling tissue. The thermocouple trace from tissue chilled at around −20°C shows a flattening of the cooling curve which is indicative of ice formation (bottom). In tissue chilled in n-hexane maintained at −70°C (top) the temperature drops rapidly, reaching −40°C in 9 sec. (From J. Chayen, L. Bitensky, and R. G. Butcher, *Practical Histochemistry*. Copyright 1973 John Wiley & Sons, Ltd. Reprinted by permission of John Wiley & Sons, Ltd.)

CHILLING, SECTIONING, AND REACTION METHODS

In order that sections may be cut from a block of tissue it must first be hardened. The conventional histochemical method is to fix the tissue, remove the tissue water, and then embed the tissue in paraffin wax. Obviously, this is of little practical use if enzyme activity is to be measured in sections of such a preparation. Chemical fixation, whether of the precipitation or nonprecipitation type (Baker, 1945), denatures protoplasm. In so doing it changes free reactive groups and may cause considerable inactivation of some if not most enzymes. The processes of dehydration, infiltration with a solvent for paraffin wax, and finally with paraffin wax itself, cause further inactivation of enzyme activity (e.g., Stafford and Atkinson, 1948). Even when fresh tissue is freeze-dried, both this procedure and the subsequent embedding in wax cause considerable inactivation of certain enzymes (Berenbom et al., 1952). Therefore, all these methods are unsuitable for measuring the biochemical response of a target cell to a hormone. The alternative way of hardening tissue is to chill it, but this also has certain pitfalls.

Chilling

It is well known that cooling tissue can lead to the formation of ice crystals. Ice formation will cause both structural and chemical damage within the tissue. Such damage will be caused either by the physical forces involved or by changes in tonicity of extracellular and intracellular fluids brought about by removal of liquid water as ice (Lovelock, 1957). The problems of ice formation in tissue have been reviewed by Chayen and Bitensky (1968). However, rapid chilling under precise conditions was shown to chill tissue to below $-40°C$ without the formation of ice. Tissue could therefore be supercooled and solidified (Lynch et al., 1966). Liquid nitrogen, used at the temperature equivalent to its boiling point ($-195°C$), is often used as the cooling agent. However, any increase in temperature, such as that caused by the immersion of a piece of tissue taken from the body or the ambient temperature of the laboratory, will cause the nitrogen to boil. The piece of tissue will therefore be surrounded by gaseous nitrogen, so effectively insulating the tissue from the chilling effect of liquid nitrogen (Moline and Glenner, 1964). On the other hand, n-hexane, although used at a higher temperature, exhibits no such drawbacks. If the n-hexane ("low in aromatic hydrocarbons" grade; boiling range $67-70°C$) is cooled to $-70°C$ by surrounding it with a slurry of CO_2 and alcohol, then the cooling rate of the tissue is faster than in liquid nitrogen. Tissue taken at room temperature will be cooled to $-40°C$ in under 9 sec without the flattening of the cooling curve indicative of ice formation (Figure 2). In order to ensure adequate chilling throughout, the block of tissue is usually kept to a size of 5 mm^3. After chilling the tissue for 30 sec to 1 min it is removed with cold forceps and stored at $-70°C$ in a dry glass tube. To absorb the excess n-hexane, tissue paper is placed in the bottom of the tube. The segment of the target organ should be used within 3-4 days of chilling to ensure minimal loss of enzyme activity.

Sectioning

Mounting
Tissue segments are mounted on cryostat chucks prior to sectioning. A drop of tap water is placed on a chuck that is cooled by being surrounded by a slurry of

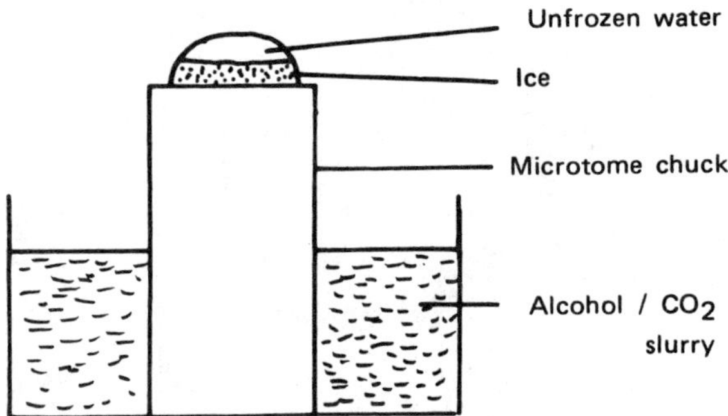

Figure 3 Mounting tissue on a cryostat chuck. The cryostat chuck is placed in a slurry of alcohol and solid CO_2. A drop of tap water is placed on top of the chuck. As the water freezes, the piece of tissue is placed in the unfrozen water, which, on freezing, anchors the tissue on the chuck. Sections should be cut only from the tissue protruding from the ice.

solid CO_2 and alcohol, and allowed to freeze. Before the water finally freezes, the block of tissue is placed in the residual film of unfrozen water, which is then allowed to freeze (Figure 3). In order to make visualization of the unfrozen water easier, a low concentration of toluidine blue (or a similar dye) is added to the water. Upon freezing, the dye changes to a lighter color, so showing up the unfrozen portion. Using this method the block of tissue is firmly anchored to the chuck. Sections should be taken only from the portion of tissue protruding from the ice, because ice damage, caused by thawing and refreezing in the residual water on the chuck, will invalidate any results obtained with that part of the tissue specimen that is embedded in the ice.

Sectioning

Sections are cut in a cryostat, and the model of choice, fitted with either a rocking or rotary microtome, is manufactured by Bright Instrument Company, Huntingdon, England. The cabinet temperature is maintained at least at $-25°C$, and the knife is further cooled to $-70°C$ by packing solid CO_2 around the haft. In this way the heat generated by the act of cutting (Thornberg and Mengers, 1957) is dissipated into the knife, which is a better conductor of heat and is at a lower temperature (Silcox et al., 1965). The sections are then "flash-dried" onto glass slides kept at room temperature. This flash-drying, over a temperature gradient approaching $100°C$, leaves the tissue water on the knife as an imprint of the section. Section thickness can be varied from 6 to 24 μm, and the speed of cutting can be controlled by the use of a motor drive. It is essential that all sections are cut at the same speed, as it has been shown that changes in the speed of cutting

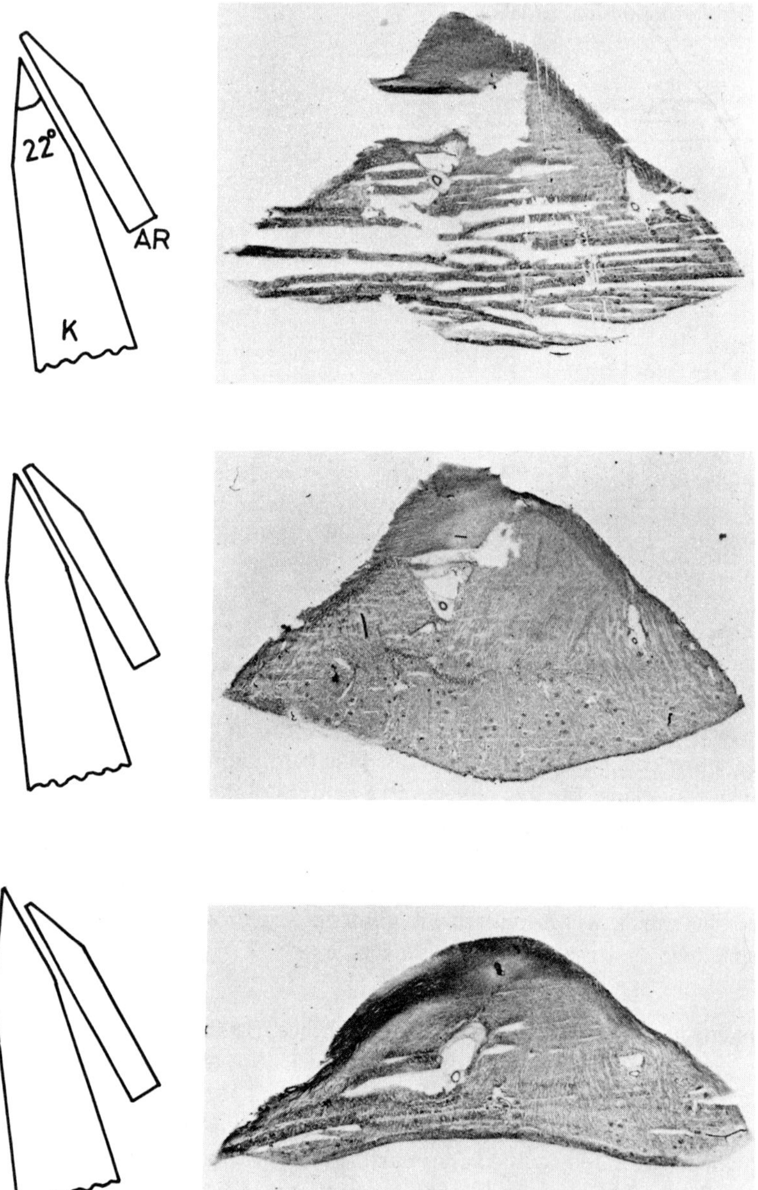

Figure 4 Section cutting. Sections should be cut on knives (K) with a tip angle of between 22° and 25°. Adjustment of the antiroll plate (AR) is critical. If the antiroll plate is too high (top) the sections will appear shattered, while if it is too low (bottom) the sections will appear crushed compared to those sections cut with an antiroll plate that is correctly adjusted so that it is level with or just above the knive (middle).

will alter the thickness of the section (Butcher, 1971a). The use of these techniques allows unfixed tissue sections to be produced with no apparent ice crystal damage. The effectiveness of these techniques has best been demonstrated by Altman and Barrnett (1975), who showed that there was no evidence of ice damage even when such sections were subsequently processed for investigation by electron microscopy.

Setting up the Cryostat

As the sections used for bioassay purposes must be of the highest quality, a brief description of the setting up of the cryostat will be given. Knives should be sharpened so that the angle at the tip of the knife is between 22° and 25° (Figure 4). The sharpness of the knife is critical, and it is often advisable to use a fresh area of the knife for each assay. The knife should be kept clear of ice particles as these will damage the sections. On rocking microtomes it is necessary to ensure that, on the upward stroke, the block of tissue clears the knife before the advance mechanism begins to operate. This is especially necessary in cases where sections are thicker than 12 μm; otherwise, the block is pressed into the back of the knife.

As critical as the adjustment of the knife is the positioning of the antiroll plate. This device ensures that the cut sections remain flat against the knife, as opposed to curling up, after being cut. If the antiroll plate is placed too high then the block of tissue will hit the antiroll plate before the knife, while if it is too low, the section will curl up and not pass beneath the antiroll plate (Figure 4). Ideally, the antiroll plate should be placed level with, or just above, the tip of the knife. Sections cut in a cryostat with a badly positioned antiroll plate will either appear shattered (antiroll plate too high) or crushed (antiroll plate too low). Similarly, sections cut with a knife that is too warm will appear corrugated and will be smaller than the face of the segment being cut. On the other hand, sections that appear shattered and are larger than the face of the segment may have been cut with a knife that was too cold.

Reaction Methods

Sections cut in the way outlined above can now be used for either histological or cytochemical analysis. Common histological staining methods (e.g., hematoxylin and eosin, or the methyl green method for DNA) for investigating tissue components can now be employed. Furthermore, accurate analysis can be made of biochemical activity within the section and this can be related to the histology of a serial section. This is because the biochemical activity should be unaltered by the process of chilling and sectioning. Precise localization of activity by trapping the reaction product at its site of generation is of great importance, so a short discussion of the methods of achieving this will be given.

Trapping Techniques

The aim of all the techniques developed for the measurement of enzyme activity
in the cytochemical bioassay systems is the precise localization of the reaction
product in a form which absorbs light sufficiently that it can be seen and measured. Precipitation or "trapping" reactions fall into three categories.

Postcoupling Techniques. In these methods the enzyme hydrolyzes an artificial substance, cleaving off an invisible, insoluble residue. Such substrates are
the naphthol AS-BI derivatives (Burstone, 1962). The reaction can be performed at the pH at which enzyme activity is optimal. The development of an
azo dye to yield sufficient color, by coupling the naphthol AS-BI residue to a
diazonium salt, is then done at the pH at which the coupling reaction is maximal. An example of this type of method is the reaction for N-acetyl-β-gluco-
saminidase activity (Robertson, 1980).

Tetrazolium Salts. In this type of reaction an artificial soluble substance,
such as a tetrazolium salt, is converted into a very insoluble chromophore.
Upon reduction, by the activity of dehydrogenases, the tetrazolium salt is converted into a highly colored and insoluble formazan (Altman, 1969a; 1972;
1976). An example of this type of reaction is that for glucose-6-phosphate
dehydrogenase activity (Chayen et al., 1973) used in the cytochemical bioassay
of PTH (Chambers et al., 1978a).

In this reaction, glucose-6-phosphate dehydrogenase oxidizes the substrate
glucose-6-phosphate to 6-phosphogluconolactone, which is spontaneously converted to 6-phosphogluconate. The resulting reducing equivalents are transferred
to the coenzyme, nicotinamide adenine dinucleotide phosphate ($NADP^+$), converting it to the reduced form (NADPH). These reducing equivalents would
normally then be allowed to pass down the microsomal respiratory pathway
until they reduce the tetrazolium salt to a formazan (Figure 5). However, this
would not be a measure of the activity of the enzyme but rather a combination
of enzyme activity and respiratory pathway efficiency. The addition of the intermediate hydrogen acceptor, phenazine methosulfate (PMS), ensures that the reducing equivalents are passed directly to the tetrazole, so that the amount of
formazan produced is in direct relation to enzyme activity. The tetrazolium salt
normally used for this reaction is neotetrazolium chloride, which has to be purified by a Soxhlet extraction prior to use (Altman, 1976). This tetrazolium salt
has a redox potential approaching that of oxygen, which will compete with the
tetrazole for the reducing equivalents. Therefore, the reaction medium must be
saturated with nitrogen, to remove any residual oxygen, prior to use. In the reaction medium used for the bioassay of PTH, potassium cyanide (10^{-2} M) is added
in order to chelate any metal ions that may be present; such ions have been
shown to be inhibitory to the enzyme (Glock and McLean, 1953). As glucose-
6-phosphate dehydrogenase is a soluble cytoplasmic enzyme (Glock and Mc-

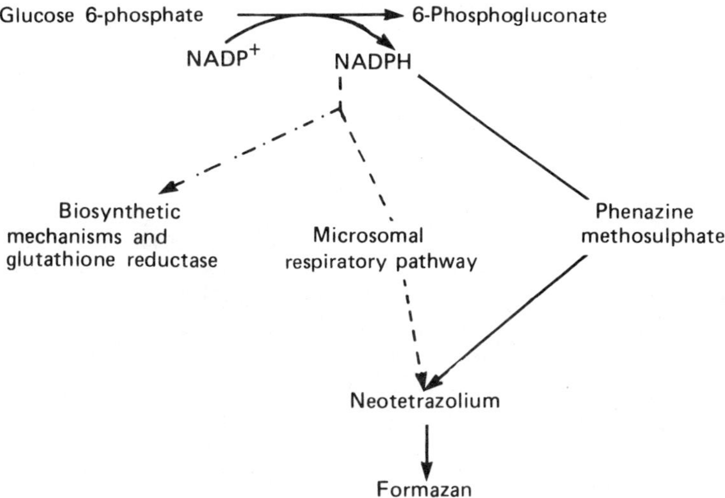

Figure 5 Measurement of glucose-6-phosphate dehydrogenase (G6PD) activity. The substrate, glucose-6-phosphate, is oxidized to 6-phosphogluconolactone, which is spontaneously converted to 6-phosphogluconate. The reducing equivalents produced in the reaction reduce the coenzyme, NADP+, to NADPH. The use of an intermediate hydrogen acceptor, phenazine methosulfate, allows the reducing equivalents to be passed directly to neotetrazolium, so bypassing the microsomal respiratory pathway. Upon reduction, the soluble neotetrazolium is reduced to form an insoluble colored formazan.

Lean, 1953), a colloid stabilizer is included in the medium to prevent the loss of enzyme activity that would normally occur if the section were immersed in an aqueous medium. The use of colloid stabilizers will be discussed later.

Simultaneous Capture Reactions. This type of reaction is that in which the enzyme liberates a soluble reaction product which is precipitated by a trapping agent included in the reaction medium. These fall into two categories, depending on whether the precipitated reaction product is colored. The first is typified by the use of diazonium salts for trapping either true naphthol, or a derivative of naphthol, as in the reaction for naphthylamidase activity. In this reaction, the substrate is a naphthol derivative (such as leucyl-β-naphthylamide) and the reaction medium contains a diazonium salt, such as tetrazotized dianisidine (Fast Blue B). The result is an azo dye caused by the coupling of the naphthylamine, released by enzymatic hydrolysis, to the tetrazotized dianisidine.

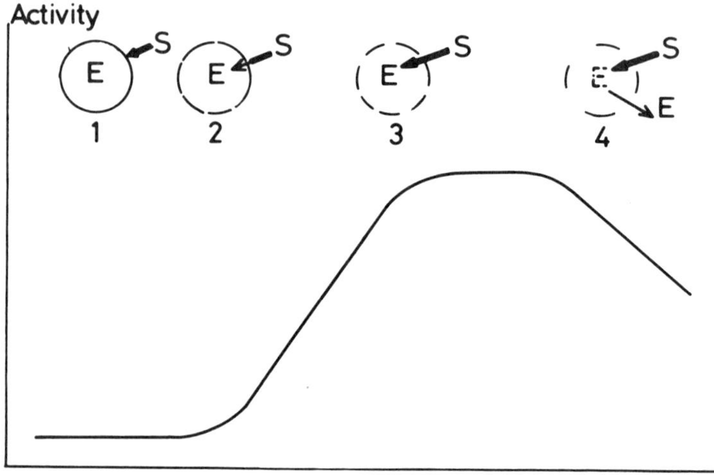

Figure 6   Measurement of lysosomal enzyme activity. State 1: The hydrophilic substrate will not be able to penetrate an intact lysosome, so no enzyme activity can be demonstrated. State 2: Upon labilization of the lysosomal membrane, some substrate will be able to enter the lysosome, so allowing some enzyme activity to be measured. State 3: Further labilization of the membrane will allow unimpeded entry of substrate so disclosing the "total" enzyme activity within the lysosome. State 4: At this point the labilization has proceeded to such a degree that the enzyme is no longer retained within the lysosomal membrane and disperses into the reaction medium, resulting in an apparent decrease in enzyme activity.

Such a method is used in the bioassay of TSH (Bitensky et al., 1974). In this assay it is not the activity of the lysosomal enzyme per se that is being measured. Rather, it is a measure of the increased permeability of the lysosomal membrane, in response to TSH, that permits a higher rate of entry of substrate into the lysosome which in turn results in a higher enzyme activity. Such increased labilization of the lysosomal membrane can be demonstrated by the use of a relatively hydrophilic substrate, such as leucyl-β-naphthylamide, which will not readily diffuse into the lysosome (Bitensky and Chayen, 1977). As the lysosomal membrane becomes more labile, so more substrate is allowed to enter with the consequential increase in enzyme activity. If required, it is possible to measure the "total" activity of the lysosomal enzyme by preincubating the sections in 0.05 M acetate buffer, pH 5.0, at 37°C prior to the reaction. Increasing time of acid preincubation will cause labilization of the lysosomal membrane until it is labilized to such a degree that substrate entry is not impeded. Further preincubation in acidic conditions will cause further labilization, which will allow the enzyme itself to leach out of the lysosomes with a consequential diminution of activity as the enzyme becomes dispersed in the reaction medium (Figure 6).

The second type of simultaneous capture reaction involves the precipitation of the product of enzymatic activity by a metal ion, such as lead, calcium, or cobalt (e.g., Danielli, 1953). For instance, in the reaction for carbonic anhydrase activity (Loveridge, 1977; 1978) as used in the bioassays of gastrin (Loveridge, 1977; Loveridge et al., 1974; 1978; Loveridge, Hoile, et al., 1980; Hoile and Loveridge, 1976; 1977), cobalt hydroxide is precipitated. These metal salts are usually colorless and so have to be converted to a colored end product, usually the sulfide. Care should be taken when using these metal trapping agents, for two reasons.

1. Most metal salts will also become attached to the tissue in a nonspecific way. Therefore, before conversion to the sulfide, this spurious attachment must be removed by washing.
2. Some enzymes are inhibited by heavy metals, and care should be taken in the choice of a particular trapping agent. Recently a new compound, lead ammonium citrate acetate (LACA), has been developed in which the inhibitory effects of lead are overcome by complexing it with ammonium, citrate, and acetate ions without diminishing the efficacy of the lead as a trapping agent (Chayen et al., 1981). This compound has been used to measure the activity of (Na^+ $-K^+$)-stimulated adenosine triphosphatase used as the end point in the assay of arginine vasopressin (Baylis et al., 1980) (see also Chapter 10).

With all types of precipitation reactions, care must be taken to ensure that the reaction product is precipitated as closely as possible to the site of reaction. This is achieved by ensuring that the enzyme will produce as much reaction prduct as possible (i.e., it is given optimal conditions for maximal activity), and that the trapping agent is present in a high enough concentration, so that the solubility product of the intended precipitate is rapidly exceeded (Danielli, 1953).

The Use of Colloid Stabilizers

Most of the enzyme activity bound within subcellular organelles will be demonstrated solely by immersing the section in the reaction medium. For instance, the activity of mitochondrial enzymes such as succinate dehydrogenase can be quantitatively examined by the addition of substrate and a tetrazolium salt to a solution buffered to the correct pH. Needless to say the conditions, such as substrate concentrations, should be optimal for the particular enzyme being studied.

In contrast, the "soluble enzymes" (i.e., those not bound to a subcellular organelle) will dissolve out and become dispersed in the reaction medium as soon as the section is immersed in an aqueous solution at a pH close to neutrality. Therefore, the activity of enzymes, such as the dehydrogenases of the pentose shunt pathway, cannot be demonstrated without the use of additional precautions. In order to prevent this solubilization, the section must be protected with some form of stabilizer. Two commercially available colloid stabilizers have been used successfully: either particular grades of polyvinyl alcohol (Altman and Chayen, 1965; Altman, 1971; Henderson et al., 1978) obtained from Wacker Chemie Ltd., Walton-on-Thames, or a polypeptide obtained by the partial degradation of collagen, Polypep 5115 (Sigma) (Butcher, 1971b). With the former, Altman (1969b) was able to measure the same amount of glucose-6-phosphate and 6-phosphogluconate dehydrogenase activity (μmole hydrogen per unit volume of tissue per unit time) as was found using conventional biochemical procedures applied to samples of the same tissue.

To economize on the use of expensive reagents the volume of reaction medium used is kept as small as possible. By encircling the tissue section, on the slide, with a Perspex ring and pipetting the medium into the cavity, as little as 0.5 ml of reaction medium is required for each section.

SECTION ASSAYS

The techniques described up to this point have been concerned with the use of quantitative cytochemistry for segment assays. In general, however, the segment assays have a very limited throughput of samples. For instance, in the case of both ACTH and TSH it is only possible to assay a single sample (at two dilutions) in each assay, due to the limiting size of the target organ. In order to improve this throughput, some of the assay systems have been modified so that instead of exposing segments of the target organ to the hormone, sections of that organ can be used instead. In the case of ACTH, these procedures have increased the number of samples able to be measured by a single worker from around 3 a week with the segment assay to 30 a week with the section procedure. Generally the techniques remain the same, although certain changes have been necessary and these will be considered for each part of the basic procedure.

Maintenance Culture

In general this stage of the procedure is the same as for segment assays with the exception that larger segments of the target organ are used. Thus, where each adrenal gland or thyroid gland would be cut into three for the segment assay, they are only halved for the section assay.

In the case of gastrin-like activity, the pH of the culture medium has been left at pH 7.6 rather than being reduced to pH 7.0 as in the segment assay procedure (Loveridge, Hoile, et al., 1980). However, in certain cases it has been found that fasting the animals can replace the maintenance culture (Hoile, 1979; Hoile and Loveridge, 1976; 1977; Klaff, 1981) (see also Chapter 9). Thus for this assay the tissue is chilled directly upon removal from the fasted animal. Similarly, the difficulties encountered with the culture of ovarian segments has been overcome by selecting animals at the correct stage of the estrous cycle so that maintenance culture is not required (Buckingham et al., 1979a,b; Buckingham and Hodges, 1981).

One of the interesting features of the section assay system is that in the majority of cases it has been found necessary to expose the tissue segments to a low concentration of hormone prior to chilling. The concentration of hormone used is usually one-tenth of the lowest concentration used in the segment assay. This "priming" procedure was shown to improve the sensitivity of the ACTH section assay (Alaghband-Zadeh et al., 1974) and the section assay for gastrin-like activity (Loveridge, Hoile, et al., 1980) where culture is used. In the case of the LH bioassay (Buckingham and Hodges, 1981) the priming was done after removal of the ovary from the animal. However, it was not found necessary to prime the segments of thyroid in the section assay for thyroid stimulators (Chayen et al., 1980). The time of exposure to the priming dose is the same as for the hormone exposure in the particular segment assay (e.g., 4 min for ACTH). After the culture period, and priming if necessary, the segments are chilled as normal.

Chilling and Sectioning

After chilling, the segments are sectioned at a thickness sufficient to encompass a cell. Thus for ACTH the section thickness is 20 μm as opposed to 12 μm in the segment assay, while for the TSH and gastrin section assays, the sections are 12 and 18 μm thick, respectively. The tissue must be used within 2-3 days of chilling, as the ability of the cells to respond to the hormone decreases even with storage at -70°C. For the same reasons, sections should be used within 2 hr of being cut.

Exposure to Hormone and Reaction Procedures

Apparatus

Whereas in the segment assay system the sections were reacted directly for the biochemical response, in the section assay procedure they first have to be exposed to hormone or to a dilution of the sample to be measured. In order to achieve this, special sets of apparatus have been designed. In the case of ACTH, TSH, and LH, the sections are held vertically while being exposed to hormone, each pair of sections being treated with either a concentration of the

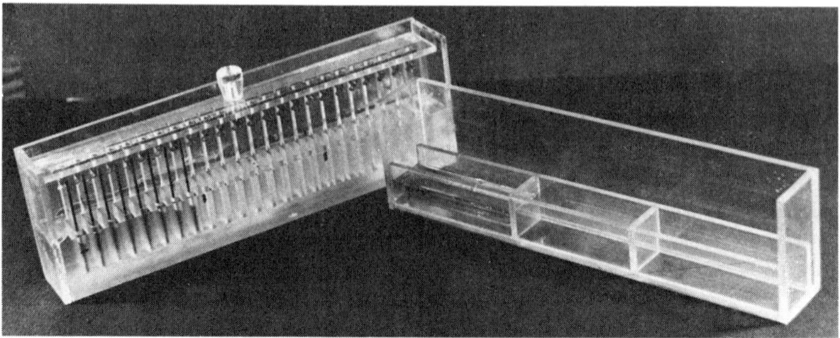

Figure 7   Section assay apparatus. For the assays of ACTH, TSH, and LH the
sections are held vertically in a rack and immersed, in duplicate, into separate
compartments, each containing a concentration of the standard hormone preparation or a dilution of the unknown. At the end of the exposure time the transfer of the rack to a second trough simultaneously immerses the sections in the
reaction medium.

hormone standard or a dilution of the unknown (Figure 7). The sections can
then be transferred simultaneously to the reaction medium. Because of the need
to restrict the depth of reaction medium in the cytochemical method for carbonic anhydrase activity (Loveridge, 1978), the sections for the assay of gastrin-
like activity must be kept horizontal during exposure to hormone (see Chapter
9). Each section is exposed to hormone or unknown, delivered from syringes
placed directly above the section. At the end of the exposure time, the trough
is flooded with the reaction medium for carbonic anhydrase activity.

The Need for Tissue Stabilizers

As discussed previously, the immersion of an unfixed tissue section in an aqueous medium at a pH close to neutrality will result in a substantial loss of soluble
material (Altman and Chayen, 1965). Immersion of the sections in T8 medium,
while exposing them to the hormone, would result in the loss of enzyme activity
or the particular moiety used as the end point of the assay. However, if the section is completely stabilized as it is, for example, in the reaction medium for
glucose-6-phosphate dehydrogenase activity, the cell membrane is also so stabilized that it does not respond to the attachment of the hormone. Therefore, a
delicate compromise must be reached whereby enough retention of cellular material is achieved but the cell can still respond to the hormone. In the case of
ACTH and LH, 5% Polypep 5115 (Sigma) is used, while in the gastrin and TSH
assays, which do not depend on strictly "soluble" enzymes, a concentration of

gum tragacanth (0.02 and 0.005%, respectively: Symourgh International Inc., 295 Fifth Avenue, New York, N.Y. 10016) is used to help maintain the integrity of the sections.

Time and pH of Exposure to Hormone

The times of exposure, to obtain a maximal response to a hormone, are much shorter in the section assay system than for the segment assays. In general the time of exposure is related to the time of exposure used in the segment assay procedure. Thus the exposure times for ACTH, gastrin, LH, and TSH are 60, 75, 80, and 90 sec, respectively (compared to times of 4, 5, 4, and 7 min, respectively, for the corresponding segment assays). In those assays where ascorbate depletion is used as the end point, 10^{-3} M sodium ascorbate is added to the carrier medium.

Although the time of exposure to hormone is much shorter for the section assay systems, changes of pH of the carrier medium must still be avoided. In the case of the gastrin assay, this is achieved by replacing the T8 medium by 0.1 M Hepes buffer, pH 7.0. In the case of LH and of ACTH, the addition of the 5% Polypep increases the buffering capacity of the medium so that the pH of the medium remains stable. However, in the TSH assay, some of the bicarbonate is removed by the addition of 1 N hydrochloride acid; the carbon dioxide content of the medium is subsequently equilibrated with that of the atmosphere (Chayen et al., 1980). This ensures that the pH remains stable throughout the time of exposure.

Reaction Methods

The methods for measuring the biochemical response to the hormones are generally the same as in the respective segment assay. The only changes that are made are to the time of the reaction, which is shorter in those cases where the thickness of the sections has been increased or if the reaction is done at $37°$C as opposed to room temperature.

MEASUREMENT

Up to this point in the assay procedure no attempt has been made to isolate the cells responsive to a particular hormone from the rest of the cells within the tissue section. As has been mentioned previously, this isolation is achieved during the process of measurement. Only the biochemical activity within the target cell is measured, thus ensuring full measurement of the response without any "dilution" of that response by the surrounding nonresponsive cells (Chayen, 1978a; Chayen and Bitensky, 1980). Hence it is possible, by optically isolating the cell, to measure enzymatic activity or the amount of a particular reactive moiety only in those cells responding to changes in their environment but without prior disruption of the tissue matrix. This process is achieved by scanning and integrating microdensitometry.

Spectrophotometry

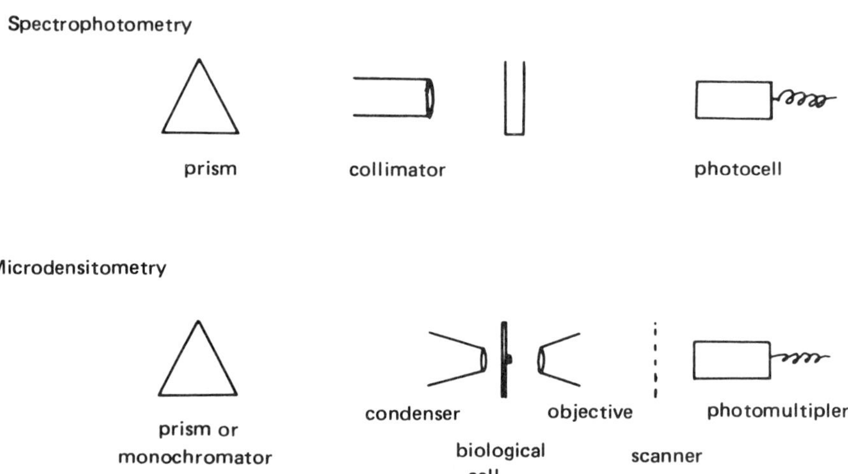

Figure 8 A schematic comparison of microdensiometry and spectrophotometry.

Scanning and Integrating Microdensitometry

The problem of measuring the intensity of color in cases where it is inhomogeneously distributed are well known. Gomori (1952) showed that an error of over 50% was incurred if, instead of measuring an evenly distributed density, the same density was concentrated into several areas making up only 10% of the whole area measured. Similarly, Chayen (1978b) calculated that if a chromophore with an extinction of 0.097 (which would transmit 80% of the light incident upon it) were precipitated so as to occupy only half the field, a simple measurement would result in an 11.3% error. However, if the chromophore was only distributed in 10% of the field, then the error would rise to 58%. Therefore, it was theoretically possible that two cells, one with a greater amount of chromophore distributed in small areas and one with a lower amount of chromophore more evenly distributed, would give similar readings. These errors have been fully discussed by Chayen and Denby (1968) and by Bitensky (1980).

This problem of inhomogeneity was overcome by Deeley (1955; Deeley et al., 1954; 1957), who developed the scanning and integrating microdensitometer. By adapting techniques used by Caspersson (1947) for the microscopic measurement of the absorption of nucleic acids at 265 nm, Deeley showed that it was possible to accurately measure a colored precipitate within a cell. He divided the total area to be measured into areas which were smaller than the optical resolution of the microscope; each such area is then optically homogeneous. The absorption of such areas could then be measured by means of a photomultiplier and the results integrated to give the total absorption over the whole area

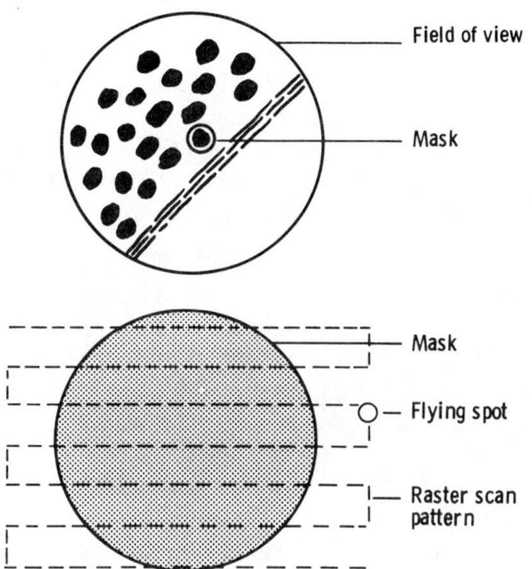

Figure 9 Scanning and integrating microdensitometry. The area or cell to be measured is selected and covered by the mask (top). The mask is scanned in a raster pattern by the flying spot of selected wavelength (bottom).

being measured. A schematic comparison of microdensitometry and spectrophotometry is shown in Figure 8. Initially the microdensitometer was applied to the measurement of the Feulgen reaction for DNA; Deeley (1955) showed that the accuracy of such measurements was ±3%. Carrying the idea a stage further it seemed reasonable to use other stoichiometric reactions to measure the cellular content of other reactive groups, such as arginine (McLeish et al., 1957). More recently, the technique has been extended to the measurement of enzyme activity (Chayen, 1978a; Bitensky, 1980).

Although microdensitometers are commercially available from several sources (Altman, 1975), the instrument that is optically acceptable (Bitensky, 1980) and that is therefore used for the cytochemical bioassays, is the Vickers M85 scanning and integrating microdensitometer (Vickers Instruments Ltd., Haxby Road, York, England). With this instrument, the field to be measured is selected by means of a masking system (Figure 9A) and then scanned in a raster pattern by a flying spot of a selected wavelength (Figure 9B). The measuring wavelength is usually the absorption maximum of the chromophore, and this can be previously determined by measuring the absorption at selected wavelengths. However, the tetrazolium reactions are measured at the isobestic point (585 nm) of

the absorption due to the two formazans produced (Butcher and Altman, 1973).
At each point in the raster scan pattern the absorption is measured, and all the
measurements are then integrated to give a measure of the total absorption within the mask. The figures displayed by the machine are standardized to units of
absolute extinction by reference to a standard calibration graph (Bitensky,
1980). The results so obtained, expressed as mean integrated extinction, make
possible the comparison of results obtained from different machines.

Practical Considerations

Mounting Medium

Before any section can be examined under a microscope, the sections must first
be covered. If no mounting medium is used, the space between the cover glass
and the slide is filled with a mixture of tissue section and air which will result
in scatter of the incident light caused by the different refractive indices of the
materials through which the light passes. Such scatter of light out of the microscope system will result in a loss of transmitted light which will be falsely inter-
preted as absorption. This can be reduced to negligible proportions by the use
of a mounting medium which has a refractive index similar to that of the tissue.

The type of mounting medium used will depend upon the chromophore being
measured. If the chromophore is soluble in water (even at a low rate), then the
section must be mounted in a nonaqueous medium, such as Styrolite (R. A.
Lamb Ltd.) or DPX. On the other hand, some chromophores (such as the formazans) are soluble in organic solvents and so should be mounted in an aqueous
medium, such as Farrants' medium.

Sections generally should be mounted only a few hours before use to avoid
the possibility of diffusion of the precipitated chromophore away from the site
of reaction.

The Vickers M85 Microdensitometer

This microdensitometer is the machine most commonly used for cytochemical
bioassays. Basically it consists of a scanning system and a spectrophotometer
built around a microscope. Therefore the setting up of the machine before use
should be done in three stages.

Setting up the Microscope.   The optical system of the microdensitometer is
designed to be set up for Köhler illumination. The section to be measured is
placed on the stage, and the objective to be used is selected. If a low-power objective is used ($\times$10), then the top lens of the condenser must first be removed.
The binocular eyepieces should be adjusted for the correct interocular distance
of the operator and each eyepiece focused individually. This is achieved by focusing upon the mask which is superimposed on the field of view. After focusing the objective, the condenser is centered by first closing down the aperture

iris and focusing the diaphragm in the plane of the section. The image of the diaphragm should be symmetrical when the condenser is raised or lowered. If this is not the case, then either there is dirt in the condenser carriage or the light is traveling obliquely for other reasons. The condenser should therefore be removed and cleaned. The condenser should now be centered and the aperture iris opened. To focus the condenser, the field iris is stopped down and the condenser refocused to give an image of the diaphragm in the plane of the section. This diaphragm can be centered if necessary and the field iris diaphragm opened so that it just fills the field. The focus of the condenser is critical, as in the measuring mode the microscope system is inverted so that the objective is used as a condenser and the condenser as an objective.

If a ×100 objective is being used, then both the top of the condenser and the slide must be oiled prior to contact with the slide and the objective, respectively. Therefore, oil should be applied to the top of the condenser before the slide is placed on the stage. The condenser should be racked down so that there is no initial contact between the slide and the oiled condenser. Having located the section with a low-power objective, the top of the slide should then be oiled. The ×100 objective can now be used and focused. To ensure sufficient light, the condenser should be raised but not to the point where contact is made between the oiled condenser and the slide until the condenser is being focused. The procedure for centering and focusing the condenser is the same as described previously, but it should be noted that good contact between the oiled slide and the objective and between the oiled condenser and the bottom of the slide should be made. Once the system is prepared, the contact between the oiled surfaces should not be broken, as this may introduce air bubbles which will affect the recorded density even though they may not be directly in the light path.

Setting up the Scanning System. At this point the microdensitometer is adjusted so that the correct area of the section is scanned by a flying spot of the selected wavelength. To do this the size of mask (ranging from an effective diameter of 660 μm, with the largest mask and a ×10 objective, to 1.5 μm, with the smallest mask and a ×100 objective) is selected and placed in the center of the field. The mask diameter should be such that it is just filled by the area of cell to be measured. The size of the raster scan frame, which should be as small as possible while still scanning the mask, is then selected. The mask should be positioned in the center of the scan pattern. In order to visualize the scanning spot, it is best to set the scanning wavelength to around 550 nm. The scanning spot should then be immobilized by switching to the manual mode and the spot centered within the mask. A spot size of 1, the scanning wavelength to be used (usually the absorption maximum), and a slit width that is the smallest possible which allows the use of spot size 1 at the selected wavelength, are now selected. The position of the stationary spot in the mask should be checked occasionally and adjusted if necessary.

Setting up the Measuring System.   With the mask on a clear field and the scanning spot immobilized, the machine should be turned to "scanning." The position of the halogen lamp should then be adjusted so that the maximum amount of light is incident upon the specimen. This involves the adjustment of the vertical and horizontal controls so that the needle on the absorption meter is maximally deflected to the right. The set 0 control should then be adjusted so that the absorption is around 0.06. The machine should not be set to zero, as in some instances the machines are not adjusted to be linear at low levels of absorption in order that any extraneous "noise" is eliminated. By blocking the light path to the photomultiplier, using an intermediate stop in the filter tray, the 100% absorption level can be set by adjustment of the infinity control. The gating level is then adjusted to 5 μa so that the machine starts integrating the densities measured by the flying spot at a point when the spot is halfway inside the mask. Integration stops when the spot is halfway out of the mask at the end of each scan line. The scanning spot is then set to automatic and the "integrate" button pressed to give a reading of the density of the blank field. The machine is then ready for use.

Measurement of Mean Integrated Extinction

Measurement of the density of the chromophore in the target cell, or zone, is made by moving the microscope table so that the mask covers the particular zone or target cell within the section. The selection of a particular cell or zone is done in relation to the histology of the section. It would be pointless to make measurements either in areas that are known to be unresponsive to the hormone being assayed (i.e., the adrenal medulla in the case of ACTH) or in any area of the section which appears to have been damaged by the preparative procedures. The relative absorption in 10 selected areas is measured in each duplicate section. The reading from the blank field is then subtracted from the mean value of the readings taken within the section, and the resultant relative absorption is converted to mean integrated extinction (MIE). This is necessary because the reading displayed by the machine is in arbitrary units and varies with the diameter of the mask, the sensitivity of the photomultiplier, the scan frame size, and the wavelength being used. Thus, the relative absorption is a measure of the total absorption within the mask as measured with the particular machine settings being used. Furthermore, the relative absorption for the same mask size will vary from machine to machine depending on the sensitivity of particular photomultipliers. Conversion to mean integrated extinction involves the relation of the relative absorption obtained from the section to the relative absorption of a density of 1.0 with the same machine settings, Thus,

$$\text{MIE} = \frac{\text{relative absorption of the chromophore}}{\text{relative absorption of a density of 1.0}}$$

and the MIE of a particular section should be the same whatever machine is being used, so allowing comparison of the results obtained with different microdensitometers. For convenience, the MIE is usually multiplied by 100 to avoid continual use of a decimal point.

In some instances it is necessary to do additional measurements in order to get a more accurate measure of the chromophore or of the level of enzyme activity. For instance, in the assays of ACTH and LH, a considerable degree of light scatter, which will contribute to the total absorption, occurs in the sections of adrenal and ovary. Thus, to compensate for this, similar readings are made both at the absorption maximum of the chromophore (680 nm) and the absorption minimum (480 nm) and converted to mean integrated extinction. The subtraction of MIE at 480 nm from that at 680 nm allows the degree of light scatter to be accounted for in the final readings.

In the assay of gastrin-like activity, another problem arises in that the cobalt salt used as a trapping agent is itself bound, nonspecifically, to the tissue. To compensate for this, readings are made in the muscle layer, where carbonic anhydrase activity is low or absent (Maren, 1967), as well as in the parietal cells. By subtracting the mean of the readings in the muscle layer from that obtained from the parietal cells, the nonspecific adsorption of cobalt is accounted for.

INTERPRETATION

In order that valid results are obtained from an assay system the properties of that system and any possible means of interference must be fully understood.

Properties of the Cytochemical Bioassay System

Biphasic Time Response

A major feature of the cytochemical bioassay (CBA) system that has to be taken into account is the biphasic nature of the responses. In all the CBA systems so far developed, the response has been biphasic with time. Thus, in the assay of AVP, Baylis et al. (1980, see also Chapter 10) showed that $(Na^+ -K^+)$-ATPase activity was increased by the hormone up to a maximum at 5 min. This was followed by a return to a basal level of activity and a second increase in activity which reached a peak at 17.5 min. Varying the concentration of the hormone did not significantly alter the timing of the second peak of activity. Similar twin phases of responses to hormone have been shown for PTH (Chambers et al., 1976). In the segment assay of ACTH the second phase of response was dependent upon the concentration of hormone (Chayen et al., 1974). Thus, hormone concentrations of 50 and 500 pg/ml showed twin responses but lower concentrations only showed a single response. In the section assay system of the same hormone, the twin responses were again evident but the second response was much more sustained than the first (Chayen et al., 1976).

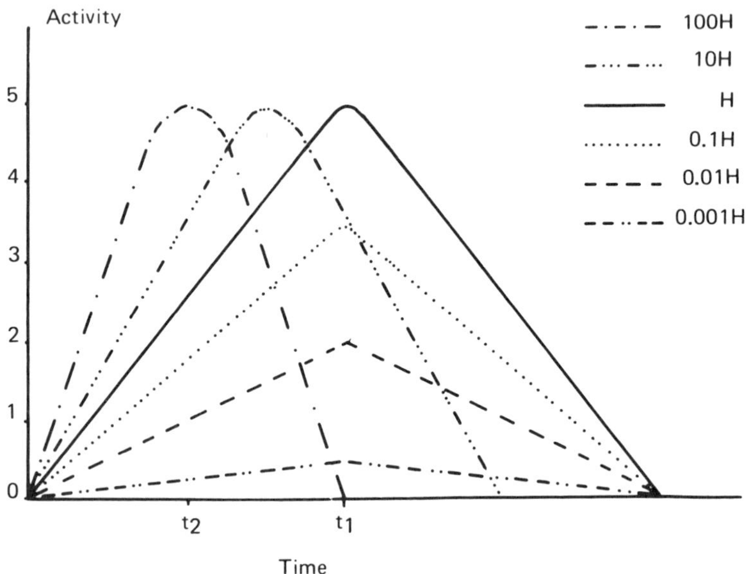

Figure 10 Hypothetical time courses of the responses to varying concentrations of hormone. At concentrations of 0.001 H to H, the time courses of the response are the same but the magnitude of the responses is dependent upon the concentration of hormone. Higher concentrations of hormone (10 and 100 H) elicit a response of the same magnitude but at earlier times.

With the exception of the assay for AVP, the assay systems so far developed have been based on the primary response to the hormone. The speed of this response, especially with section assays, indicates the need to be exact about the timing of exposure of the section or segment to the hormone.

Unlike the secondary response in the assay for AVP, it would seem that high concentrations of hormone cause a quicker primary response. Thus in the segment assay of ACTH (Chayen et al., 1974: Figure 8), extrapolation of the time-response curves for different concentrations of the hormone show that 500 pg/ml gave a maximum response at 2.5 min, 50 pg/ml gave a response of the same magnitude at 4 min, while concentrations of 5 and 0.05 pg/ml gave responses of different magnitude but at the same time of 5 min. Similarly, Loveridge et al. (1979) showed that in the segment assay for thyroid stimulators, concentrations of TSH of 10 and 1 μU/ml gave peak responses of the same magnitude at 3 and 6 min, respectively. In the same system, Bitensky et al. (1974) had previously shown that concentrations of 4×10^{-2} and 4×10^{-4} μU/ml gave peak responses of differing magnitudes at the same time of 7 min.

Interestingly, Ealey et al. (1981) showed that in the section assay of thyroid stimulators the time courses of 10^{-1} and 10^{-3} μU/ml of TSH were distinctly different but gave the same magnitude of response. Thus, there may be certain differences between the responses in the segment and section assay procedures, at least for TSH.

Therefore it would appear that, in segment assays at least, high concentrations of hormone elicit their effect by shortening the time necessary to achieve maximal stimulation. Beyond a certain dilution, the time to achieve maximal stimulation is unaltered by further dilution, but the responses show dose relationship (Figure 10). It is obvious that the assays should be done only over the range of concentration at which there is a dose response, with no change in the time of response. It should be understood, therefore, that where the discrimination between two stimulators relies upon a separation of the time at which they elicit their maximal effect, the discrimination may be upset if one or other of the stimulators is present in inordinately high concentrations. Such a situation could, in theory, occur in the assay of thyroid stimulators (Bitensky et al., 1974; Petersen et al., 1975) if the concentration or potency of thyroid-stimulating immunoglobulins (TSI) were high enough to cause a peak response before or at the same time as the peak response to TSH (Loveridge et al., 1979). In this situation the discrimination could be achieved by the use of a specific antibody to the polypeptide hormone, so determining whether the effect is due to the hormone or the immunoglobulins.

Biphasic Dose Response

Although the biphasic nature of the time response to the hormone presents interesting conceptual information in terms of the mode of action of polypeptide hormones, it does not affect the assay system if its effects are taken into account. Conceptually, at least, the standard calibration graph of any assay system could be based either on the time taken to produce a maximal response for each hormone concentration or on the magnitude of the response to each hormone concentration at a single time of exposure. It is usually more convenient and conventional to use a single time of exposure and measure the magnitude of the response. With the cytochemical bioassay system the biphasic time courses can result in biphasic dose responses. Thus, with the hypothetical situation described in Figure 10, if the time t_1 was used as the time of exposure, then the dose-response graph would be biphasic with a linear increase up to a concentration of H and a decrease in response as the concentration was increased to 100 H (Figure 11A). If, on the other hand, time t_2 was chosen, then the increase would be maximal at 100 H (Figure 11B), although in all probability higher concentrations would show a decrease in activity. However, it should be noted that the slope of the dose-response graph at time t_2 is less than that at time t_1. Such

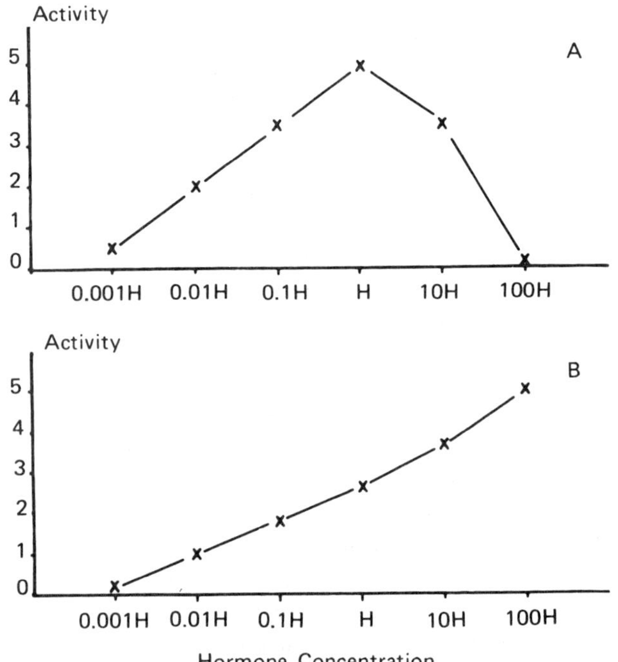

Figure 11   The effect on dose-response graphs if the times of exposure of t_1 and t_2 from Figure 10 were used in an assay. At time t_1 (A) the dose response is biphasic. However, at time t_2 (B), the response to hormone is not biphasic but the slope of the response is considerably less.

biphasic dose responses (Figure 11A) have been described for the cytochemical bioassay systems of ACTH (Chayen et al., 1974), gastrin (Loveridge, Hoile, et al., 1980), and thyroid stimulators (Ealey et al., 1981).

"False-Negative" and "False-Positive" Results

The preceding discussion on the biphasic nature of the time and dose responses should have made clear the possibility of obtaining false-negative results if the concentration of hormone in a given sample is too high. In common with all other assays it is essential, therefore, that the sample is diluted until it elicits a response parallel to the standard graph. Single-point determinations should be avoided, as no reliance can be placed upon such estimations. If a response is obtained that is the reverse of that of the standard graph, it is quite possible that the concentration of hormone within the plasma is too high and therefore higher dilutions must be used in subsequent assay.

In general, false-positive results have not been reported for these assay systems, although in the case of the gastrin section assay described by Hoile and Loveridge (1976; 1977) the response to graded concentrations of the hormone was not always a simple stimulation of carbonic anhydrase over the basal level of activity. Thus, occasionally the level of activity in sections exposed to carrier medium alone was higher than that in sections exposed to some of the standard hormone concentrations. If single-point estimations had been used for a sample containing little or no gastrin, then a false level of gastrin-like activity might have been obtained. However, measuring the sample at two dilutions would have given no change in activity between different concentrations of the sample, which would have been indicative of the absence of gastrin. This problem with the gastrin section assay system has now been overcome (Loveridge et al., 1978; Loveridge, Hoile, et al., 1980) (Chaper 9).

Thus, in summary, it is essential that assays are done with at least two dilutions of the unknown, and preferably three if there is some doubt about the approximate concentration of hormone in the sample.

Interference

When dealing with any bioassay system, it should be remembered that the assay is measuring a functional capacity rather than the number of molecules present. For instance, in certain cases of central hypothyroidism, while patients might be secreting immunoreactive TSH, the biological activity of that TSH is reduced (Faglia et al., 1979). Conversely, it may well be that the apparent hormone activity present in a given sample may not be due solely to a particular hormone. The biological response of the target cell will be the sum of the activity of all the agents present, whether stimulatory or inhibitory, that affect the biochemical response at the particular time of exposure being used. A certain degree of specificity is conferred upon the system by the choice of a single biochemical parameter and a single time of exposure as well as the cell type being used. However, this does not totally rule out interference by other hormones or agents which may have similar effects or modulate the response at some intermediate point between the process of hormone binding to a receptor and the biochemical parameter being used as an end point.

Effect of Other Hormones

In general, the assays so far developed have proven to be remarkably specific in their response to a particular hormone. For instance, in the assay of PTH, the effect of six other hormones was tested, none having any significant effect on the end point of the system, namely, glucose-6-phosphate dehydrogenase activity (Goltzman et al., 1980). However, it should always be borne in mind that while a particular hormone may not stimulate a target cell to the same degree as the hormone under examination, it might affect the response of the cell to that hormone. Thus, in the case of gastrin, cholecystokinin (CCK, which has a sequence homology with gastrin at the biologically active C-terminus), while being 1000-fold less active on its own, markedly inhibits the parietal cell response to gastrin if present at the same time (Loveridge, Hoile, et al., 1980). However, this interference is manifested by a lack of parallelism between dilutions of the standard gastrin preparation and the unknown. Similar cases of cross-reactivity between hormones with a sequence homology has occurred with ACTH and β-melanocyte-stimulating hormone (βMSH), although the latter was 10^5 times less active (Holdaway et al., 1974). In this case, the effect of simultaneous exposure was not examined.

In developing new assay systems, therefore, it should be noted that the specificity of the response should be examined not only by testing the solitary effect of the other hormones likely to have a similar action but also by a combination of the other hormones, with the specific hormone that is to be assayed.

Other Forms of Interference

While it is possible, when establishing an assay system, to test the effect of other hormones or agents which might affect the response, there is always the possibility of encountering an assay result indicative of interference by unknown moieties or agents. Present experience has indicated that the type of interference falls into one of several categories.

1.   Where the unknown moiety alters the time course of the response to the hormone under test: This will usually be manifested by a deviation from parallelism, possibly to the extent of a reversed slope of activity induced by the unknown sample compared to that of the standard curve. An example of this is the effect of acetylcholine and histamine on the parietal cell response to gastrin with the segment assay system. Both these agents shortened the time necessary to achieve maximal stimulation of carbonic anhydrase activity, this being dependent upon the concentration (Loveridge, 1981; Loveridge et al., 1978). The possibility of such interference can be tested for simply by looking at a time-activity response of one or more dilutions of the unknown.

2.   Where the sample contains a high proportion of inactive hormone which results in a lower than expected estimate of biological activity; in general such a sample will elicit a parallel response, but the possibility of the presence of inactive hormone can be tested by a simple recovery experiment. This entails the addition of exogenous hormone to the sample prior to assay. If the inactive hormone is not binding to the receptor, then total recovery of the exogenous hormone should be achieved. On the other hand, if binding between inactive hormone and receptor does occur, then full recovery will not be achieved until

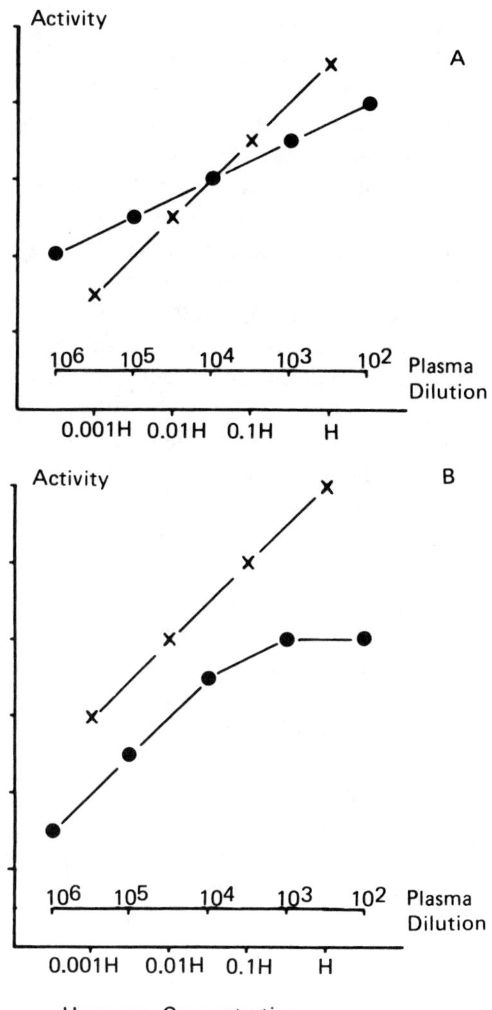

Figure 12 Hypothetical results to show some possible aberrant responses. Response to the dilutions of unknown (● ——— ●); response to concentrations of the standard (x ——— x). (A) The effect of a partial agonist which elicits a response not parallel to that of the standard. (B) The effect of diluting out an inhibitor, the presence of which results in a nonparallel response at low dilutions. But as the dilution of the unknown sample increases, the response becomes parallel to that of the standard hormone preparation.

enough exogenous hormone has been added to overcome the effect of the inactive hormone. The possibility of such a situation has arisen in a study of pseudohypoparathyroidism type 1 (PSP1), where the discrepancy between immunoreactive PTH and biologically active PTH has been shown to be much greater than in primary hyperparathyroidism (Nagant et al., 1981a). Addition of exogenous PTH to plasma from patients with PSP1 did not result in the same degree of recovery as in normal subjects (Loveridge, Fischer, et al., 1982).

3.   Where the sample contains a moiety that acts as a partial agonist, such as the interaction between CCK and gastrin already described, the sample will exhibit a nonparallel response at all dilutions (Figure 12A). Again, the presence of a partial agonist could be detected by a recovery experiment where the full response to the exogenous hormone will not be elicited until the effect of the partial agonist is overcome.

4.   In situations where the sample contains an antagonist to the hormone (e.g., an antibody) the response will probably be nonparallel at low dilutions but eventually will give a parallel response as the dilutions are increased (Figure 12B). At the high dilutions, presumably, dissociation of the hormone-antagonist complex occurs, so allowing full expression of the hormonal activity. Recovery of exogenous hormone will again not be complete until the effect of any excess antagonist is overcome.

5.   Loveridge, Bitensky, et al. (1980) showed that plasma from patients with pernicious anemia elicited a dose response which was not parallel at low dilutions but upon further dilutions the response became parallel (Figure 12B). Recovery of added gastrin was negligible except at high dilutions of the sample. A distinction from the situation described in (4) was made by culturing the segments of target organ in the presence of a dilution of plasma from either normal subjects or patients with pernicious anemia. After the culture period, the segments were exposed to exogenous gastrin in the absence of the plasma. Segments cultured in the presence of normal plasma showed a normal response to gastrin, but this was inhibited in the segments cultured in the presence of plasma from a patient with pernicious anemia. Thus, by showing that the response to exogenous gastrin was inhibited by the plasma, it was possible to conclude that the antagonist was active against the target cell and not against the hormone.

ACKNOWLEDGMENTS

I should like to acknowledge the general financial support of the Arthritis and Rheumatism Council for Research. Thanks go to Miss D. Stewart for typing the manuscript and the helpful advice of Dr. J. Chayen and Dr. L. Bitensky in the preparation of this chapter is gratefully acknowledged.

Appendix The Composition of Trowell's T8 Culture Medium[a]

Component	Concentration (mg/liter)
Amino acids	
L-Arginine HCl	21
L-Cysteine HCl	47
L-Histidine	10
L-Isoleucine	26
L-Leucine	26
L-Lysine	36
DL-Methionine	15
L-Phenylalanine	33
L-Threonine	48
L-Tryptophan	4
L-Tyrosine	18
L-Valine	23
Inorganic salts	
Calcium chloride	220
Potassium chloride	450
Magnesium sulfate·$7H_2O$	250
Sodium bicarbonate	2820
Sodium chloride	6100
Sodium dihydrogen orthophosphate	398
Vitamins	
p-Aminobenzoic acid	35
Thiamine HCl	17
Other components	
Glucose	4000
Insulin	50
Phenol red	10

[a]The pH of the medium in the presence of 95% O_2:5% CO_2 is 7.6.

REFERENCES

Alaghband-Zadeh, J., Daly, J. R., Bitensky, L., and Chayen, J. (1974). The cytochemical section assay for corticotrophin. *Clin. Endocrinol. (Oxf)*, 3:319-327.

Allgrove, J., Chayen, J., and O'Riordan, J. L. H. (1983). The cytochemical bioassay of parathyroid hormone: Further experience. *J. Immunoassay*, 4:1-19.

Altman, F. P. (1969a). The use of eight different tetrazolium salts for a quantitative study of pentose shunt dehydrogenation. *Histochemie*, 19:363-374.

Altman, F. P. (1969b). A comparison of dehydrogenase activities in tissue homogenates and tissue sections. *Biochem. J.*, 114:13P-14P.

Altman, F. P. (1971). The use of a new grade of polyvinyl alcohol for stabilising tissue sections during histochemical incubations. *Histochemie*, 28: 236-242.

Altman, F. P. (1972). Quantitative dehydrogenase histochemistry with special reference to the pentose shunt dehydrogenases. *Prog. Histochem. Cytochem.*, 4:225-273.

Altman, F. P. (1975). Quantitation in histochemistry: A review of some commercially available microdensitometers. *Histochem. J.*, 7:375-395.

Altman, F. P. (1976). Tetrazolium salts and formazans. *Prog. Histochem. Cytochem.*, 9:4-56.

Altman, F. P., and Barrnett, R. J. (1975). The ultrastructural localisation of enzyme activity in unfixed tissue sections. *Histochemistry*, 41:179-183.

Altman, F. P., and Chayen, J. (1965). Retention of nitrogenous material in unfixed sections during incubation for histochemical demonstration of enzymes. *Nature* 207:1205-1206.

Baker, J. R. (1945). *Cytological Technique.* Methuen, London.

Baylis, P. H., Pitchfork, J., Chayen, J., and Bitensky, L. (1980). A cytochemical bioassay for arginine vasopressin: Preliminary studies. *J. Immunoassay*, 1:399-411.

Bayliss, W. M., and Starling, E. H. (1902a). On the causation of the so-called 'Peripheral reflex secretion' of the pancreas. *Proc. R. Soc.*, 69:352-353.

Bayliss, W. M., and Starling, E. H. (1902b). The mechanism of pancreatic secretion. *J. Physiol. (Lond.)*, 28:325-353.

Berenbom, M., Yokoyama, H. O., and Stowell, R. E. (1952). Chemical and enzymatic changes in liver following freeze-drying and acetone fixation. *Proc. Soc. Exp. Biol. Med.*, 81:125-128.

Bitensky, L. (1980). Microdensitometry. In *Trends in Enzyme Histochemistry and Cytochemistry,* D. Evered and M. O'Connor (eds.). (Ciba Foundation Symposium 73). Excerpta Medica, Amsterdam, pp. 181-202.

Bitensky, L., Alaghband-Zadeh, J., and Chayen, J. (1974). Studies on thyroid stimulating hormone and the long acting thyroid stimulating hormone. *Clin. Endocrinol. (Oxf)*, 3:363-374.

Bitensky, L., and Chayen, J. (1977). Histochemical methods for the study of lysosomes. In *Lysosomes, a Laboratory Handbook,* 2nd ed., Dingle, J. T. (ed.). North-Holland, Amsterdam, pp. 209-243.

Buckingham, J. C., Chayen, J., Hodges, J. R., Robertson, W. R., and Weisz, J. (1979a). A cytochemical section assay for the determination of luteinizing hormone. *J. Endocrinol.*, 81:160P.

Buckingham, J. C., Chayen, J., Hodges, J. R., Robertson, W. R., and Weisz, J. (1979b). A cytochemical section assay for the determination of luteinizing hormone in human and rat plasma. *Acta Endocrinol. (Kbh) (Suppl.)*, 225: 131.

Buckingham, J. C., and Hodges, J. R. (1981). A cytochemical bioassay method for the determination of luteinizing hormone in biological fluids and tissues. *Br. J. Pharmacol.* 73:111-118.

Burstone, M. S. (1962). *Enzyme Histochemistry and Its Applications to the Study of Neoplasms.* Academic, New York.

Butcher, R. G. (1971a). The chemical determination of section thickness. *Histochemie,* 28:131-136.

Butcher, R. G. (1971b). Tissue stabilisation during histochemical reactions: The use of collagen polypeptides. *Histochemie,* 28:231-235.

Butcher, R. G., and Altman, F. P. (1973). Studies on the reduction of tetrazolium salts. II. The measurement of the half-reduced and fully reduced formazans of neotetrazolium chloride in tissue sections. *Histochemistry,* 37: 351-363.

Caspersson, T. (1947). Relations between nucleic acid and protein synthesis. *Symp. Soc. Exp. Biol.,* 1:127-151.

Chambers, D. J., Dunham, J., Zanelli, J. M., Parsons, J. A., Bitensky, L., and Chayen, J. (1978a). A sensitive bioassay of parathyroid hormone in plasma. *Clin. Endocrinol. (Oxf),* 9:375-379.

Chambers, D. J., Schäfer, H., Laugharn, J. A., Jr., Johnstone, J., Zanelli, J. M., Parsons, J. A., Bitensky, L., and Chayen, J. (1978b). Dose-related activation by PTH of specific enzymes in different regions of the kidney. In *Endocrinology of Calcium Metabolism,* D. H. Copp and R. V. Talmage (eds.). Excerpta Medica, Amsterdam, pp. 216-220.

Chambers, D. J., Zanelli, J. M., Parsons, J. A., and Chayen, J. (1976). Cytochemical responses of the guinea pig kidney cortex to low concentrations of bovine parathyroid hormone (0.001-0.1 pg/ml). *J. Endocrinol.,* 71:87P.

Chayen, J. (1978a). The cytochemical approach to hormone assay. *Int. Rev. Cytol.,* 53:333-396.

Chayen, J. (1978b). Microdensitometry. In *Biochemical Mechanisms of Liver Injury,* T. F. Slater (ed.). Academic, London, pp. 257-291.

Chayen, J., and Bitensky, L. (1968). Multiphase biochemistry. In *The Biological Basis of Medicine, Vol, I,* E. E. Bittar and N. Bittar (eds.). Academic, London, pp. 337-368.

Chayen, J., and Bitensky, L. (1980). The application of cytochemistry to the highly sensitive bioassay of polypeptide hormones. *Pathol. Res. Pract.,* 170: 39-49.

Chayen, J., Bitensky, L., and Butcher, R. G. (1973). *Practical Histochemistry.* Wiley, London.

Chayen, J., Bitensky, L., Chambers, D. J., Loveridge, N., and Daly, J. R. (1974). Studies on the mechanism of cytochemical bioassays. *Clin. Endocrinol. (Oxf),* 3:349-360.

Chayen, J., Daly, J. R., Loveridge, N., and Bitensky, L. (1976). The cytochemical bioassay of hormones. *Recent Prog. Horm. Res.,* 32:33-79.

Chayen, J., and Denby, E. F. (1968). *Biophysical Technique as Applied to Cell Biology.* Methuen, London.

Chayen, J., Frost, G. T. B., Dodds, R. A., Bitensky, L., Pitchfork, J., Baylis, P. H., and Barrnett, R. J. (1981). The use of a hidden metal-capture reagent for the measurement of Na^+-K^+-ATPase activity: A new concept in cytochemistry. *Histochemistry,* 71:533-542.

Chayen, J., Gilbert, D. M., Robertson, W. R., Bitensky, L., and Besser, G. M. (1980). A cytochemical section assay for thyrotrophin. *J. Immunoassay,* 1:1-13.

Cuatrecasas, P. (1971). Perturbation of the insulin receptor of isolated fat cells with proteolytic enzymes: Direct measurement of insulin-receptor interaction. *J. Biol. Chem.,* 246:6522-6531.

Cumming, H. (1970). *Virology—Tissue Culture,* Laboratory Aids Series, J. T. Baker (ed.). Butterworths, London.

Danielli, J. F. (1953). *Cytochemistry: A Critical Approach.* Wiley, New York.

Deeley, E. M. (1955). An integrating microdensitometer for biological cells. *J. Sci. Instrum.,* 32:263-267.

Deeley, E. M., Davies, H. G., and Chayen, J. (1957). The DNA content of cells in the root of *Vicia faba. Exp. Cell Res.,* 12:582-591.

Deeley, E. M., Richards, B. M., Walker, P. M. B., and Davies, H. G. (1954). Measurements of Feulgen stain during the cell-cycle with a new photo-electric scanning device. *Exp. Cell Res.,* 6:569-572.

Ealey, P. A. (1979). *The Validation of the Cytochemical Bioassay of Thyrotrophin and Its Application to Selected Clinical Problems.* Ph.D. thesis, University of London.

Ealey, P. A., Marshall, N. J., and Ekins, R. P. (1981). Time-related thyroid stimulation by thyrotropin and thyroid stimulating antibodies, as measured by the cytochemical section bioassay. *J. Clin. Endocrinol. Metab.,* 52:483-487.

Faglia, G., Bitensky, L., Pinchera, A., Ferrari, C., Paracchi, A., Beck-Peccoz, P. Ambrosi, B., and Spada, A. (1979). Thyrotropin secretion in patients with central hypothyroidism: Evidence for reduced biological activity of immunoreactive thyrotropin. *J. Clin. Endocrinol. Metab.,* 48:989-998.

Fenton, S., Somers, S., and Heath, D. A. (1978). Preliminary studies with the sensitive cytochemical assay for parathyroid hormone. *Clin. Endocrinol. (Oxf),* 9:381-384.

Glock, G. E., and McLean, P. (1953). Further studies on the properties and assay of glucose 6-phosphate dehydrogenase and 6-phosphogluconate dehydrogenase in rat liver. *Biochem. J.,* 55:400-408.

Goltzman, D., Henderson, B., and Loveridge, N. (1980). Cytochemical bioassay of parathyroid hormone: Characteristics of the assay and analysis of circulating forms. *J. Clin. Invest.,* 65:1309-1317.

Gomori, G. (1952). *Microscopic Histochemistry.* University Press, Chicago.

Henderson, B., Loveridge, N., and Robertson, W. R. (1978). A quantitative study of the effects of different grades of polyvinyl alcohol on the activities of certain enzymes in unfixed tissue sections. *Histochem. J.,* 10:453-463.

Hoile, R. W. (1979). *The Biological Activity of Gastrin in Peptic Ulceration.* M.S. thesis, University of London.

Hoile, R. W., and Loveridge, N. (1976). Preliminary studies on a cytochemical section bioassay for gastrin. *J. Endocrinol.,* 71:87-88P.

Hoile, R. W., and Loveridge, N. (1977). Early results from the cytochemical assay for gastrin. *Br. J. Surg.,* 64:293.

Holdaway, I. M., Rees, L. H., Ratcliffe, J. G., Besser, G. M., and Kramer, R. M. (1974). Validation of the redox cytochemical assay for corticotrophin. *Clin. Endocrinol. (Oxf),* 3:329-334.

Klaff, L. (1981). *Quantitative Cytochemical Studies of Acid Secretagogue Effects on the Carbonic Anhydrase Activity of Gastric Parietal Cell Sections.* Ph.D. thesis, University of Cape Town.

Liotta, A., and Krieger, D. T. (1975). A sensitive bioassay for the determination of human plasma ACTH levels. *J. Clin. Endocrinol. Metab.,* 40:268-277.

Lovelock, J. E. (1957). The denaturation of lipid-protein complexes as a cause of damage by freezing. *Proc. R. Soc. Lond. [Biol.],* 147:427-433.

Loveridge, N. (1977). *The Development of a Cytochemical Bioassay for Gastrin-like Activity.* M.Phil. thesis, Brunel University.

Loveridge, N. (1978). A quantitative cytochemical method for measuring carbonic anhydrase activity. *Histochem. J.,* 10:361-372.

Loveridge, N. (1981). *Studies on the Mode of Action of Polypeptide Hormones.* Ph.D. thesis, Brunel University.

Loveridge, N., Bitensky, L., Chayen, J., Hausamen, T.-U., Fischer, J. M., Taylor, K. B., Gardner, J. D., Bottazzo, G. F., and Doniach, D. (1980). Inhibition of parietal cell function by human gamma-globulin containing gastric parietal cell antibodies. *Clin. Exp. Immunol.,* 41:264-270.

Loveridge, N., Bloom, S. R., Welbourn, R. B., and Chayen, J. (1974). Quantitative cytochemical estimation of the effect of pentagastrin (0.005-5 pg/ml) and of plasma gastrin on the guinea-pig fundus in vitro. *Clin. Endocrinol. (Oxf),* 3:389-396.

Loveridge, N., Fischer, J. A., Nagant de Deuxchaisnes, C., Dambacher, M. A., Tschopp, F., Werder, E., Devogelaer, J-P., De Meyer, R., Bitensky, L., and Chayen, J. (1982). Inhibition of cytochemical bioactivity of parathyroid hormone by plasma in pseudohypoparathyroidism Type I. *J. Clin. Endocrinol. Metab.,* 54:1274-1275.

Loveridge, N., Hoile, R. W., Johnson, A. G., and Chayen, J. (1978). The cytochemical measurement of gastrin-like activity. *J. Endocrinol.,* 77:40-41P.

Loveridge, N., Hoile, R. W., Johnson, A. G., Gardner, J. D., and Chayen, J.

(1980). The cytochemical section-bioassay of gastrin-like activity. *J. Immunoassay*, 1:195-209.

Loveridge, N., Zakarija, M., Bitensky, L., and McKenzie, J. M. (1979). The cytochemical bioassay for thyroid stimulating antibody of Graves' disease: Further experience. *J. Clin. Endocrinol. Metab.*, 49:610-615.

Lynch, R., Bitensky, L., and Chayen, J. (1966). On the possibility of super cooling in tissues. *J. Roy. Microsc. Soc.*, 85:213-222.

Maren, T. H. (1967). Carbonic anhydrase: Chemistry, physiology and inhibition. *Physiol. Rev.*, 47:595-781.

McLeish, J., Bell, L. G. E., La Cour, L. F., and Chayen, J. (1957). The quantitative cytochemical estimation of arginine. *Exp. Cell Res.*, 12:120-125.

Moline, S. W., and Glenner, G. G. (1964). Ultrarapid tissue freezing in liquid nitrogen. *J. Histochem. Cytochem.*, 12:777-783.

Nagant de Deuxchaisnes, C., Fischer, J. A., Dambacher, M. A., Devogelaer, J.-P., Arber, C. E., Zanelli, J. M., Parsons, J. A., Loveridge, N., Bitensky, L., and Chayen, J. (1981). Dissociation of bioactive and immunoreactive parathyroid hormone in pseudohypoparathyroidism type I. *J. Clin. Endocrinol. Metab.*, 53:1105-1109.

Paul, J. (1975). *Cell and Tissue Culture*, 5th ed. Churchill-Livingstone, Edinburgh.

Petersen, V. B., Smith, B. R., and Hall, R. (1975). A study of thyroid stimulating activity in human serum with the highly sensitive cytochemical bioassay. *J. Clin. Endocrinol. Metab.*, 41:199-202.

Rees, L. H., Holdaway, I. M., Kramer, R. M., McNeilly, A. S., and Chard, T. (1973). A new bioassay for luteinizing hormone. *Nature*, 244:232-234.

Robertson,W.R.(1980). A quantitative study of N-acetyl-β-glucosaminidase activity in unfixed tissue sections of the guinea-pig thyroid gland. *Histochem.J.*,12:87-96.

Sayers, G. (1977). Bioassay of ACTH using isolated cortex cells. *Ann. N.Y. Acad. Sci.*, 297:220-241.

Silcox, A. A., Poulter, L. W., Bitensky, L., and Chayen, J. (1965). An examination of some factors affecting histological preservation in frozen sections of unfixed tissue. *J. Roy. Microsc. Soc.*, 84:559-564.

Stafford, R. O., and Atkinson, W. B. (1948). Effect of acetone and alcohol fixation and paraffin embedding on activity of acid and alkaline phosphatases in rat tissues. *Science*, 107:279-280.

Starling, E. H. (1905). The chemical correlation of the functions of the body. *Lancet*, ii:339-341.

Takeuchi, K., Spier, G. R., and Johnson, L. R.(1979). Mucosal gastrin receptor. I. Assay standardization and fulfillment of receptor criteria. *Am. J. Physiol.*, 237:E284-E294.

Thornberg, W., and Mengers, P. E. (1957). An analysis of frozen section techniques. I. Sectioning fresh frozen tissue. *J. Histochem. Cytochem.*, 5:47-52.

Trowell, O. A. (1959). The culture of mature organs in a synthetic medium. *Exp. Cell Res.*, 16:118-147.

4

Adrenocorticotropic Hormone

W. H. C. Walker / McMaster University Medical Centre, Hamilton,
Ontario, Canada

ADRENOCORTICOTROPIN: STRUCTURE AND FUNCTION

Chemical Structure and Synthesis

Human adrenocorticotropic hormone (ACTH) is a polypeptide of molecular
weight 5250, composed of 39 amino acid residues in a single chain with no disulfide cross-linkages. Its biological activity resides in the sequence of its first
24 N-terminal amino acid residues, and this sequence is identical in all species
that have been studied. The same 24-amino acid sequence is available in synthetic form (Synacthen, Tetra-Cosactryn, or Cortrosyn) and is widely used in
clinical practice. Minor changes in this sequence may have profound effects; for
example, oxidation of the methionine at position 4 can cause a marked reduction in activity.

The C-terminal part of the ACTH molecule varies in beef, sheep, pig, and
human at residues 31 and 33 (Riniker et al., 1972). Human corticotropin has
been completely synthesized (Bajusz et al., 1967), and several analogs have been
prepared which have higher biological activity than the natural hormone, probably because of their relative resistance to peptidase cleavage and longer circulating half-life. The nomenclature of ACTH precursors, analogs and fragments is
complex, and usage generally follows the proposals of Li (1959) and of the
IUPAC-IUB Commission of Biochemical Nomenclature (1967).

ACTH arises in the pituitary from a large glycoprotein precursor, pro-adrenocorticotropin-endorphin, the nucleotide sequence for which has been defined
(Nakanishi et al., 1979). Proteolytic cleavage yields β-lipotropin with 91 amino

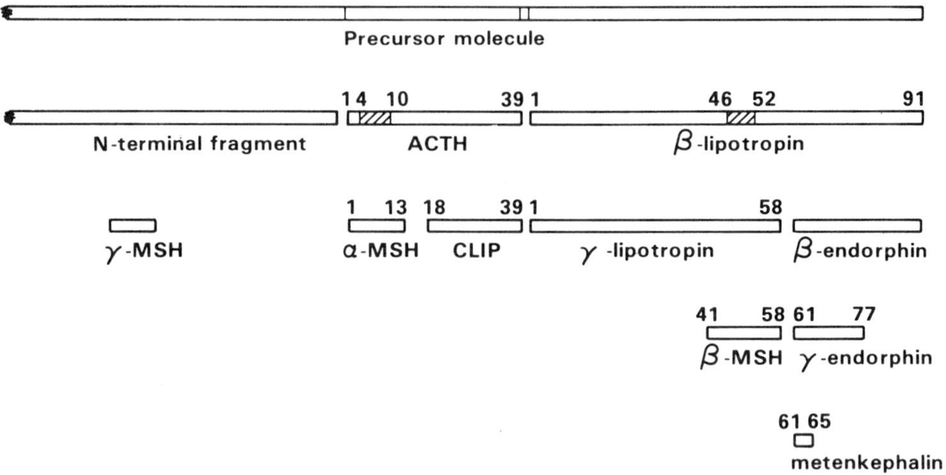

Figure 1   Structural relationships of ACTH-related peptides. The common heptapeptide core occurs in positions 4-10 in ACTH and 46-52 in β-lipotropin.

acid residues and an ACTH precursor which in turn is then cleaved to form ACTH and a larger fragment termed the 16K fragment (Eipper and Mains, 1980; Allen et al., 1980). The 7-amino acid sequence, -Met-Glu-His-Phe-Arg-Trp-Gly-, occurs at residues 4-10 of ACTH and also at residues 46-52 of β-lipotropin. Some of the β-lipotropin is cleaved to produce γ-lipotropin (residues 1-58), β-endorphin (residues 61-91), γ-endorphin (residues 61-77), α-endorphin (residues 61-76), and metenkephalin (residues 61-65) (Figure 1).

β-Lipotropin is released into the blood at the same time as ACTH, their concentrations varying similarly under physiological stimuli and in pathological states, such as Nelson's syndrome, Addison's disease, and the ectopic ACTH syndrome (Kreiger et al., 1979; Gray and Ratcliffe, 1979). β-Melanocyte-stimulating hormone does not occur in the human, earlier reports to the contrary being apparently due to artifactual degradation of β-lipotropin. A cleavage product of the 16K fragment, termed melanocyte-stimulating hormone, has been shown to activate cholesterol ester hydrolase (Pederson et al., 1980), but its physiological role is uncertain.

Pituitary synthesis, storage, and release of ACTH is influenced by stress, by circulating levels of corticosteroids, and by a 24 hr cyclic rhythm which responds slowly to night-day changes and to sleep patterns. All these regulatory effects are believed to be mediated by an hypothalamic corticotropin releasing factor (CRF).

Ectopic production of ACTH, β-lipotropin, and related fragments may occur in four major classes of tumors: oat-cell carcinoma of the bronchus, ovarian carcinoma, pheochromocytoma, and endocrine tumors of the foregut, such as carcinoid of bronchus or thymus, medullary carcinoma of the thyroid, and islet cell tumors of the pancreas (Azzopardi and Williams, 1968; Rosai et al., 1976). The proteolytic cleavage sites may differ between malignant tumors and normal pituitary (Rees et al., 1977; Orth et al., 1973), accounting for the presence of "big ACTH" in tumor extracts and the venous effluent of such tumors.

Effects on Target Cells

The sequence of events leading to corticosteroid secretion following ACTH binding to specific cell surface receptors is rapid, an increase in secretion occurring within 1 min of ACTH stimulation. ACTH activates adenyl and guanyl cyclases, resulting in the formation of adenosine $3',5'$-cyclic monophosphate (cAMP) (Grahame-Smith et al., 1967) and guanosine $3',5'$-cyclic monophosphate (cGMP) (Perchellet and Sharma, 1980), and the increase of cyclic nucleotides occurs at least as rapidly as the release of corticosteroid. Other short-term effects include activation of $5'$-nucleotidase (Hilf et al., 1961; Chambers and Chayen, 1976), ornithine decarboxylase (Byus and Russell, 1975; Levine et al., 1975), 5-ene,3β steroid dehydrogenase (Loveridge and Robertson, 1978), and glucose-6-phosphate dehydrogenase (Chayen et al., 1976). In addition, there is depletion of intracellular ascorbate within 1 min, followed by a brief increase and a second prolonged depletion (Chayen et al., 1976).

Other effects of ACTH include increases of glucose oxidation, phosphate turnover, protein turnover, cholesterol uptake, and synthesis of RNA and DNA. Synthesis of a short-lived protein has been described (Koritz et al., 1977). Long-term exposure to ACTH leads to increased weight of the adrenal gland, this being the trophic effect for which the hormone was named. Increase in adrenal blood flow occurs in response only to supraphysiological levels of ACTH.

Ascorbate is present in unstimulated adrenal tissue at a concentration of 1-5 mg/g adrenal tissue (Elton et al., 1959; Colurso, 1979). Maximal stimulation with ACTH leads to a loss of up to 40% intracellular ascorbate in vivo (Sayers et al., 1944; 1946; 1948), and in vitro (Colurso, 1979). The loss varies with animal species and stimulus but falls far short of total depletion. There is rapid appearance of ascorbate in the adrenal venous blood after ACTH stimulation (Liddle et al., 1962), and this is associated with inhibition of transport of ascorbate from blood into adrenal cells (Sharma et al., 1963; 1964). Corticosteroids also produce an inhibitory effect on ascorbate uptake (De Nicola et al., 1968).

The metabolic pathway leading from cholesterol esters to corticosteroids involves a series of enzyme-mediated steps that occur partly in the endoplasmic

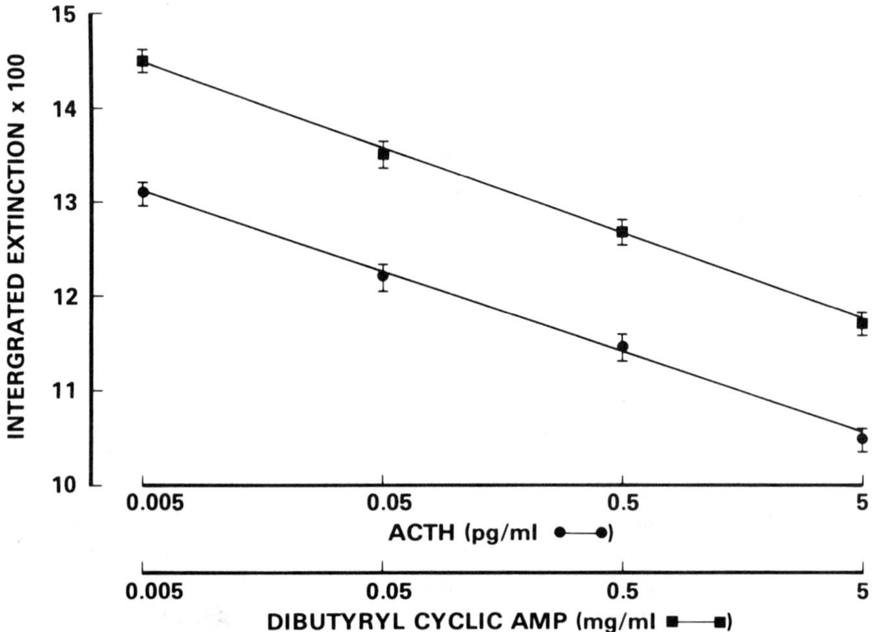

Figure 2   Dose-response curve, mean ± standard error for ACTH, Third International Working Standard, and for dibutryl cAMP. The curve for dibutyryl cGMP was indistinguishable from that for dibutyryl cAMP.

reticulum but start and finish in the mitochondria. Cholesterol esters stored in cells of the zona fasciculata and zona reticularis are hydrolyzed. The free cholesterol enters the mitochondria and is there converted to pregnenolone by a group of enzymes termed *desmolase*. The reaction requires cytochrome P_{450}, NADPH, and molecular oxygen. The three next steps occur in the endoplasmic reticulum and involve steroid 17β-hydrogenase, 5-ene,3β-hydroxysteroid dehydrogenase, and steroid 21-hydroxylase. The hydroxylase enzymes require microsomal cytochrome P_{450}, NADPH, and molecular oxygen. The dehydrogenase is NAD^+ linked. The final step leading to cortisol synthesis involves mitochondrial steroid 11β-hydroxylase. It requires cytochrome P_{450}, NADPH, molecular oxygen and Mg^{2+}. The level of mitochondrial corticosteroid exerts feedback inhibition on the desmolase enzymes (Nelson, 1980).

Causal relationships between the observed molecular effects of ACTH and the regulation of corticosteroid biosynthesis are not clearly established. Dibutyryl cAMP stimulates corticosteroid secretion with a time course similar to that seen with ACTH (Urquhart and Li, 1969), and we have found a decrease in intracellular ascorbate after a 1 min exposure of guinea pig adrenal segments to dibutyryl cAMP or to dibutyryl cGMP with dose-response relationships parallel to that resulting from ACTH stimulation (Figure 2). Both 5-ene,3β-hydroxysteroid dehydrogenase and glucose-6-phosphate dehydrogenase, although potential regulators, appear not to be involved in rate-limiting steps (Hilf, 1965). The hydroxylation reactions in corticosteroid biosynthesis have much higher capacities than are required for maximal physiological steroidogenesis (Hechter et al., 1951). The activity of the desmolase enzymes is influenced by ACTH (Pedersen et al., 1980) as well as being inhibited by corticosteroids.

Chayen et al. (1976) postulated that a rate-limiting factor in steroidogenesis may be availability of molecular oxygen as determined by the intracellular concentration of ascorbate. The demonstration by Hodges and Hotston (1970) that scorbutic guinea pigs secrete corticosteroids at maximal rate has been cited in support of this hypothesis. The biphasic nature of ascorbate loss following continuous ACTH stimulation suggests that the role of ascorbate in steroidogenesis is likely to be complex. There is insufficient experimental evidence to identify the role of changes in intracellular ascorbate as being either a cause or a consequence of increased steroidogenesis. Available evidence does, however, indicate a close correlation between ascorbate levels and steroidogenesis, and this is the basis of both in vivo (Sayers et al., 1948) and cytochemical in vitro bioassays. By contrast, the tropic effect of ACTH can be dissociated from steroidogenesis, this being seen most clearly in individuals with desmolase deficiency who develop adrenal hyperplasia as a result of enhanced pituitary secretion of ACTH. Similarly, experiments with actinomycin (Ferguson and Morita, 1965) demonstrate blockage of ACTH-induced RNA synthesis with persistence of the steroidogenic effect.

ASSAYS AVAILABLE FOR ADRENOCORTICOTROPIN

ACTH, like many other hormones, may be measured in three ways: by reaction with specific antibody in immunoassay (Yalow et al., 1964; Demura et al., 1966; Landon and Greenwood, 1968), by reaction with cell surface receptors in receptor assay (Taunton et al., 1969; Lefkowitz et al., 1970; Wolfsen et al., 1972), and by interaction with living cells in vivo (Sayers et al., 1948; Lipscomb and Nelson, 1962) or in vitro (Saffran and Schally, 1955; Kloppenborg et al., 1968; Sayers et al., 1971; Chayen et al., 1971, 1972; Lowry et al., 1973).

The amino acid sequences involved in reactions with antibody are often not those involved in receptor recognition, and the specificities of immunoassay and receptor assay therefore differ. Antibodies to ACTH are often directed against the C-terminal species-dependent part of the molecule rather than against the N-terminal biologically active moiety (Besser et al., 1971), although antibodies with specificities directed to parts of the N-terminal sequence may also be prepared. Some of the precursors and degradation fragments of ACTH that occur in plasma cross-react with antibodies and lead to radioimmunoassay levels that are higher than receptor assay levels. A similar effect results when biological activity is lost without cleavage of the molecule (Nicholson et al., 1976).

Antibodies and receptors for ACTH have similar equilibrium constants (K) of the order of 10^{11} liters/mol, and sensitivity of both immunoassay and receptor assay, being limited to 0.2 K^{-1} (Walker and Keane, 1977), is therefore about 2 pmol/liter (10 pg/ml). Bioassay has a potential, not always attained, for much higher sensitivity because of its cascade multiplication of the initial binding stimulus. The greater complexity of bioassay systems can introduce variables affecting delivery and breakdown of hormone with consequent difference of assay results between one assay system and another.

In early bioassays using whole animals, sensitivity was lost by delayed release of test material from the site of injection, by dilution throughout the body fluids and by breakdown of hormone while in transit. The Lipscombe and Nelson (1962) bioassay overcame these problems by retrograde injection of test material into the adrenal vein, leading to sensitivities of 20 pmol/liter (100 pg/ml), sufficient to detect pathologically elevated but not normal ACTH levels. Isolated cell bioassays (Sayers et al., 1971; Lowry et al., 1973) further improved the on-site delivery of hormone and for the first time realized the potential enhancement of sensitivity over receptor assay with detection limits of 0.2 pmol/ml. The cytochemical bioassay (Chayen et al., 1971; 1976), using thin tissue slices in maintenance organ culture, carried this enhancement of sensitivity a major step forward, with detection limits of 0.004 pmol/liter (0.02 pg/ml). This technique combines delivery of hormone direct to target cells with retention of the integrity of the adrenal tissue and the ability to measure the response exclusively in the cells of the zona reticularis where ACTH-regulated ascorbate depletion is maximal.

The measured response in all ACTH in vivo bioassays has been either adrenal steroidogenesis (Lipscomb and Nelson, 1962) or adrenal ascorbate depletion (Sayers et al., 1948). The role of ascorbate metabolism in steroidogenesis is still unclear, but no evidence has yet emerged to discount the general validity of the close correlation between ascorbate depletion and steroidogenesis in response to ACTH stimulation (Rees et al., 1973). In vitro assays have also employed adenosine 3′,5′-cyclic monophosphate and a variety of induced enzymes as end points of ACTH activity. In measuring any response it is necessary to take into account the time taken to attain the maximal response and the duration of the response. Chayen (1980) has argued cogently for using the early rate of change of response rather than the long-term integrated response which must be affected by variations in the rate of hormone degradation.

All assays involve comparison of unknown with standard, and the resultant activity assigned to the unknown will be of general applicability, independent of assay conditions, only if unknown and standard relate to all the assay variables in a quantitatively similar fashion. The major variables are those of delivery, reaction, and persistence. Delivery problems may be reduced by ensuring that the test material is applied to the target tissue as directly as possible; persistence of effect may be accommodated by measuring the early stage of response. Differences in reaction kinetics, for instance when unknown and standard differ in their receptor-binding equilibrium constants, may be revealed by assaying both unknown and standards at several different dilutions. Only if there is a similar proportionality of effect in both unknown and standard dilutions, manifested by "parallelism" of responses on a dose-response curve, can the unknown material have a unique activity assigned to it. If the responses are not parallel, the assigned activity will depend on the arbitrarily chosen dilution of unknown and, more generally, will differ as assay conditions are changed or a different assay is used. It should be noted that parallelism may occur when test and standard are different, as with cyclic nucleotides and ACTH. Parellism is necessary but not sufficient to demonstrate identity; it is sufficient within any given assay system to unambiguously assign a concentration in the arbitrary equivalent units of the standard.

THE CYTOCHEMICAL BIOASSAY

ACTH was the first hormone to be assayed by the cytochemical bioassay technique. The guinea pig has proven to be a highly suitable species for this assay since it is small and inexpensive and has adrenals large enough to divide into two or three segments, yet small enough to be able to mount an entire cross-section on a microscope slide for quantitative microdensitometry. Under favorable conditions, more than 100 sections may be obtained from each guinea pig adrenal. By contrast, the number of sections from the mouse adrenal is limited to 15 and from the rat adrenal to 50, and in both these species the zona reticularis is less well demarcated than in the guinea pig. The assay is based on the depletion of ascorbate in cells of the zona reticularis, demonstrated by reduction of ferricyanide in the presence of ferric ions to yield the insoluble dye, ferro-ferricyanide, known as Prussian blue.

The original cytochemical bioassay was described in 1971 by Chayen and his associates (Chayen et al., 1971; 1972; Daly et al., 1972). The technique was limited to the assay of one unknown sample for each guinea pig, the two adrenals yielding six segments which were incubated with each of four dilutions of the standard, and two of the unknown. The routine applicability of the assay was greatly facilitated by development of techniques that permitted sectioning of the adrenal segments with retention of viability and responsiveness of the

Table 1 Cytochemical Bioassay for ACTH

5 hr maintenance culture in vitro of adrenals from one animal	Recover from surgery Remove endogenous hormone Restoring resting metabolism Provide within-animal comparisons
Prime with 0.5 fg/ml ACTH 4 min	Enhance responsiveness
Section at 20 μm	Retain intact whole cells within section
React with standards/unknowns 60 sec	Create dose-response curve based on ascorbate depletion
Stain for ascorbate	Provide chromogen for scanning
Scanning and integrating microdensitometry	Measure ascorbate depletion in cells of zona reticularis

adrenal cells and subsequent exposure of the sections to standards and unknowns (Alaghband-Zadeh, 1974; Alaghband-Zadeh et al., 1974). As a result, the section bioassay can accommodate up to 40 unknown samples, and this is a realistic weekly throughput for one technologist.

Section Bioassay

The section bioassay consists of the following steps: maintenance organ culture, priming, chilling, sectioning, exposure to standards, staining, and measurement (Table 1).

Culture

T8-Ascorbate: To 50 ml Trowell's T8 culture medium (Trowell, 1959) (Gibco Laboratories, Grand Island Biological Company, Grand Island, N.Y. 14072, USA) at 37°C, 10 mg sodium ascorbate (Sigma Chemical Co., St. Louis, Mo.) is added to give a final concentration of 10^{-3} mol/liter ascorbate and the pH of the medium adjusted to 7.6 with sodium hydroxide. A male or female guinea pig, Hartley strain, 350-550 g, which has been maintained unstressed on a normal diet for 1 week, is killed by asphyxiation with nitrogen or by a blow on the head. The adrenals are gently removed, trimmed of connective and adipose tissue, and each is cut transversely into two or three equal segments.

The segments are separately maintained in culture medium for 5 hr as described in detail in Chapter 3. The culture period allows the adrenal segments to recover from their in vivo exposure to ACTH and from the trauma of surgery. At the same time, the segments take up ascorbate from the medium, thereby enhancing their responsiveness to ACTH.

Priming
At the end of the culture period, the culture medium is aspirated and replaced
by a similar volume of fresh Trowell's-ascorbate medium containing 0.5 fg
ACTH per milliliter medium. The chamber is again closed and the segments left
in contact with the priming ACTH solution for 4 min at 37°C. The priming dose,
although below the detection limit of the assay, has been found to stabilize the
responsiveness of the tissue to subsequent graded standard concentrations of
ACTH.

Chilling and Sectioning
The tissue is chilled and sectioned (20 μm) as described in full in Chapter 3.

Exposure to ACTH
The slides are placed in pairs back-to-back into the holders in the lid of the segmented incubation vessel. Each chamber of the vessel is filled with one of a
graded series of concentrations of standard or of the unknowns in 10 ml volumes
of T8-ascorbate medium with added 5 g per 100 ml Polypep 5115 (Sigma Chemical Co., St. Louis, Mo.), with the pH adjusted to 7.6. Polypep 5115 is a hydrolyzed collagen which acts as a colloid stabilizer, preventing the major loss of
soluble cellular material that would otherwise occur when sections are immersed in aqueous solution.

ACTH standard, either human reference material or synthetic cosyntropin
(α1-24-corticotropin; Cortrosyn, 0.25 mg ampules, Organon, Inc., West Orange,
N.J. USA) is diluted in 0.05 N HCl to a concentration of 50 ng ACTH per
milliliter. In this form, it can be stored in plastic vials in 0.3 ml aliquots at –70°C
for at least 3 months. Working dilutions of standards and plasmas are prepared
in Trowell's T8-ascorbate with Polypep 5115. The dilutions are made in plastic
tubes, using plastic disposable pipette tips to avoid contact with glass and consequent adsorptive loss of ACTH.

A 24-chamber incubation vessel (Figure 7, Chapter 3) has a capacity for four
standards at 5, 50, 500, and 5000 fg/ml together with 10 unknown samples each
at two dilutions, usually 1 in 100 and 1 in 1000. When the vessel and its contents have been brought to 37°C, the slides, already held in place in the lid, are
inverted into their respective chambers and exposed to ACTH for 1 min.

The lid with its slides is then transferred to a vessel containing the ferricya-
nide-ferric chloride solution at room temperature.

Staining
The stain is prepared immediately before use by mixing 3 volumes of 1.35%
ferric chloride ($FeCl_3 \cdot 6H_2O$) and 1 volume of 0.1% potassium ferricyanide.
Sections are stained for 15 min, the staining solution being replaced with fresh
solution at 5 and 10 min. Slides are then rinsed for 5 min in tap water, air

dried, cleared in xylol, and mounted in DePex or Permount. They are stored in the dark to prevent fading of the stain.

Microdensitometry

The intensity of the Prussian blue stain in the zona reticularis is measured by scanning and integrating microdensitometry at 680 nm with a ×10 objective and an A-5 mask size of diameter 100 μm and a spot size of 2 μm (spot size 1). At this magnification, 10 fields encompass most of the zona reticularis in sections from an average-sized guinea pig adrenal. The staining of the zona reticularis is relatively uniform, and absorbance integrated over these large fields is satisfactory. Is is, however, essential that cells of the medulla and of the zona fasciculata are not inadvertently included in the areas scanned. Nonspecific absorbance is measured by taking consecutive readings of 10 fields of the zona reticularis at 480 nm. The difference of the means of readings at 680 and 480 nm represents the absorbance due to Prussian blue, and this is used to prepare a dose-response curve.

Dose-Response Curve

There is an inverse linear dose-response relationship between the integrated absorbance due to ascorbate staining and the logarithm of the ACTH dose over the range of the assay standards. On occasion, the highest standard, 5 pg/ml, may exceed the linear response of the assay. The arbitrary integrated absorbance units generated by the scanning microdensitometer may for purposes of comparison be converted to absolute absorbance units by the use of a calibration graph prepared using neutral density filters (Bitensky et al., 1973).

Segment Bioassay

In the original cytochemical bioassay for ACTH (Chayen et al., 1972), the adrenal segments were exposed directly to either the graded standards or unknowns, thus replacing the 4 min priming stage after the 5 hr tissue culture period. Sectioning and staining then proceeded as described above. When a laboratory is setting up the cytochemical bioassay for the first time, it is reasonable to demonstrate tissue responsiveness by using the segment assay sequence. The section bioassay then requires control of only two added steps, maintenance of tissue responsiveness throughout the stages of section cutting, and the reaction of the section with ACTH in the presence of colloid stabilizer.

VALIDATION OF THE CYTOCHEMICAL BIOASSAY

The cytochemical bioassay procedure for ACTH has been subject to extensive investigation, defining the effects of varying assay conditions, and exploring the reactions occurring with ACTH and its analogs. Results have been compared

Table 2 The Effect of pH on the Response of Zona
Reticularis Cells to Varying ACTH Concentrations[a]

pH	ACTH Concentration (pg/ml)				
	0.027	0.27	2.7	27	270
7.0	7.2	3.0	*1.6*	3.8	8.9
7.4	1.1	6.5	4.2	*3.5*	7.1
7.6	6.7	4.7	3.4	*2.0*	3.6

[a]The maximum response is in italics.
Source: Chayen et al. (1972).

with those obtained by other bioassays and by radioimmunoassay using standards and biological samples from a variety of physiological studies defining pituitary responsiveness in health and disease.

Assay Conditions

Incubation

The Trowell's medium contains a bicarbonate-carbonic acid buffer system which is in equilibrium with pCO_2 in the gas phase. The control of culture conditions, including pH, has been shown to be critical for maintenance of viability and responsiveness of the adrenal cells (Anthony et al., 1979). Chayen et al. (1972) showed that changing the pH of the medium over the range 7.0-7.6 influenced the concentration of ACTH at which a maximum response occurred (Table 2).

When adrenal tissue is incubated in the presence of ascorbate, cells of the zona reticularis, but not those of the zona fasciculata, have been shown (Chayen et al., 1972) to take up ascorbate, reaching a maximum at 5 hr. At the same time, in the presence of 10^{-3} mol/liter ascorbate, the activities of glucose-6-phosphate dehydrogenase and 6-phosphogluconate dehydrogenase in cells of the zona reticularis return to the levels observed in adrenal biopsy samples. This contrasts with a loss of about half the activity of these enzymes after 5 hr culture in the absence of ascorbate and only partial recovery in the presence of $10^{-10^{-4}}$ mol/liter ascorbate (Chayen et al., 1976).

The culture medium is added up to the level of the lens tissue, which acts as a wick. If tissue is submerged in culture medium it fails to respond adequately. Tissue that has been kept in maintenance culture for 3 hr or less fails to respond to ACTH (Chayen et al., 1976). The 5 hr culture period appears to be essential for recovery of the tissue from the trauma and anoxia of surgical removal and allows time for disappearance of the effects of endogenous hormones.

Priming

Tissue that had not been primed had impaired responsiveness in the section bioassay with a 10-fold increase in the limit of detection (Alaghband-Zadeh, 1974).

Section Cutting

Following tissue culture and priming, tissue may be stored at $-70°C$ for up to 3 days (Alaghband-Zadeh, 1974). It is essential that sections are of uniform thickness and that the manipulations involved do not impair tissue responsiveness to ACTH. Use of a motor-driven cryostat removes the variability of cutting velocity that is the main cause of uneven sectioning. Tissue responsiveness is maintained only if meticulous care is taken to avoid transient thawing of the segments during mounting and cutting. It is important to keep the liquid water used for mounting confined to a small layer of the segment that will be furthest away from the cutting surface.

Section thickness influences the magnitude of response to ACTH, being optimal at 20 μm, which exceeds the diameter of the zona reticularis cells. The response is small and variable at 10 μm thickness, increases up to 20 μm thickness, and shows no increase thereafter (Alaghband-Zadeh, 1974).

Exposure to ACTH

The air-dried sections may be held for 1 hr at room temperature in a dessicator or 3 hr at $-20°C$ before exposure to standards and unknowns (Alaghband-Zadeh et al., 1974).

The use of the colloid stabilizer Polypep 5115 in the aqueous standards preserves cellular integrity as judged by phase contrast microscopy (Chayen et al., 1976). The 5% concentration used represents a compromise. At lower concentrations, the response to ACTH is irregular due to loss of cell solutes by diffusion; at higher concentrations, the response is reduced, probably due to inhibition of the binding and subsequent plasma membrane events by the colloid stabilizer (Alaghband-Zadeh, 1974).

The time course of ascorbate depletion in the zona reticularis following ATCH stimulation is complex. There is a rapid initial fall in ascorbate concentration which continues for 60 sec; this is the response parameter used in the cytochemical section bioassay. A short period of increasing ascorbate concentration ensues, and this in turn is followed by a second and more prolonged fall (Chayen et al., 1976). Similar changes are seen in the segment bioassay (Chayen et al., 1974). During the same period, the ascorbate concentration in the zona fasciculata shows little change or even increases (Chayen et al., 1972).

The duration of the initial decrease in ascorbate concentration may be no more than 20 sec if standards are made up in Trowell's T8 medium without added ascorbate, and the inclusion of 10^{-3} mol/liter ascorbate in the medium at this stage prolongs the initial fall to permit a 60 sec incubation period (Alaghband-Zadeh, 1974).

It is known that within less than 1 min of injeciton of ACTH in vivo, ascorbate appears in the adrenal venous blood (Liddle et al., 1962). Colurso (1979) explored the fate of ^{14}C-ascorbate included in the 5-hr culture medium. She found that at the end of the culture period the adrenal contained 2.5-3.0 mg ascorbate per gram tissue, a level similar to that found in vivo. After treatment of sections with standards or unknowns for 1 min, there was a 7-10% loss of ascorbate. The amount of ^{14}C-ascorbate released into the culture medium was not directly proportional to that lost from the zona reticularis as measured by Prussian blue staining.

The early loss of ascorbate in the zona reticularis is linearly correlated with the amount of corticosteroid secreted into the culture medium (Chayen et al., 1974), and both bear a linear relationship to the logarithm of the ACTH dose used for stimulation over the 1000-fold range used in the assay.

Staining

The Prussian blue reaction was originally introduced as a stain for sulfhydryl groups, but Loveridge et al. (1975) showed that the reactivity of the stain toward different reducing substances could be modified by the relative concentrations of ferricyanide ions and ferric ions in the mixture. They tested the reactivity of the stain to ascorbate, glutathione, and cysteine dissolved in gelatin disks which were sectioned at 10 μm. When ferric ions exceeded ferricyanide ions in the ratio of 50:1, ascorbate reacted rapidly and preferentially, exceeding the molar response of glutathione and cysteine by factors of 7 and 2, respectively. When ferric and ferricyanide ions were present in nearly equimolar amounts, glutathione and cysteine were more reactive than ascorbate. Sections of guinea pig adrenal reacted strongly with the 50:1 molar ratio stain, and gave a linear response to logarithmic increments of ACTH stimulation. The equimolar mixture produced little reaction, suggesting that glutathione and cysteine are probably not determinants of the measured response in the assay. These authors also used an alcoholic silver nitrate stain selective for ascorbate and obtained a similar linear response to logarithmic increments of ACTH.

In the early descriptions of the cytochemical bioassay, it was noted that there was some unevenness of staining of the zona reticularis. Alaghband-Zadeh et al. (1972) found that a more even staining resulted if the ferric ammonium sulfate in the original stain was replaced by ferric chloride. This modification also yielded a more intense stain, and the slope of the dose-response curve of the assay was increased.

Standards

ACTH potency is measured in International Units (U), defined since 1962 by The Third International Standard for Corticotrophin. This material is of porcine origin. It is provided in vials containing 5 subcutaneous U, and the USP

Corticotrophin Reference Standard is derived from the same material. Confusion may arise because this material also has a potency designation of 1.5 intravenous U. The difference arose because of a requirement to maintain comparability with previous international reference material even though bioassays using the subcutaneous route of administration had given way to assays using intravenous administration. Ten picograms of α1-24-ACTH is equivalent to approximately one intravenous microunit (μU) of the third international ACTH standard. It has been shown that α1-24-ACTH is equipotent on a molar basis with α1-39-ACTH in the isolated cell bioassay (Sayers, 1977) and in the cytochemical bioassay (Chayen et al., 1972; 1976).

Stability of ACTH in Plasma

ACTH in plasma rapidly loses its biological activity in vitro, a 90% loss being described when plasma samples were left for 1 hr at room temperature (Holdaway et al., 1974). Besser et al. (1971), using the Lipscomb-Nelson bioassay, found in vitro disappearance half-times in plasma of 30-197 min at 37°C. To minimize destruction by proteolytic enzymes, heparinized blood specimens must be transported on ice and centrifuged at 4°C within 5 min. Plasma should be stored at –70°C until assayed (Daly, Loveridge, et al., 1974). Jubiz and Nolan (1978) showed that N-ethylmaleimide at a concentration of 1.35 mg/ml greatly delayed ACTH inactivation, as measured by immunoassay, when added to evacuated blood collection tubes before blood was drawn, but this additive has not been evaluated in bioassay systems.

Assay Characteristics

Specificity

Specificity may be described as the extent to which the assay erroneously responds to substances other than the target hormone. Such cross-reactivity is assessed by selective removal of the target hormone prior to assay and by testing the response to known related and potentially reactive materials in the assay. Further evidence for specificity for the target hormone comes from comparisons with other assay techniques, although it has to be recognized that these are seldom definitive, being themselves subject to methodological variables.

Antibody neutralization studies using a rabbit anti-ACTH antibody have been described (Holdaway et al., 1974; Holdaway, Rees, et al., 1973). After incubation with a rabbit anti-ACTH antibody, plasmas from two patients, one with hypopituitarism and the other with ectopic ACTH syndrome, both showed decreases of bioassayable ACTH of 99%. Incubation with nonimmune rabbit serum under similar conditions showed a loss of bioassayable ACTH of 50 and 83% in the two patients, reflecting the instability of ACTH in plasma. The residual plasma ACTH bioactivity after incubation with antibody was 10 fg/ml in the hypopituitary patient and 3 pg/ml in the patient with ectopic ACTH syndrome, indicating an exceedingly low level of cross-reacting material under the two extremes of hormone deficit and excess.

Cross-reactivity studies (Holdaway et al., 1974) have shown less than 0.01% molar reactivity for synthetic bovine β-MSH, synthetic α-MSH, human luteinizing hormone (LH), ovine prolactin, and a C-terminal α18-39-ACTH fragment. Cyclic AMP in concentrations up to 5 mg/ml is inactive, but its dibutyryl analog, which readily penetrates the cell membrane, is active in the cytochemical bioassay and generates a dose-response curve parallel to that of ACTH.

In a direct comparison between the cytochemical bioassay, the Lipscomb and Nelson bioassay, and two radioimmunoassays (Rees et al., 1973), the Third International Standard was added to horse serum to give concentrations of 0.5, 5, 50, and 200 U/ml. Within the range of sensitivity of the assays, there was complete agreement among all four assay systems.

Sensitivity

A concentration of 5 fg ACTH per milliliter incubation medium regularly produces an ascorbate depletion distinguishable from zero (Daly, Loveridge, et al., 1974). Since plasma at a dilution of 1 in 10 causes no loss of parallelism in the assay, it is possible to detect 50 fg ACTH per milliliter plasma. At this detection level, a plasma volume of 1.1 ml is required to provide the requisite 1 in 10 and 1 in 100 dilutions. Under the more usual operating conditions of the assay, dilutions of 1 in 100 and 1 in 1000 are used, and a detection limit of 0.5 pg/ml is achieved using 11 μl plasma.

The extraordinary sensitivity of the cytochemical bioassay is exemplified by the work of Chambers and Chayen (1976) using membrane-bound 5'-nucleotidase as a marker of response. Significant activation of this enzyme was found with an ACTH concentration of 5×10^{-20} g/ml, although the slope of the dose-response line was too shallow for this to be used as a routine assay.

Precision

The fiducial limits (P = 0.95) of the assay extend to 87-115% (Daly, Loveridge, et al., 1974). These workers reported that results calculated from each of the two dilutions of 100 consecutive unknown samples deviated from the mean of the two results by more than 15% only 5 times with plasma ACTH concentrations ranging from 4 to 320 pg/ml.

In a detailed study of assay precision, Anthony et al. (1979) found a variability of section thickness of ±4% (1 coefficient of variation). The variability between sections stained with Prussian blue was ±3-6%, while variability on repeated scanning of a single specimen area was 1-2%. The overall index of precision (λ) over a 4-point calibration curve was 0.10-0.21. Similar values for assay variability have been reported by Chayen et al. (1976) and by Daly et al. (1977).

Accuracy

If accuracy is defined as the extent to which the target hormone is being measured by the assay, this may be evaluated by recovery experiments. Addition of 250 pg/ml ACTH to a plasma initially containing 15 pg/ml showed recoveries of 236 and 200 pg/ml when assayed by the segment assay on 2 consecutive days (Chayen et al., 1972). In another report, additions of ACTH at 50 and 100 pg/ml to a plasma containing 69 pg/ml ACTH resulted, in the section assay, in mean recoveries of 102 and 101%, respectively (Alaghband-Zadeh, 1974).

Comparison of Bioassay with Radioimmunoassay

Apart from the study made on hormone added to horse serum (Rees et al., 1973), where the results of cytochemical bioassay and two radioimmunoassays were in complete agreement (see above), direct comparisons between radioimmunoassay and cytochemical bioassay for ACTH reveal variable diffferences, the immunoassay results always being the greater. The differences are least in the resting state, increase in the stimulated state, and are greatest in the post-stimulation period.

Using immunoassay and cytochemical bioassay, Fleisher et al. (1974) compared ACTH results in 18 subjects subjected to insulin-induced hypoglycemia. Immunoassay results at rest were some 5% above those obtained by bioassay; at 60 min, the immunoassay results were 60% above bioassay results. Similar findings comparing an in vivo bioassay with immunoassay results were reported by Besser et al. (1971) and by Matsuyama et al. (1972), and using an isolated adrenal cell assay by Liotta and Krieger (1975). In experiments on rats, Anthony et al. (1979) found that the immunoreactive/bioactive ratio decreased on repeated noise stress. This may be due to preferential release of bioactive ACTH from the pituitary when it contains more newly synthesized hormone.

A large part of the difference between immunoassay and bioassay is explained by the prolongation of the half-life of disappearance of immunoreactive ACTH compared with that of bioactive ACTH (Krieger and Allen, 1975). There are also artifacts related to plasma interference in some radioimmunoassays (Moldow and Yalow, 1980). Major discrepancies between immunoassay and bioassay have been described in plasma samples from patients with ectopic production of ACTH-like molecules from nonendocrine tumors (Bloomfield et al., 1977; Rees et al., 1977).

PHYSIOLOGICAL STUDIES AND CLINICAL INVESTIGATIONS

The great sensitivity of the cytochemical bioassay for ACTH permits assay on small sample volumes from children and repeated sampling at short time intervals under controlled conditions of stimulation or suppression when the repeated

10 ml blood volumes required for immunoassay would lead to unacceptable blood loss. The sensitivity also allows investigation of subjects with low or subnormal ACTH concentrations as in hypopituitarism. Investigations of the pituitary adrenal axis in rats under conditions of stimulation and suppression have been made (Buckingham and Hodges, 1974; 1975; Anthony et al., 1979). Tissue extracts from lung and pituitary have been assayed (Holdaway, Bloomfield, et al., 1973; Hodges and Vellucci, 1975; Bloomfield et al., 1977), and the assay has been used in the two-stage measurement of corticotropin releasing factor (Gillham et al., 1975; Buckingham and Hodges, 1976).

The biological half-life of ACTH has been evaluated by Rees et al. (1973), who injected 100 mg cortisol intravenously into a volunteer at rest and monitored the change in the circulating level of ACTH. After 4 hr the ACTH had fallen from 43 to 0.035 pg/ml. The rate of disappearance indicated a biological half-life of a normal level of endogenous ACTH of 10.4 min. Daly, Fleisher, Chambers, et al. (1974) in a similar study found a half-life of 7 min. These workers also injected α1-24-ACTH into human volunteers following dexamethasone suppression. The peak plasma levels of ACTH indicated a volume of distribution roughly equal to the blood volume. The biological half-life of α1-24-ACTH was approximately 9 min. By contrast, Tanaka et al. (1978) found a mean half-life for immunoreactive ACTH of 40 min. A substituted α1-18-ACTH was shown by Daly, Fleisher, Chambers, et al. (1974) to have a prolonged half-life in vivo. Reader et al. (1976) showed short-term oscillations of ACTH levels in human volunteers over the range 17-62 pg/ml when plasma samples were assayed at 15 min intervals over 2 hr, starting at 0930 hr. Infusion of cortisol sufficient to raise plasma levels to 20-25 g per 100 ml resulted in a gradual decline of ACTH levels to 3-7 pg/ml with diminution of spontaneous oscillations. In hospital patients, Daly, Fleisher, Chambers, et al. (1974) showed a circadian rhythm with a mean ACTH plasma level at 0800 hr of 70 pg/ml and at 2000 hr of 24 pg/ml, corresponding to mean cortisol levels of 18 μg per 100 ml at 0800 hr and 7 μg per 100 ml at 2000 hr.

Holdaway, Rees, et al. (1973) measured plasma ACTH in 10 μl heel-prick blood samples from 14 healthy babies aged 7-10 days and in blood samples of healthy adults. The mean ACTH level in the babies was 63 pg/ml with a range of 13-137 pg/ml and in the adults, 30 pg/ml, with a range of 8-50 pg/ml. These workers also measured the ACTH levels in six patients with hypopituitarism and in a patient with an autonomous adrenal tumor secreting cortisol. All patients had ACTH levels of less than 1 pg/ml. Three of the hypopituitary subjects showed small but measurable increases in plasma ACTH in response to insulin-induced hypoglycemia.

Daly, Fleisher, Glass, et al. (1974) evaluated the pituitary responsiveness under hypoglycemic stress in three groups of patients with rheumatoid arthritis who had been treated with long-term ACTH, long-term corticosteroid, and no hormone therapy, respectively. In the ACTH-treated group the stress-related secretion of endogenous ACTH was delayed but the total amount of ACTH secreted was normal. In the corticosteroid-treated group, both the rate of secretion and the total amount of ACTH secreted were reduced.

FUTURE CLINICAL APPLICATIONS

As the technique of cytochemical bioassay becomes more widely available, it may be expected to find increasing application in conjunction with immunoassay in defining those situations where divergence between immunoreactive and bioactive hormone occurs, especially in ectopic production of hormone by tumors and in disorders of biosynthesis of hormone. The exquisite sensitivity of the bioassay permits continued definition of pituitary-adrenal status in the neonate and in situations of deficient ACTH production.

REFERENCES

Alaghband-Zadeh, J. (1974). Development of a section-bioassay for the routine assay of corticotrophin. *Ann. Clin. Biochem.*, 11:43-48.

Alaghband-Zadeh, J., Daly, J. R., Bitensky, L., and Chayen, J. (1974). The cytochemical section assay for corticotrophin. *Clin. Endocrinol. (Oxf)*, 3:319-327.

Alaghband-Zadeh, J., Daly, J. R., Tunbridge, R. D. G., Loveridge, N., and Chayen, J. (1972). Methodological refinements in the redox bioassay for adrenocorticotrophin. *J. Endocrinol.*, 58:xix.

Allen, R. G., Orwoll, E., Kendall, J. W., Herbert, E., and Paxton, H. (1980). The distribution of forms of adrenocorticotropin and β-endorphin in normal, tumorous and autopsy human anterior pituitary tissue. *J. Clin. Endocrinol. Metab.*, 51:376-380.

Anthony, A., Brister, N. W., and Colurso, G. J. (1979). Cytochemical bioassay and radioimmunoassay of ACTH in noise stressed rats. *J. Histochem. Cytochem.*, 27:1380-1381.

Azzopardi, J. G., and Williams, E. D. (1968). Pathology of "nonendocrine" tumors associated with Cushing's Syndrome. *Cancer*, 22:274-286.

Bajusz, S., Medzihradszky, K., Paulay, Z., and Lang, Z. (1967). Total-Synthese des meuschlichen Cortictropins. *Acta Chim. Acad. Sci. Hung.*, 52:335-341.

Besser, G. M., Orth, D. N., Nicholson, W. E., Byyny, R. L., Abe, K., and Woodham, J. P. (1971). Dissociation of the disappearance of bioactive and radioimmunoreactive ACTH from plasma to man. *J. Clin. Endocrinol. Metab.*, 32:595-603.

Bitensky, L., Butcher, R. G., and Chayen, J. (1973). Quantitative cytochemistry in the study of lysosomal function. In *Lysosomas in Biology and Pathology*, Vol. 3, J. T. Dingle (ed.). North Holland, Amsterdam, pp. 465-510.

Bloomfield, G. A., Holdaway, I. M., Corrin, B., Ratcliffe, J. G., Rees, G. M., Ellison, M., and Rees, L. H. (1977). Lung tumors and ACTH production. *Clin. Endocrinol. (Oxf)*, 6:95-104.

Buckingham, J. C., and Hodges, J. R. (1974). Interrelatinships of pituitary and plasma corticotrophin and plasma corticosterone in adrenalectomized and stressed, adrenalectomized rates. *J. Endocrinol.*, 63:213-222.

Buckingham, J. C., and Hodges, J. R. (1975). Hypothalamopituitary-adrenal activity in the rat after treatment with betamethasone. *J. Endocrinol.*, 65: 14P.

Buckingham, J. C., and Hodges, J. R. (1976). Effects of hypothalamic extracts and corticosteroids on pituitary adrenocorticotrophic activity in vitro. *J. Endocrinol.*, 69:34P.

Byus, C. V., and Russell, D. H. (1975). Ornithine decarboxylase activity: Control by cyclic nucleotides. *Science*, 187:650-652.

Chambers, D. J., and Chayen, J. (1976). The response of a plasma membrane enzyme to very low concentrations of corticotrophin in the cytochemical section bioassay system. *J. Endocrinol.*, 68:24P.

Chayen, J. (1980). The cytochemical bioassay of polypeptide hormones. In *Monographs on Endocrinology*, Vol. 17. F. Grass, A. Labhart, T. Mann, and J. Zander (eds.). Springer Verlag, Berlin.

Chayen, J., Bitensky, L., Chambers, D. J., Loveridge, N., and Daly, J. R. (1974). Studies on the mechanisms of cytochemical bioassays. *Clin. Endocrinol. (Oxf)* 3:349-360.

Chayen, J., Daly, J. R., Loveridge, N., and Bitensky, L. (1976). The cytochemical bioassay of hormones. *Recent Prog. Horm. Res.*, 32:33-79.

Chayen, J., Loveridge, N., and Daly, J. R. (1971). The measurable effect of low concentrations (pg/ml) of ACTH on reducing groups of adrenal cortex maintained in organ cultures. *Clin. Sci.*, 41:2P.

Chayen, J., Loveridge, N., and Daly, J. R. (1972). A sensitive bioassay for adrenocorticotrophic hormone in human plasma. *Clin. Endocrinol. (Oxf).* 1: 219-233.

Colurso, G. J. (1979). Biochemical and Histochemical Correlates of Pituitary Adrenocortical Activation in Rats Exposed to Noise Stress. Ph.D. Thesis, Pennsylvania State University, p. 109.

Daly, J. R., Alaghband-Zadeh, J., Loveridge, N., and Chayen, J. (1977). The cytochemical bioassay of corticotropin (ACTH). *Ann. N.Y. Acad. Sci.*, 297:242-259.

Daly, J. R., Loveridge, N., Bitensky, L., and Chayen, J. (1972). Early experience with a highly sensitive bioasssay for ACTH. *Ann. Clin. Biochem.*, 9:81-84.

Daly, J. R., Loveridge, N., Bitensky, L., and Chayen, J. (1974). The cytochemical bioassay of corticotrophin. *Clin. Endocrinol. (Oxf)*, 3:311-318.

Daly, J. R., Fleisher, M. R., Chambers, D. J., Bitensky, L., and Chayen, J. (1979). Application of the cytochemical bioassay for corticotrophin to clinical and physiological studies in man. *Clin. Endocrinol. (Oxf)*, 3:335-345.

Daly, J. R., Fleisher, M. R., Glass, D., Chambers, D. J., Bitensky, L., and Chayen, J. (1974). Comparison of effects of long-term corticotrophin and corticosteroid treatment on responses of plasma growth hormone, ACTH, and corticosteroid to hypoglycemia. *Br. Med. J.*, 2:521-524.

Demura, H., West, C. D., Nugent, C. A., Nakagawa, K., and Tyler, F. H. (1966). A sensitive radioimmunoassay for plasma ACTH levels. *J. Clin. Endocrinol. Metab.*, 26:1297-1302.

De Nicola, A. F., Clayman, M., and Johnstone, R. M. (1968). Hormonal control of ascorbic acid transport in rat adrenal glands. *Endocrinology*, 82:436-446.

Eipper, B. A., and Mains, R. E. (1980). Structure and biosynthesis of pro-adrenocorticotropin-endorphin and related peptides. *Endocrinol. Rev.*, 1:1-27.

Elton, R. L., Zarrow, I. G., and Zarrow, M. X. (1959). Depletion of adrenal ascorbic acid and cholesterol. *Endocrinology*, 65:152-157.

Ferguson, J. J., Jr., and Morita, Y. (1965). RNA synthesis and adrenocorticotropin responsiveness. *Biochim. Biophys. Acta*, 87:348-350.

Fleisher, M. R., Glass, D., Bitensky, O., Chayen, J., and Daly, J. R. (1974). Plasma corticotrophin levels during insulin-hypoglycemia: Comparison of radioimmunoassay and cytochemical bioassay. *Clin. Endocrinol. (Oxf)*, 3:203-208.

Gilham, B., Jones, M. T., Hillhouse, E. W., and Burden, J. (1975). Preliminary observations on the nature of corticotrophin-releasing hormone from the rat hypothalamus in vitro. *J. Endocrinol.*, 65:12P-13P.

Grahame-Smith, D. G., Butcher, R. W., Ney, R. L., and Sutherland, E. W. (1967). Adrenosine 3′,5′-monophosphate as the intracellular mediator of the action of adrenocorticotropic hormone on the adrenal cortex. *J. Biol. Chem.* 242:5535-5541.

Gray, C. E., and Ratcliffe, J. G. (1979). Clinical evaluation of a radioimmunoassay for β-MSH-related peptides (lipotrophins) in human plasma. *Clin. Endocrinol. (Oxf)*, 10:163-172.

Hechter, O., Zaffaroni, A., Jacobson, R. P., Levy, H., Jeanloz, R. W., Schenker, V., and Pinous, G. (1951). The nature and the biogenesis of teh adrenal secretory product. *Recent Prog. Horm. Res.*, 6:215-241.

Hilf, R. (1965). The mechanisms of ACTH. *N. Engl. J. Med.*, 273:798-811.

Hilf, R., Breuer, C., and Borman, A. (1961). Adrenal adenine nucleotide metab-
olizing enzymes: Alterations induced by ACTH treatment. *Arch. Biochem.
Biophys.*, 94:319-327.

Hodges, J. R., and Hotston, R. T. (1970). Ascorbic acid deficiency and pituitary
adrenocortical activity in the guinea pig. *Br. J. Pharmacol.*, 40:740-746.

Hodges, J. R., and Vellucci, S. V. (1975). The effect of reserpine on hyopthal-
amo-pituitary-adrenocortical function in the rat. *Br. J. Pharmacol.*, 53:555-
561.

Holdaway, I. M., Bloomfield, G., Ratcliff, J. C., Hinson, K-W. F., Rees, G. M.,
and Rees, L. H. (1973). Adrenocorticotrophin levels in normal and neoplastic
lung tissue. In *Endocrinology,* S. Taylor (ed.). Heinemain, London, pp. 309-
315.

Holdaway, I. M., Rees, L. H., and Landon, J. (1973). Circulating corticotrophin
levels in severe hypopituitarism and in the neonate. *Lanced* ii:1170-1172.

Holdaway, I. M., Rees, L. H., Ratcliffe, J. G., Besser, G. M., and Kramer, R. M.
(1974). Validation of the redox cytochemical assay for corticotrophin. *Clin.
Endocrinol. (Oxf),* 3:329-334.

IUPAC-IUB Commission of Biochemical Nomenclature (1967). Rules for naming
synthetic modifications of natural peptides. *Biochemistry,* 6:362-364.

Jubiz, W., and Nolan, G. (1978). N-ethylmaleimide prevents destruction of corticotropin (ACTH) in plasma. *Clin. Chem.,* 24:826-827.

Kloppenborg, P. W. C., Island, D. P., Liddle, G. W., Michelakis, A. M., and Nicholson, W. E. (1968). A method of preparing adrenal cell suspensions and its
applicability to the in vitro study of adrenal metabolism. *Endocrinology,* 82:
1053-1058.

Koritz, S., Bhargava, G., and Schwartz, E. (1977). ACTH action on adrenal
steroidogenesis. *Ann. N.Y. Acad. Sci.,* 297:329-335.

Krieger, D. T., and Allen, W. (1975). Relationship of bioassayable and immunoassayable plasma ACTH and cortisol concentrations in normal subjects and
in patients with Cushing's disease. *J. Clin. Endocrinol. Metab.,* 10:675-687.

Krieger, D. T., Liotta, A. S., Suda, T., Goodgold, A., and Condon, E. (1979).
Human plasma immunoreactive lipotropin and adrenocorticotropin in normal
subjects and in patients with pituitary-adrenal disease. *J. Clin. Endocrinol.,
Metab.,* 48:566-576.

Landon, J., and Greenwood, F. C. (1968). Homologous radioimmunoassay for
plasma levels of corticotrophin in man. *Lancet,* i:273-276.

Lefkowitz, R. J., Roth, J., and Pastan, I. (1970). Radioreceptor assay of adrenocorticotropic hormone: New approach to assay of polypeptide hormones
in plasma. *Science,* 170:633-635.

Levine, J. H., Nicholson, W. E., Peytremann, A., and Orth, D. N. (1975). The
mechanism of ACTH stimulation of adrenal ornithine decarboxylase activity.
Endocrinology, 97:136-144.

Li, C. H. (1959). Proposed system of terminology for preparations of adrenocorticotropic hormone. *Science,* 129:969-970.

Liddle, G. W., Island, D., and C. K. Meador (1962). Normal and abnormal regulation of corticotrophin secretion in man. *Recent Prog. Horm. Res.,* 18:125-153.

Liotta, A., and Krieger, D. T. (1975). A sensitive bioassay for the determination of human plasma ACTH levels. *J. C.in. Endocrinol. Metab.,* 40:268-277.

Lipscomb, H. S., and Nelson, D. H. (1962). A sensitive biologic assay for ACTH. *Endocrinology,* 71:13-23.

Loveridge, N., Alaghband-Zadeh, J., Daly, J. R., and Chayen, J. (1975). The nature of the redox change measured in the cytochemical bioassay of corticotrophin. *J. Endocrinol.,* 67:28P.

Loveridge, N., and Robertson, W. R. (1978). Stimulation of adrenal 5-ene, 3 hydroxysteroid dehydrogenase by corticotrophin in vitro. *J. Endocrinol.,* 78:457-458.

Lowry, P. J., McMartin, C., and Peters, J. (1973). Properties of a simplified bioassay for adrenocorticotrophic activity using the steroidogenic response of isolated adrenal cells. *J. Endocrinol.,* 59:43-55.

Matsuyama, H., Harada, G., Ruhmann-Wennhold, A., Nelson, D. H., and West, C. D. (1972). A comparison of bioassay and radioimmunoassay for plasma corticotropin in man. *J. Clin. Endocrinol. Metab.,* 34:713-717.

Moldow, R. L., and Yalow, R. S. (1980). Artifacts in the radioimmunoassay of ACTH in tissue extracts and plasma. *Horm. Metab. Res.,* 12:105-110.

Nakanishi, S., Inoue, A., Kita, A., Nakamura, M., Chang, A. C. Y., Cohen, S. N., and Numa, S. (1979). Nucleotide sequence of cloned cDNA for bovine corticotropin-β-lipotropin precursor. *Nature,* 278:423-427.

Nelson, D. H. (1980). *The Adrenal Cortex: Physiological Function and Disease.* Saunders, Philadelphia, pp. 16-18.

Nicholson, W. E., Liddle, R. A., and Puett, D. (1976). Corticotropin: Plasma clearance, catabolism and biotransformations. *Endocrinology (Suppl.),* 98:59.

Orth, D. N., Nicholson, W. E., Mitchell, W. M., Island, D. P., and Liddle, G. W. (1973). Biologic and immunologic characterization and physical separation of ACTH and ACTH fragments in the extopic ACTH syndrome. *J. Clin. Invest.,* 52:1756-1769.

Pedersen, R. C., Brownie, A. C., and Ling, N. (1980). Pro-adrenocorticotropin/endorphin-derived peptides: Coordinate action on adrenal steroidogenesis. *Science,* 208:1044-1045.

Perchellet, J.-P.. and Sharma, R. K. (1980). Mediatory role of calcium and guanosine 3′,5′-monophosphate in adrenocorticotrophin-induced steroidogenesis by adrenal cells. *Science,* 203:1258-1259.

Reader, S. C. J., Alaghband-Zadeh, J., Carter, G. D., and Daly, J. R. (1976). Observations on the feedback regulation of adrenocorticotrophin secretion. *J. Endocrinol.*, 71:57P.

Rees, L. H., Bloomfield, G. A., Gilkes, J. J. H., Jeffcoate, W. J., and Besser, G. M. (1977). ACTH as a tumor marker. *Ann. N.Y. Acad. Sci.*, 297:603-620.

Rees, L. H., Ratcliffe, J. G., Besser, G. M., Kramer, R., and Landon, J. (1973). Comparison of the redox assay for ACTH with previous assays. *Nature (New Biol.)*, 241:84-85.

Riniker, B., Sieber, P., Rittel, W., and Zuber, H. (1972). Revised amino-acid sequences for procine and human adrenocorticotrophic hormone. *Nature (New Biol.)*, 235:114-115.

Rosai, J., Levine, G., Weber, W. R., and Higa, E. (1976). Carcinoid tumors and oat cell carcinomas of the thymus. *Pathol. Ann.*, 11:201-226.

Saffran, M., and Schally, A. V. (1955). In vitro bioassay of corticotrophin. *Endocrinology*, 56:523-532.

Sayers, G. (1977). Bioassay of ACTH using isolated cortex cells. *Ann. N.Y. Acad. Sci.*, 297:220-241.

Sayers, G., Sayers, M. A., Lewis, H. L., and Long, C. N. H. (1944). Effect of adrenocorticotropic hormone on ascorbic acid and cholesterol content of the adrenal. *Proc. Soc. Exp. Biol. Med.*, 55:238-239.

Sayers, G., Sayers, M. A., Liang, T. Y., and Long, C. N. H. (1946). Effect of pituitary adrenocorticotropic hormone on the cholesterol and ascorbic acid content of the adrenal of the rat and guinea pig. *Endocrinology*, 38: 1-9.

Sayers, G., Swallow, R. W., and Giordano, N. D. (1971). An improved technique for the preparation of isolated rat adrenal cells: A sensitive accurate and specific method for the assay of ACTH. *Endocrinology*, 88:1063-1068.

Sayers, M. A., Sayers, G., and Woodbury, L. A. (1948). The assay of adreno-corticotrophic hormone by the adrenal ascorbic acid depletion method. *Endocrinology*, 42:379-393.

Sharma, S. K., Johnstone, R. M., and Quastel, J. H. (1963). Active transport of ascorbic acid in adrenal cortex and brain cortex in vitro and the effects of ACTH and steroids. *Can. J. Biochem. Physiol.*, 41:597-604.

Sharma, S. K., Johnstone, R. M., and Quastel, J. H. (1964). Corticosteroids and ascorbic acid transport in adrenal cortex in vitro. *Biochem. J.* 92:564-573.

Tanaka, K., Nicholson, W. E., and Orth, D. N. (1978). Diurnal rhythm and disappearance half-time of endogenous plasma immunoreactive β-MSH (LPH) and ACTH in man. *J. Clin. Endocrinol. Metab.*, 46:883-890.

Taunton, O. D., Roth, J., and Pastan, I. (1969). Studies on the adrenocortico-tropic hormone-activated adenyl cyclase of a functional adrenal tumor. *J. Biol. Chem.*, 244:247-253.

Urquhart, J., and Li, C. C. (1969). Dynamic testing and modelling of adreno-
  cortical secretory function. *Ann. N.Y. Acad. Sci.*, 156:756-778.
Walker, W. H. C., and Keane, P. M. (1977). Theoretical aspects of radioimmuno-
  assay. In *Handbook of Radioimmunoassay*, G. Abraham (ed.). Dekker, New
  York, p. 94.
Wolfsen, A. R., McIntyre, H. B., and Odell, W. B. (1972). Adrenocorticotropin
  measurement by competitive binding receptor assay. *J. Clin. Endocrinol.
  Metab.*, 34:684-689.
Yalow, R. S., Glick, S. M., Roth, J., and Berson, S. A. (1964). Radioimmuno-
  assay of human plasma ACTH. *J. Clin. Endocrinol. Metab.*, 24:1219-1225.

5

Thyroid-Stimulating Hormone

Klaus-Dieter Döhler, Alexander von zur Mühlen, Thomas O. F. Wagner,
Christoph Lucke,* Hans K. Weitzel,‡ and Takuma Hashimoto† / Medizinische
Hochschule Hannover, Hannover, Federal Republic of Germany

D. Emrich / Universität Göttingen, Göttingen, Federal Republic of Germany

NATURE OF THYROID-STIMULATING HORMONE, ITS ORIGIN, RELEASE, AND BIOLOGICAL FUNCTION

Thyroid-stimulating hormone (TSH) is a glycoprotein hormone, produced by, and released from, the anterior pituitary gland. Its production and release are controlled by the brain, mainly by the hypothalamus. Nerve cells in the hypothalamus produce thyrotrophin-releasing hormone (TRH) and release it into the hypothalamo-pituitary portal system where it is transported to the pituitary gland. TRH is a tripeptide, which can be made synthetically.

TSH is composed of two subunits, an α subunit, which is biologically inactive, and a β subunit, which determines the biological function of TSH. The α subunit is common to both TSH and to two gonadotrophic pituitary hormones, follicle-stimulating hormone (FSH) and luteinizing hormone (LH). The main function of TSH is stimulation of thyroid gland growth and activity, which includes iodide accumulation and thyroid hormone synthesis and release. Thyroid hormones are mainly involved in metabolic activities of the organism. They also control the release of TSH from the pituitary gland by means of inhibitory feedback mechanism on the pituitary gland and hypothalamus. Eleva-

---

Present affiliation: Hagenhof-Klinik, Langenhagen, Federal Republic of Germany.
†*Present affiliation*: Kanazawa University Hospital. Kanazawa. Japan.
‡*Present affiliation*: Steglitz Clinic, Free University of Berlin, Berlin, Federal Republic of Germany.

tion of thyroid hormone levels in the circulation will result in inhibition of TSH release. Reduced TSH release will then result in reduced thyroid hormone production and release from the thyroid gland, which result in lower blood levels of thyroid hormones, and, subsequently, less inhibition of TSH release. Interference with this feedback mechanism is the reason for most clinical disturbances of the hypothalamo-pituitary-thyroid system. Determination of TSH and thyroid hormone levels in the plasma of patients with several pituitary and thyroid diseases is, therefore, helpful in diagnosis of the disease. Moreover, TSH determination contributes tremendously to a better understanding of the regulation of thyroid function under physiological and pathophysiological conditions.

DEVELOPMENT AND NATURE OF ASSAY SYSTEMS FOR THE DETERMINATION OF TSH

Classic bioassays were not sensitive enough to determine thyroid-stimulating hormone levels in plasma of normal euthyroid or of hyperthyroid patients. Sensitivity limits of several classic bioassays ranged from 30 to 50 μU/ml (McKenzie, 1958), 25 μU/ml (Brown and Munro, 1967), or from 5 to 10 μU/ml (Chapman et al., 1976). Only after the radioimmunoassay (RIA) had been developed could normal TSH levels be measured (Odell et al., 1967). However, even RIA is not sensitive enough (the sensitivity limit of commonly used RIA for determination of TSH is close to 1 μU/ml plasma) to detect decreased plasma TSH levels like those found in patients with hyperthyroid Graves' disease.

A few years ago, a highly sensitive cytochemical bioassay (CBA) technique was developed (Chayen et al., 1976) for quantitative determination of a variety of hormones, immunoglobulins, and growth-promoting or growth-inhibiting factors in the blood circulation (as described in other chapters of this book). The basic principle of this method is the alteration in biochemical activity of target organ cells by the respective hormones or growth factors of interest. In regard to TSH this method is based on the ability of this hormone to stimulate the endocytosis of colloid by thyroid follicle cells. The colloid vesicles fuse with lysosomes, and the permeability of the lysosomal membranes is increased. Lysosomes are then able to take up added substrate in quantities directly proportional to their permeability. Intralysosomal enzymes, then, hydrolyze the substrate. This mechanism can be made visible and is used for the quantitative determination of previously added TSH. This technique made it possible to detect concentrations of human TSH as low as 5×10^{-5} μU/ml medium (Petersen et al., 1975), which, in conventional radioimmunoassays, are undetectable. Another advantage of this assay over the RIA is its ability to measure biological activities instead of immunological activities. This ability of the CBA is advantageous in clinical conditions where the biological activity of the released TSH may be impaired (see Clinical Applications).

METHODOLOGY OF THE CYTOCHEMICAL BIOASSAY FOR TSH

The Segment Assay

As described by Bitensky et al. (1974), a guinea pig (250-300 g) is killed by asphyxiation with nitrogen and the thyroid gland is rapidly removed by dissection. The gland is trimmed of adhering fat and connective tissue and divided into six segments. Each segment is placed individually in nonproliferative organ culture and maintained with Trowell's T8 medium (Trowell, 1959) at 37°C and pH 7.6 for 5 hr in an atmosphere of 95% O_2:5% CO_2 (for full details see Chapter 3). The medium is then replaced by fresh Trowell's T8 medium containing one of a graded series of concentrations of the standard preparation of TSH (i.e., 10^{-1}-10^{-4} μU MRC Research Standard A per milliliter medium) or the plasma to be assayed, usually diluted 100- and 1000-fold. The purpose of this treatment is to achieve graded labilization of lysosomal membranes in thyroid follicle cells. Maximal reaction is achieved after 7 min of incubation (Bitensky et al., 1974), after which the segments are chilled in n-hexane at –70°C and are stored at this temperature.

The segments are cut at 10 μm in a cryostat (Bright FS/CS/M/LT) at a cabinet temperature of below –25°C and with the knife cooled to –70°C with solid CO_2. The sections are then flash-dried onto warm slides (room temperature). The sections are then reacted for 7.5 min in a substrate solution (pH 6.5; 37°C) containing 8 mg leucine-β-naphthylamide (Serva), 10 mg Fast Blue B (Serva), 1 ml 0.02 M KCN, and 8 ml 0.9% NaCl in every 10 ml of 0.1 M acetate buffer. During this incubation the substrate, leucyl-β-naphthylamide, enters the lysosomes of thyroid follicle cells where it is hydrolyzed by lysosomal naphthylamidases. The reaction product, naphthylamine, is coupled to a diazonium salt (Fast Blue B) to produce a colored precipitate. By the subsequent treatment with 0.1 M $CuSO_4$ the diazonium-naphthylamine complex is chelated, resulting in color intensification.

The slides are then mounted in Farrants' medium (pH 6.5), and the density of the chromogenic reaction product is measured in individual thyroid follicle cells by means of a scanning and integrating microdensitometer (Vickers M85) using a ×100 oil-immersion objective. Readings are taken at a wavelength of 550 nm. The mean of 10-20 readings is computed from each section and is used as reference point for the standard curve or for the plasma samples, respectively. All determinations are done in duplicate.

Although the advantages of this assay (sensitivity and bioactivity) are quite obvious, the major disadvantage is the small number of samples which can be determined within one assay. A guinea pig thyroid gland can only be cut into maximally six segments, without causing major destruction of gland tissue. Thus, only six hormone determinations can be made. As four determinations are reserved to establish the standard curve, only one unknown plasma sample at two dilutions can be measured in each assay.

The Section Assay

For better exploitation of the cytochemical bioassay, efforts were made in London (Gilbert et al., 1977; Chayen et al., 1980; Ealey et al., 1980) and in Hannover (von zur Mühlen et al., 1978) to develop a cytochemical section assay for TSH, which is based on the principle that incubation with TSH is performed on thyroid sections, rather than on thyroid segments.

Guinea pig thyroid glands are bisected and incubated for 5 hr in Trowell's T8 medium, as described for the segment assay. Subsequently the thyroid segments are chilled in n-hexane at $-70°C$ and are stored at this temperature until the following day. The thyroid segments are cut at 14 μm in a cryostat at a cabinet temperature below $-25°C$ and with the knife cooled up to $-70°C$ with CO_2-ice. The thyroid sections are flash-dried onto warm slides (room temperature).

To allow the simultaneous handling of many slides, a simple apparatus was constructed of Perspex according to the design of Alaghband-Zadeh et al. (1974). The apparatus consists of a trough, which is divided into 22 compartments. The lid of the trough has 22 metal clips, into each of which two slides, containing thyroid sections, can be inserted, back-to-back. When the lid is in place, each pair of slides dips into one of the 22 compartments which are filled with incubation medium. The incubation medium (pH 7.6) is composed of Trowell's T8 medium, which contains 0.05 M sodium acetate and 0.02% gum tragacanth (Sigma) as colloid stabilizer. The first four compartments contain, in addition, one of a graded series of concentrations of a TSH standard preparation (see segment assay for details). The next two compartments contain two dilutions of one plasma sample to be assayed, and the next eight pairs of compartments contain two dilutions of another eight plasma samples. With this device nine samples of plasma, each at two dilutions, can be measured within one assay.

Using 14 μm thick thyroid sections, the optimal incubation time has been 90 sec (at $37°C$). After 90 sec the slides, containing the thyroid sections, are removed and immediately immersed into a second trough, which contains the leucyl-β-naphthylamide substrate medium as described for the segment assay. After 6 min of incubation the slides are washed briefly in 0.9% NaCl solution. The subsequent procedures are identical with those described for the segment assay.

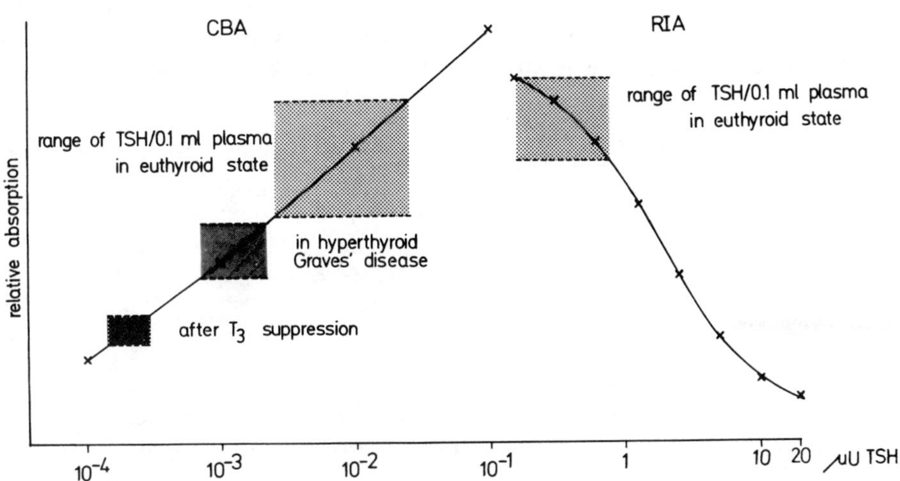

Figure 1   Comparison of a typical cytochemical bioassay (CBA) standard curve
(left) with a typical radioimmunoassay (RIA) standard curve (right). The abscissa indicates the logarithmic concentrations of TSH in microunits (μU) of MRC
Research Standard A. The CBA standard curve ranges from 10^{-4} to 10^{-1} μU
TSH, the RIA standard curve ranges from approximately 10^{-1} to 20 μU TSH.
By using 100 μl of plasma, the TSH range of euthyroid persons is covered by
the lightly dotted areas on each standard curve. In the RIA, euthyroid TSH
levels range from approximately 1 to 8 μU/ml (Lemarchand-Beraud, 1976;
Odell et al., 1967). Lower TSH levels cannot be detected. In the CBA, by using
100 μl plasma samples, which are diluted 100-fold before going into the assay,
it is not only possible to measure the complete range of euthyroid TSH levels
[approximately 0.3-3.2 μU/ml, measured by Petersen et al. (1975), Döhler
et al. (1978), and Hashimoto et al. (1979)], but also the low TSH levels of
hyperthyroid patients (Petersen et al., 1975; Döhler et al., 1978; and Hashimoto et al., 1979) and even lower levels (0.015-0.072 μU/ml, as indicated)
after suppression of TSH release by treatment of normal volunteers with T$_3$
for 7 consecutive days (Petersen et al., 1975). (Modified from Döhler et al.,
1978).

Table 1   Determination of TSH in Normal Human Plasma By CBA
After Precipitative Treatment with Various Concentrations of a Specific
Antibody to Human TSH[a]

Antibody Concentration	Amount of TSH (μU/ml ± SE)
None	1.45 ± 0.35
1:400,000	0.09 ± 0.03
1:100,000	0.037 ± 0.02
1: 50,000	< 0.004

[a]Four determinations were performed for each treatment.

The principles of the two cytochemical bioassays described here, the segment assay and the section assay, are based on the same cytochemical reactions. The section assay, however, has a much larger capacity than the segment assay.

EVALUATION OF THE CYTOCHEMICAL BIOASSAY FOR TSH

Sensitivity

A typical standard curve for the CBA is presented in Figure 1, where it is compared with a typical RIA standard curve. Whereas the RIA standard curve could only register TSH values between approximately 0.1 and 20 μU, the CBA standard curve could detect TSH values as low as 10^{-4} μU. With 100 μl plasma samples the RIA was barely able to measure TSH concentrations below 1 μU/ml. The CBA, however, was able to measure not only the whole range of euthyroid TSH levels [approximately 0.3-3.2 μU/ml measured by Petersen et al. (1975) and Döhler et al. (1978)], but also the low levels in hyperthyroid patients (Petersen et al., 1975; Döhler et al., 1978; Hashimoto et al., 1979) and even lower levels after suppression of TSH release (Petersen et al., 1975) by treatment of normal volunteers with tri-iodothyronine (T$_3$) for 7 consecutive days (see Figure 1).

Specificity

Plasma aliquots from a normal euthyroid volunteer were treated with various concentrations of a specific antibody to human TSH (Kabi). The TSH-antibody complex was precipitated by a double-antibody solid-phase method (Dasp from Organon) and the TSH, remaining in the supernatant, was determined. With increasing antibody concentrations decreasing levels of TSH remained in the supernatant (Table 1). At an antibody concentration of 1:50,000 plasma TSH was completely inactivated and became undetectable in the supernatant. This indicates that the thyroid-stimulating effect of normal human plasma is likely to be due to TSH, and not due to other possible thyroid-stimulating substances.

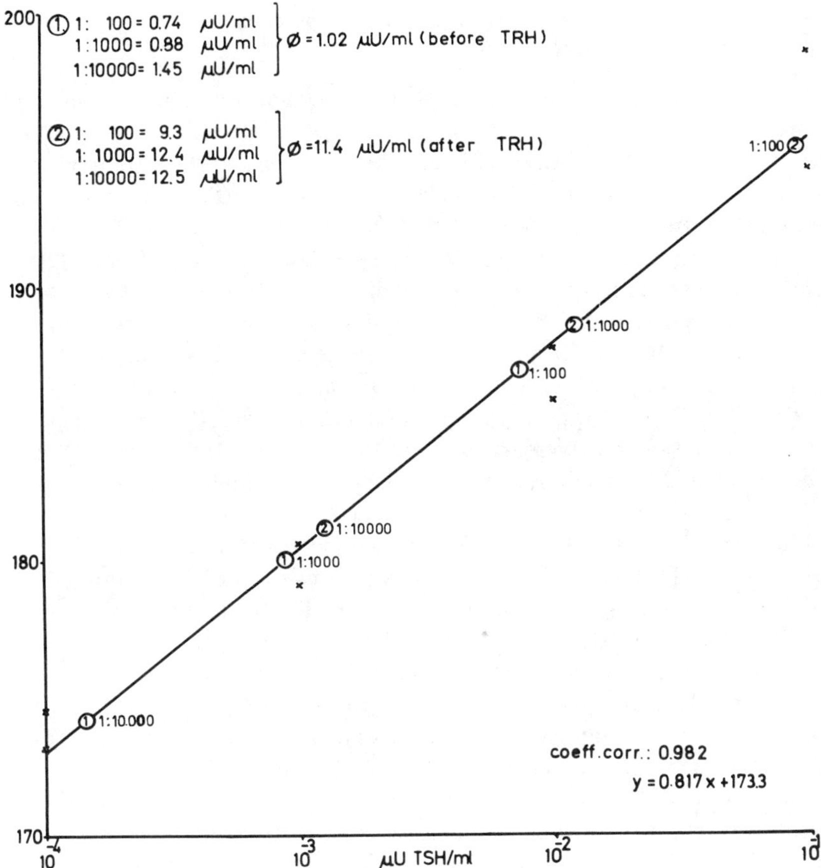

Figure 2   Demonstration of a typical standard curve in a cytochemical bioassay (section assay) for human TSH. A positive linear correlation was regularly observed between the concentration of TSH administered over the range of 10^{-4}-10^{-1} μU/ml medium and the response (relative absorption) of thyroid follicle cells. Coefficient of correlation for this curve was 0.982, the slope y was 0.187x. TSH concentrations in plasma obtained from a normal euthyroid volunteer before (plasma 1) and 30 min after injection of 200 μg TRH (plasma 2) are demonstrated at three dilutions (1:100, 1:1000, and 1:10,000). The numerical values of the actually measured TSH levels are listed in the upper left corner after conversion into concentration per milliliter of original plasma. Note that there was parallelism in the middle part of the curve between the TSH values of plasma dilutions and the TSH values of standard dilutions. There was slight deviation from parallelism only at both extreme ends of the standard curve.

Is There Parallelism Between the Values of Plasma Dilutions and Values of TSH Standard Dilutions?

In a total of 10 cytochemical section assays, a positive linear correlation was regularly observed between the concentration of TSH (MRC Research Standard A) over the range of 10^{-4}-10^{-1} μU/ml of incubation medium and the response (relative absorption) of thyroid follicle cells. Correlation coefficients for the 10 standard curves were between 0.956 and 0.995.

Further certainty that the biologically active substance in plasma is identical to the substance used as standard preparation can be obtained by testing whether the values measured for the different plasma dilutions run parallel to the values measured for the different dilutions of the standard preparation. Plasma was taken from a normal euthyroid volunteer before and 30 min after IV injection of 200 μg TRH. Each plasma sample was diluted 100-, 1000-, and 10,000-fold, and all samples were measured in the same CBA (section assay). As indicated in Figure 2 the measured values of the three basal plasma dilutions were 0.74×10^{-2} (dilution 1:100), 0.88×10^{-3} (dilution 1:1000), and 1.45×10^{-4} (dilution 1:10,000) μU/ml TSH. These values were equivalent to 0.74, 0.88, and 1.45 μU TSH per milliliter of original plasma (mean 1.02 μU/ml). The three dilutions of the plasma sample obtained after TRH treatment yielded 0.93×10^{-1} (dilution 1:100), 1.24×10^{-2} (dilution 1:1000), and 1.25×10^{-3} (dilution 1:10,000) μU/ml TSH (Figure 2). These values were equivalent to 9.3, 12.4, and 12.5 μU/ml of original plasma (mean: 11.4 μU/ml).

Thus, for both plasma samples the TSH values measured in different dilutions were parallel to the values measured in different dilutions of the standard preparation, with the exception of the extreme top and the extreme bottom of the standard curve. This phenomenon of impaired parallelism at very high or very low concentrations is also well known in radioimmunoassays. In the upper part of the CBA standard curve this effect is probably due to supraoptimal lysosomal membrane permeability induced by the high TSH concentrations (Bitensky et al., 1974; Hashimoto et al., 1979).

Interassay Variation in the Cytochemical Section Assay for TSH

TSH levels in plasma of a normal euthyroid volunteer were determined at two dilutions (1:100 and 1:1000) in three different section assays. The TSH values determined in assay 1 were 1.85 and 1.95 μU/ml plasma, in assay 2 they were 1.55 and 1.90 μU/ml, and in assay 3 TSH values were 1.70 and 2.05 μU/ml plasma (Figure 3). Interassay variation of the mean values was 5.1%.

With the development of the cytochemical section assay, a highly sensitive biological method for TSH determination became available, with a reasonable

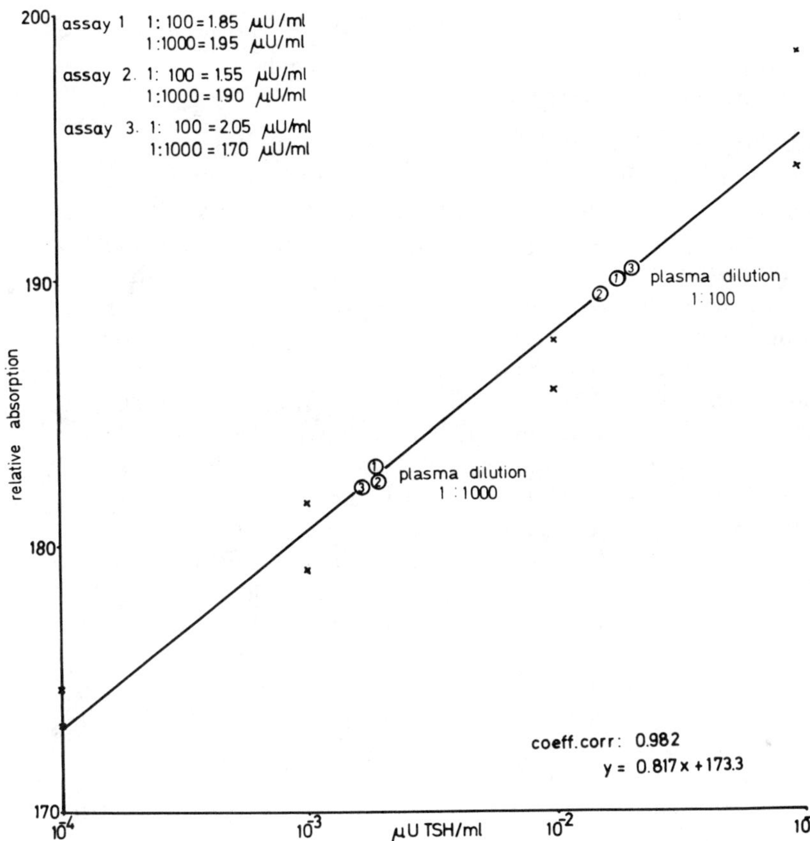

Figure 3 Demonstration of the same standard curve as in Figure 2. TSH levels were determined at two dilutions (1:100 and 1:1000) in three different cytochemical section assays (1, 2, and 3). Individual TSH values, measured in each of the three assays, are listed in the upper left corner after conversion into concentration per milliliter of original plasma.

measuring capacity. Although its capacity is still far from that of the radioimmunoassay, its high sensitivity and its ability to measure biological activities make it a valuable tool for the study of such clinical and scientific problems which cannot be solved by use of RIA techniques. Some of these clinical and scientific problems and the application of the cytochemical bioassay will be discussed below.

CLINICAL APPLICATIONS OF THE CBA FOR TSH

TSH Activity Determined in Human Plasma by CBA and by RIA Before and After TRH Administration

The TRH test is one of the most important tests today for the study of pituitary-thyroid function. In a normally functioning system, the intravenous (IV) injection of 200 μg of the thyrotropin-releasing hormone will result in rapid release of TSH from the pituitary gland, leading to significant elevation of TSH levels in the plasma after 20-30 min. Where the function of the pituitary-thyroid feedback system is disturbed, TRH treatment will show different results. If the thyroid gland is continuously hyperactive (primary hyperthyroidism), the resulting elevation of thyroid hormone levels in the blood will cause inhibition of TSH synthesis in the pituitary gland. The thyrotropic cells of the pituitary gland will be depleted of TSH and are thus unable to respond with TSH release after a single stimulation by TRH (negative TRH test). Also, in cases where the thyroid gland does not produce and release normal amounts of thyroid hormones, due to a primary defect in pituitary function (primary hypopituitarism or secondary hypothyroidism), TRH treatment will be without effect, because the target gland forTRH, the pituitary gland, is not functioning. Thus a TRH test provides information about the functional state of the pituitary gland; it does not provide information about the primary causes for pituitary defects (primary hyperthyroidism, primary hypopituitarism, or primary nonfunctioning of the hypothalamus). For proper diagnosis of pituitary-thyroid dysfunction, the TRH test alone is, therefore, not sufficient. Supplementary tests or measurements (i.e., determination of thyroid hormone levels) have to be performed.

Methodology of Supplementary Laboratory Examinations
Plasma TSH levels were determined by CBA and by RIA before, and 30 min after, IV injection of 200 μg TRH. The same reference standard (MRC Research Standard A) was used in both assay systems. Antiserum for the RIA was obtained from Kabi (Munich). In the RIA the standard curve was run in TSH-low plasma (0.45 μU/ml, determined by CBA), which had been obtained from normal volunteers after treatment with thyroxine (T_4) for several days. Human TSH-α and human TSH-β subunit RIA were performed with materials from the National Institute of Arthritis, Metabolism and Digestive Diseases (NIAMDD) in Bethesda, Maryland (USA). Human luteinizing hormone (materials from NIAMDD), human prolactin (Serono), estradiol, and progesterone were determined by RIA. Serum tri-iodothyronine was measured by RIA as described elsewhere (Hesch et al., 1972). Serum T_4 was determined by the commercial competitive protein binding test (Tetralute Miles). Tanned red-cell hemagglutination methods were used for this detection of thyroglobulin antibodies and

Table 2 Plasma TSH Levels of Four Normal Subjects Determined by CBA and by RIA Before and After IV Treatment with 200 µg TRH

| | | Plasma TSH Levels (µU/ml) | | | |
| | | Before TRH | | 30 min after TRH | |
Patient	Sex	CBA	RIA	CBA	RIA
U.O.	F	1.06	4.2	20.25	18.4
D.S.	F	1.20	2.9	8.50	9.4
D.A.	M	2.30	4.2	18.50	16.5
B.P.	F	3.20	5.3	19.00	16.0
Mean		1.94	4.15	16.56	15.08
± Standard error		±0.50	±0.49	±2.71	±1.96

thyroid microsomal antibodies (Amino et al ,.1976) using a commercial test kit (Fujizoki Pharmaceutical Co.). It has been shown previously (Amino et al., 1976) that 29% of patients with Graves' disease are thyroglobulin antibody positive (titer 1:20), and 85% are microsomal antibody positive (titer 1:160).

TRH Test in Euthyroid Subjects
In Table 2 plasma TSH concentrations are shown in four normal euthyroid volunteers before, and 30 min after, the IV injection of 200 µg TRH. Biologically active plasma TSH concentrations increased significantly (p < 0.01, paired student's t-test) in all subjects after TRH treatment. There was parallelism between the responses to the plasma dilutions and those to the reference standard dilutions, which indicates that the biologically active, thyroid gland-stimulating, substance in the plasma of normal volunteers was likely to be identical to the reference standard (human TSH). There was general agreement between results obtained by CBA or RIA at elevated plasma TSH concentrations. There was more divergence at lower concentrations, which is probably due to the limit of sensitivity of the RIA. The TSH levels, measured in plasma of euthyroid subjects before and after treatment with TRH, are in the same range as those reported by Petersen et al. (1975).

TRH Test in Patients with Pituitary Unresponsiveness
A large number of reports have been published demonstrating unresponsiveness of pituitary function to TRH stimulation in various thyroid disorders, such as treated and untreated Graves' disease (Buerklin et al., 1976; Clifton-Bligh et al., 1974; Emrich et al., 1976; Harada et al., 1975; Hesch et al., 1974; Kumahara et al., 1971; Martino et al., 1976; von zur Mühlen et al., 1974; Sanchez-Franco et al., 1974), subacute thyroiditis (Ogihara et al., 1973), nontoxic nodular goiter

(Elte et al., 1976; Emrich and Bähre, 1978; Gemsenjäger et al., 1976; Kirkegaard et al., 1977), and euthyroid Graves' disease (Chopra et al., 1973, 1974; Clifton-Bligh et al., 1974; Hesch et al., 1974). The lack of sensitivity of existing assay systems and the inability to determine biological activities have made intrepretations rather difficult. Whether the unresponsiveness of the pituitary gland to TRH stimulation was due only to the limited sensitivity of the TSH-RIA system was tested in patients with hyperthyroid Graves' disease, before and after treatment, by the use of the highly sensitive cytochemical bioassay.

TRH Test in Patients with Hyperthyroid Graves' Disease.   Biologically active TSH was detected by CBA in plasma of three patients with untreated Graves' disease. The mean concentration, however, was rather low (0.15 μU/ml) (Table 3). No increase in the apparent TSH levels was seen in the plasma samples taken after injection of 200 μg TRH. In all cases TSH levels determined by RIA were below the limit of detection ($<$ 1.25 μU/ml). Treatment of the plasma with a specific antibody to human TSH resulted in the complete loss of TSH-like activity (Table 3), which indicates that the small amount of biological thyroid-stimulating activity in patients with Graves' disease was likely to be due to TSH, and not due to other thyroid-stimulating substances. All three patients had microsomal antibody titers of 1:8000. Only one patient had detectable amounts of thyroglobulin antibodies (titer 1:25). In all cases, plasma T_3 and T_4 levels were elevated (Table 3).

We confirmed earlier reports by Petersen et al. (1975; 1976) that basal plasma TSH levels are low, but measurable by CBA, in patients with Graves' disease. We also demonstrated that there was no further increase in biologically active TSH levels in these patients after TRH administration (Table 3).

TRH Test in Patients with Graves' Disease After Treatment.   Various reports have been published concerning pituitary responsiveness to TRH in hyperthyroidism during the course of treatment (Buerklin et al., 1976; Clifton-Bligh et al., 1974; Emrich et al., 1976; Harada et al., 1975; Kumahara et al., 1971; Martino et al., 1976; von zur Mühlen et al., 1971; 1975). They have pointed out a time lag between normalization of circulating thyroid hormone levels and the recovery of responsiveness of thyrotropic pituitary cells to TRH stimulation. We determined plasma TSH levels in two patients with Graves' disease, who had been treated either by subtotal thyroidectomy (2 years previously) or by radioiodine (4 years previously). In the RIA their plasma TSH levels before and after the TRH test were below the limit of assay detection. By use of the CBA, low but detectable biologically active plasma TSH levels were measured (Table 4), which increased 4.5- to 10-fold after TRH treatment. Both patients had rather high titers of microsomal antibodies (1:2000 and 1:8000, respectively), but no detectable levels of thyroglobulin antibodies. Thyroid hormone levels in the plasma were normal (Table 4).

Table 3 TRH Test and Plasma Levels of TSH (Measured by CBA), Thyroid Hormones, and Thyroidal Antibodies in Three Patients with Untreated Graves' Disease

| Patient | Plasma TSH (μU/ml) | | | T_4 (μg per 100 ml) | T_3 (ng/ml) | Titers of microsomal antibodies | Titers of thyroglobulin antibodies |
	Before TRH	After TRH	After TSH antibody 1:50,000				
A.R.	0.19	0.15	Undetectable	16.4	3.52	1:8000	1:25
M.B.	0.34	0.39	Undetectable	12.8	3.43	1:8000	Undetectable
A.D.	0.14	0.14	Undetectable	12.4	5.55	1:8000	Undetectable

Table 4 TRH Test and Plasma Levels of TSH (Measured by CBA), Thyroid Hormones, and Thyroidal Antibodies in Two Patients with Graves' Disease After Treatment

| Patient | Treatment | Plasma TSH (μU/ml) | | T_4 (μg per 100 ml) | T_3 (ng/ml) | Titers of microsomal antibodies | Titers of thyroglobulin antibodies |
		Before TRH	After TRH				
E.L.	Subtotal thyroidectomy	0.28	1.25	8.2	1.86	1:8000	Undetectable
W.K.	Radioiodine therapy	0.18	1.80	6.9	0.79	1:2000	Undetectable

The results indicate that, despite normalization of thyroid hormone levels in the plasma, TSH levels are still rather low in patients who were formerly hyperthyroid and who had undergone anti-thyroid treatment. The results further indicate that such patients respond to TRH stimulation with an increase in biologically active TSH release. This response may, however, remain undetected by RIA, due to the limited sensitivity of the assay system. The reasons for the long-lasting TSH suppression after anti-thyroid treatment remain speculative (von zur Mühlen et al., 1975).

TRH Test in Patients with Euthyroid Goiter.   We have used the CBA technique for the investigation of patients with euthyroid goiter and a negative TRH test. These patients are euthyroid by clinical examination and by thyroid hormone concentrations in the blood but they are unresponsive to TRH stimulation (Elte et al., 1976; Emrich and Bähre, 1978; Gemsenjäger et al., 1976; Hesch et al., 1975; Kirkegaard et al., 1977). They show no signs of Graves' disease (including euthyroid Graves' disease or autonomous adenoma).

Ten euthyroid patients with goiter showed low but detectable plasma levels of biologically active TSH when measured by CBA (Table 5). Their plasma TSH levels were significantly lower than in normal subjects (p $<$ 0.001, determined by Student's t-test), and no rise in TSH concentrations was observed after TRH administration (Table 5). When determined by RIA the TSH levels were below the limit of detection in all cases ($<$ 1.25 μU/ml). Plasma levels of thyroglobulin antibodies and microsomal antibodies were undetectable in all patients. Thyroid hormone levels in the plasma were normal in all cases (Table 5).

Notwithstanding the actual mechanism of the phenomenon, we have been able to confirm that the circulating basal levels of biologically active TSH in these subjects are low and unresponsive to TRH. Emrich and Bähre (1978) demonstrated that the frequency of negative TRH tests increased with the duration of the euthyroid goiter. They further pointed out that mean concentrations of T_4 and T_3, although in the normal range, are significantly higher compared to patients with euthyroid goiter and a positive TRH test. In about 40% of them, ^{131}I-thyroid uptake could not be suppressed by thyroid hormones. The authors assumed that this functional autonomy may be due to autonomous microadenomas as described by Miller and Block (1970). Other reasons, such as defects in TSH production and in feedback mechanism between the pituitary and the thyroid gland, also have to be taken into account.

Circadian Variation of Basal Plasma TSH Levels in Euthyroid Subjects as Determined by CBA Section Assay

There is much controversy in the literature as to whether basal TSH release shows a circadian rhythm (Hershman and Pittman, 1971; Lucke et al., 1977; Odell et al., 1967; Patel et al., 1972; Utiger, 1965; van Cauter et al., 1974;

Table 5 TRH Test and Plasma Levels of TSH (Measured by CBA), Thyroid Hormones, and Thyroidal Antibodies in 10 Patients with "Euthyroid Goiter"

| Patient | Age | Plasma TSH (μU/ml) | | T_4 | T_3 | Microsomal | Thyroglobulin |
		Before TRH	After TRH	(μg per 100 ml)	(ng/ml)	antibodies	antibodies
C.J.	65	0.12	0.13	6.9	1.30	Undetectable	Undetectable
E.P.	46	0.23	0.23	5.5	1.80	Undetectable	Undetectable
E.K.	60	0.24	0.28	10.3	2.0	Undetectable	Undetectable
B.P.	71	0.34	0.33	4.5	1.20	Undetectable	Undetectable
E.S.	63	0.18	0.20	7.7	0.80	Undetectable	Undetectable
W.S.	49	0.11	0.10	8.2	1.00	Undetectable	Undetectable
E.H.	54	0.17	0.20	8.4	0.98	Undetectable	Undetectable
H.K.	67	0.22	0.17	6.6	1.06	Undetectable	Undetectable
O.T.	63	0.11	0.22	5.8	0.96	Undetectable	Undetectable
W.D.	65	0.12	0.14	6.1	0.96	Undetectable	Undetectable
Mean $\pm$ SE		0.18 $\pm$ 0.02	0.20 $\pm$ 0.02	7.0 $\pm$ 0.5	1.21 $\pm$ 0.12		

Vanhaelst et al., 1972; Webster et al., 1972; Wecke, 1973). This controversy is possibly due to the application of rather insensitive methods (including RIA) or to infrequent blood sampling. Therefore, we reinvestigated the mode of TSH release in normal euthyroid subjects throughout a 24 hr period using the highly sensitive CBA technique.

Blood samples were taken from three healthy euthyroid male volunteers at 20 min intervals as described by Lucke et al. (1977). The first sample was taken at 1400 hr. The subjects were permitted to sleep between 2300 and 0630 hr. Plasma TSH levels were determined by the cytochemical section assay.

Plasma TSH levels showed distinct variations in each subject throughout the 24 hr period (Figure 4). Subject J started out with peak plasma TSH levels between 1400 and 1500 hr. During the subsequent 2 hr, plasma TSH levels decreased rapidly. Afterward they increased again, reaching peak levels between 2100 and 220 hr. This trend was reversed subsequently, causing plasma TSH levels to drop dramatically until midnight. They remained low during the subsequent 2-3 hr, but increased continuously thereafter, reaching peak levels at the end of the blood-sampling period (1400 hr).

Subject F displayed a similar plasma TSH concentration profile to that of subject J. The two profiles were, however, not synchronous. They were shifted in such a way that subject F's profile was trailing subject J's profile by approximately 2 hr (Figure 4). Subject F showed elevated TSH levels between 1500 and 1600 hr. Subsequently plasma TSH levels dropped and reached lowest levels between 1800 and 2200 hr. There was a subsequent increase in TSH levels until midnight, but TSH levels were low again from 0200 to 0400 hr. Subsequently, subject F's plasma TSH profile showed a tendency to increase with minor intermittent variations until the end of the blood-sampling period (1400 hr).

Subject S displayed a similar plasma TSH concentration profile to those of subjects J and F, but his profile was not synchronous with the other two. The TSH profile of subject S was shifted in such a way that it trailed subject F's TSH profile by approximately 1-2 hr and subject J's profile approximately 3-4 hr (Figure 4).

The results demonstrate that basal TSH concentrations in plasma of healthy euthyroid subjects underwent distinct variations throughout a 24 hr test period. Peak levels were four- to six-fold higher than the trough levels. In each subject the TSH concentration profile demonstrated two peaks and two troughs during the 24 hr test period. The profiles were, however, not synchronous. They were shifted among individuals by 2-4 hr, which might reflect different activity rhythms among the three subjects during daily life. The disagreement among earlier studies about the existence or nonexistence of circadian variability of plasma TSH levels may have been due to the application of low-sensitivity methods (including RIA), infrequent blood sampling, and individual variability among test subjects.

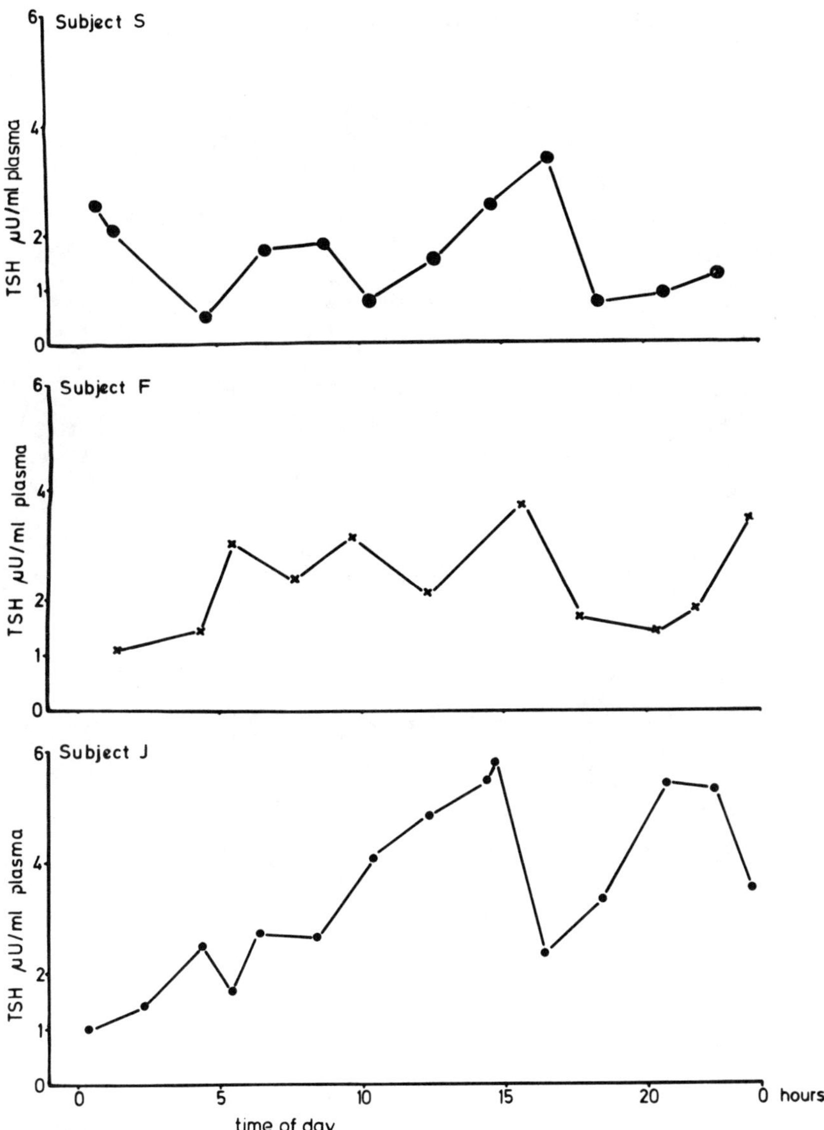

Figure 4   TSH concentrations in the plasma of three healthy male euthyroid subjects throughout a 24 hr period. TSH levels were determined by cytochemical section assay. Time of day is listed on the abscissa. For each subject the blood sampling period started at 1400 hr and lasted until 1400 hr the subsequent day. The subjects were permitted to sleep between 2300 and 0630 hr.

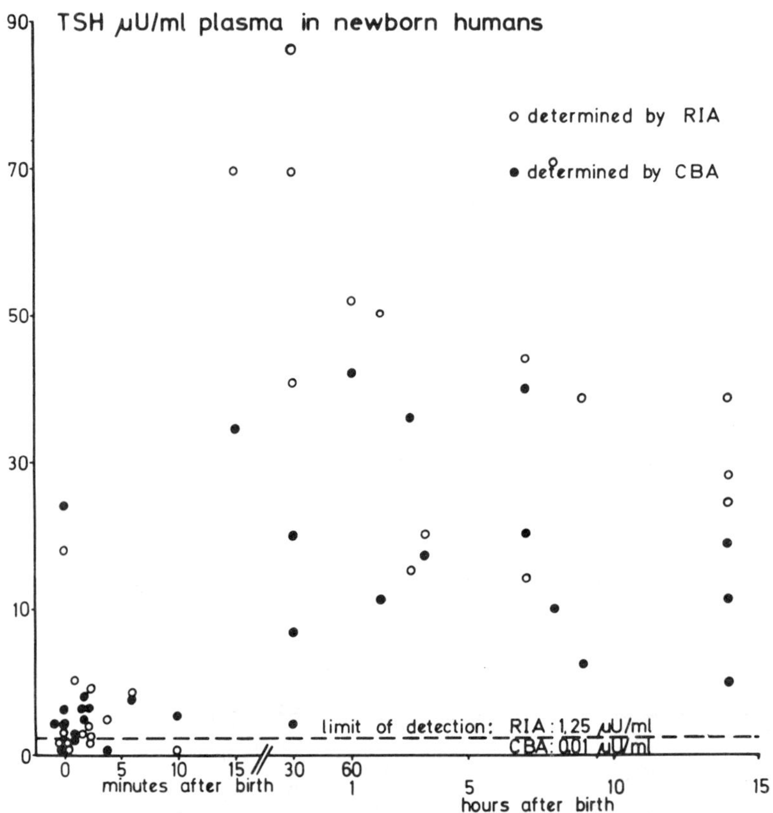

Figure 5   Plasma TSH levels of newborn children during the first 14 hr after birth, as determined by radioimmunoassay (RIA) and cytochemical bioassay (CBA). The abscissa indicates time in minutes and hours after birth. Sensitivity limits for RIA (1.25 μU/ml) and CBA (0.01 μU/ml) are listed.

Plasma Levels of TSH in Newborn Children

In some instances blood withdrawal for clinical tests may be limited to very small quantities. In such cases the availability of a highly sensitive method for the determination of clinical parameters is of great advantage. As the CBA technique for TSH determination requires less than 100 μl of plasma, we also measured plasma levels of biologically active TSH in newborn children. It is known that newborn children respond to the "trauma" of birth with increased release of immunologically active TSH (Fisher and Odell, 1969). In the study detailed below, we confirmed these observations and demonstrated, by use of CBA, that these elevated TSH levels were biologically active.

TSH levels were determined by cytochemical section assay and by RIA in plasma of 29 newborn children. One blood sample was taken from each child during the first 14 hr of life. During the first 10 min after birth, plasma TSH levels were low in 13 of 14 children (Figure 5). At 15 min after birth, plasma TSH levels started to increase tremendously. This increase was observed with both assay systems. After approximately 2-3 hr, plasma TSH levels started to decrease slowly. They were still elevated, however, at 14 hr after birth. There was no apparent sex difference in plasma TSH levels (not shown in Figure 5). Both assay systems registered similar trends in plasma TSH concentrations. Absolute TSH values, however, were generally somewhat lower when determined by cytochemical section assay as compared to RIA. Discrepancies between the measurements of the two assay systems were also observed in a few cases. This might indicate that, during the first hours of life, not only is biologically active TSH being released, but also some biologically inactive but immunologically active fragments or subunits.

The Syndrome of Inappropriate TSH Secretion

There have been several reports concerning inappropriately high TSH levels (determined by RIA) in patients with or without thyroid dysfunction (Gershengorn and Weintraub, 1975). As this "syndrome of inappropriate TSH secretion" was found in several of our patients, we investigated it further to elucidate the nature of this phenomenon.

Seven patients, aged 19-51 years, underwent thorough clinical examination; two of these patients were men, five were premenopausal women with normal menstrual cycles by history. No clinical sign of endocrine dysfunction was detected. Clinical diagnoses are given in Table 6. In all patients, plasma levels of TSH, LH, prolactin, TSH α subunits, TSH β subunits (materials from NIAMDD), T_3, T_4, 17β-estradiol, and progesterone were determined by RIA. In addition, plasma TSH levels were measured by CBA. In the RIA the antiserum cross-reacted with α subunits by 2.5% and with β subunits by 8%. The α subunit antiserum demonstrated a 4% cross-reactivity with TSH and 2.8% cross-reactivity with β subunits. The β-subunit antiserum cross-reacted with TSH by 3.4% and with α subunits by 3.5%.

All patients had dramatically elevated plasma TSH levels (up to 10,000 μU/ml; Table 6) as determined by RIA. There were no signs of thyroid dysfunction from the clinical evaluation or from the thyroid hormone levels in the plasma (Table 6). When determined by CBA, plasma levels of TSH were normal or low (Table 6). In patients with normal biologically active (but high immunologically active) plasma TSH levels, injection of TRH resulted in a further in-

Table 6 Dissociation of Biological and Immunological TSH Activity in
Plasma of Patients with Inappropriate TSH Secretion

Patient (sex, age)	Clinical Diagnosis	Date	H-TSH-RIA (μU/ml)		H-TSH-CBA (μU/ml)	
			Basal Value	30 min after 200 μg TRH	Basal Value	30 min after 200 μg TRH
A	Osteoporosis,	May 76	66	–	0.13	0.17
(M, 51)	hypotension	June 76	22	22	0.14	0.14
		June 78	26	27	0.69	–
B	Anorexia	May 76	120	120	0.28	0.32
(F, 36)	nervosa	June 76	125	–	0.15	0.27
C	–	June 78	246	–	2.6	–
(F, 42)						
D	–	Aug. 78	28	27	2.0	3.3
(F, 19)						
E	Idiopathic	Nov. 77	492	499	1.02	11.4
(F, 40)	edema					
F	–	Oct. 77	10,000	10,000	1.0	4.5
(F, 31)						
G	–	Jan. 78	1,357	–	3.55	–
(M, 30)						

crease of biologically active TSH levels. In patients with low biologically active
(but high immunologically active) plasma TSH levels, the injection of TRH was
without effect (Table 6). Plasma levels of LH, as determined by RIA, were also
elevated, exceeding 320 μU/ml in most cases, while plasma levels of prolactin
and gonadal hormones were within the normal range (not shown in the table).

The observation that plasma levels of TSH and LH were extremely high, when
determined by RIA, whereas plasma levels of thyroid and gonadal hormones
were normal, points to two possible explanations. Either the biological activities
of TSH and LH are greatly reduced, or the end organs (thyroid gland and
gonads) are insensitive to these hormones. By the use of the CBA we demonstrated that the biological activity of TSH in these patients was markedly reduced as compared to immunological activity. Plasma concentrations of α subunits, which are common to both LH and TSH, but which are biologically inactive, were elevated in all patients. Plasma concentrations of β subunits, which are
specific for TSH only, were in the normal range. We may conclude, therefore,

H-TSH-α-(ng/ml)		H-TSH-β-(ng/ml)				
Basal Value	30 min after 200 μg TRH	Basal Value	30 min after 200 μg TRH	T$_4$ (μg per 100 ml)	T$_3$ (ng/ml)	LH (mu/ml)
–	–	–	–	7.1	1.08	–
14.3	–	–	–	7.4	1.24	76.4
10.5	10.7	5.0	4.8	7.2	1.19	15.6
48.2	50.6	1.8	1.8	6.3	0.3	320
–	–	–	–	7.0	0.42	320
259	–	5.24	–	5.2	1.12	320
45.2	45.9	9.1	9.7	6.7	1.44	235
198.2	201.1	3.6	8.4	8.0	1.5	320
465.0	430.0	1.4	1.9	4.5	1.35	320
243.0	–	5.2	–	8.7	1.5	320

that in addition to normal or subnormal amounts of biologically active TSH and LH, these patients also release vast amounts of biologically inactive TSH- or LH-like material, containing α subunits, or they secrete the α subunit itself. Dissociation of biological and immunological TSH activity was also observed in a few patients with hypothalamo-pituitary disease (Faglia et al., 1979; Petersen et al., 1978). Bioimmune dissociation is also discussed in Chapter 3.

SUMMARY AND DISCUSSION

The cytochemical bioassay, originally developed by Chayen et al. (1976), has played an important role in the establishment of sensitive and specific assay procedures for the determination of polypeptide hormones. The technique has been applied to the measurement of a variety of hormones, immunoglobulins, and growth-promoting or growth-inhibiting factors (as described in other chapters of this book). The early version of the CBA segment assay for TSH as developed by Bitensky et al. (1974) has been applied by Petersen et al. (1975; 1976; 1978), by Döhler et al. (1978), by Hashimoto et al. (1979), and by Faglia et al. (1979). This technique was rather slow, but recently the more advanced and somewhat faster technique of a CBA section assay for TSH has been developed (Gilbert et al. 1977; von zur Mühlen et al., 1978; Chayen et al., 1980).

In this report we describe the application of the CBA segment assay and the CBA section assay for the determination of human TSH in health and disease. The sensitivity and precision of our assays were similar to the ones described by Bitensky et al. (1974) and Petersen et al. (1975). Whereas the RIA standard curve only covered TSH concentrations from 10^{-1} to 20 μU/ml, the CBA standard curve reached sensitivity limits at 1000-fold lower concentrations (10^{-4} μU/ml). Due to this high sensitivity the CBA is able to measure such low levels of TSH as occur in the plasma of patients with severe hyperthyroidism, or in plasma of normal subjects who had been treated with T_3 for several days (Petersen et al., 1975).

The lysosomal response in thyroid follicle cells not only occurs after stimulation by TSH, but also after activation by other thyroid stimulators (i.e., thyroid-stimulating immunoglobulins). The optimal reaction time for TSH was observed in the CBA segment assay to be 7 min; the optimal reaction time for thyroid-stimulating immunoglobulins, however, was 20 min (Bitensky et al , 1974; Petersen et al ,.1975). It was, therefore, assumed that biological activity of the two thyroid stimulators could be distinguished by choosing the appropriate incubation time .Recently, however, Loveridge et al. (1979) discovered, in the plasma of some patients with Graves' disease, thyroid-stimulating immunoglobulins which were able to increase the permeability of lysosomal membranes after a few minutes of incubation .Therefore, we have to be aware of the possibility of confusing thyroid-stimulating immunoglobulins with TSH in the cytochemical bioassay system. We regularly checked for specificity of our assay system by testing for parallelism between the responses to different dilutions of the same plasma sample with those to different dilutions of the TSH standard. As demonstrated in Figure 3, we did obtain parallelism between the values of different plasma dilutions and those of the TSH standard curve. This indicates that the measured thyroid stimulator (TSH) in plasma was likely to be similar to the human standard TSH preparation used to set up the CBA standard curve. Only in the extreme upper or lower part of the standard curve was the parallelism slightly impaired .Specificity of our assay system for TSH also became apparent after treatment of human plasma with an antibody to human TSH. Subsequent precipitation of the TSH-antibody complex resulted in inactivation of biological TSH activity in the supernatant .

There was regularly a linear relationship between the logarithm of the TSH concentration applied and the response of the thyroid follicle cells over the range of 10^{-4}-10^{-1} μU TSH per milliliter. There was also excellent correlation between individual standard concentrations From earlier studies it is known that at very high TSH concentrations the linearity will disappear and the standard curve will eventually reverse its slope (Hashimoto et al., 1979). This phenomenon, which has also been described by Bitensky et al. (1974) during prolonged acidic pretreatment of thyroid sections, is probably due to supraoptimal lysosomal membrane permeability, induced by the higher TSH concentrations. Under these conditions, lysosomal naphthylamidases apparently leak out of the lysosomes so that the amount of color developed in the subsequent incubation with the chromogenic substrate is diminished .

Repeated measurement of the TSH concentration in aliquots of the same control plasma gave good agreement in three different CBA section assays, with an interassay variation of 5.1%.

We applied the cytochemical bioassay for TSH in several clinical studies. TSH levels in euthyroid subjects before and after treatment with TRH were in the same range as that reported by Petersen et al. (1975). In addition, we demonstrated that basal TSH concentrations in plasma of healthy euthyroid subjects underwent distinct variations throughout a 24 hr test period. The disagreement in the earlier studies concerning the possible presence of circadian variation in plasma TSH levels may have been due to the application of methods of low sensitivity (including RIA), infrequent blood sampling, and individual variability among test subjects .

As reported by Petersen et al. (1975; 1976), basal plasma TSH levels are low, but measurable by CBA in patients with Graves' disease. We have confirmed these results and have shown additionally that there was no further increase in biologically active TSH levels in these patients after TRH administration. Treatment with a specific antibody to human TSH resulted in the complete loss of thyroid-stimulating activity. This indicates that the low TSH-like activity observed in the plasma of patients with Graves' disease was indeed likely to be due to low circulating levels of TSH, and not to any other biologically active substance. By the use of the CBA we could demonstrate increased TSH release to TRH stimulation in two formerly hyperthyroid patients 2 and 4 years after treatment, respectively. This increase in plasma TSH levels remained undetected in the RIA, as it barely reached the lowest limit of detection by this method.

We have used the CBA technique to investigate patients with euthyroid goiter and a negative TRH test. These patients are euthyroid as assessed by clinical examination and by thyroid hormone concentrations in the blood. By the use of the CBA we determined that the circulating basal levels of biologically active TSH in these patients are low and we confirmed earlier results, obtained by RIA, that the pituitary gland in these patients is unresponsive to TRH.

Newborn children respond to the trauma of birth with increased release of immunologically active TSH (Fisher and Odell, 1969). We confirmed these observations and demonstrated, by the use of the CBA, that these elevated TSH levels were biologically active.

There have been reports concerning inappropriately high TSH levels, as determined by RIA, in plasma of patients with or without thyroid dysfunction (Gershengorn and Weintraub, 1975). Possible explanations for these observations are either lack of biological activity of TSH, or end-organ insensitivity. By the use of the CBA, we demonstrated that the biological activity of TSH was greatly reduced in these patients, which was, at least in part, due to the release of vast amounts of biologically inactive TSH α-subunits.

ACKNOWLEDGMENTS

We want to record our gratitude to Dr. L. Bitensky, Dr. J. Chayen, and Dr. D. Gilbert for teaching us the cytochemical bioassay technique and for their always useful discussions. Furthermore, we want to thank Ms. J. Poernomo, Ms. H. Schwarze, and Ms. M. Th. Suchy (Hannover) for excellent technical assistance and Ms. N. Lotwin (Los Angeles) and Ms. M. König (Hannover) for secretarial assistance. The supply of TSH-low plasma by Dr. F. W. Erhardt and Professor P. C. Scriba (Munich) is gratefully acknowledged. We are also grateful to the Deutsche Forschungsgemeinschaft for their financial support and to the Alexander-von-Humboldt-Foundation for providing a research fellowship. We thank Professor R. Hall and the Medical Research Council for providing biologically active human TSH Research Standard A and the NIAMDD (Bethesda) for providing purified human TSH for iodination and RIA kits for TSH α and β subunits.

REFERENCES

Alaghband-Zadeh, J., Daly, J. R., Bitensky, L., and Chayen, J. (1974). The cytochemical section assay for corticotrophin. *Clin. Endocrinol. (Oxf)*, 3:389-396.

Amino, N., Hagen, S. R., Yamada, N., and Refetoff, S. (1976). Measurement of circulating thyroid microsomal antibodies by the tanned red cell hemagglutination technique: Its usefulness in the diagnosis of autoimmune thyroid disease. *Clin. Endocrinol. (Oxf)*, 5:115-125.

Bitensky, L., Alaghband-Zadeh, J., and Chayen, J. (1974). Studies on thyroid stimulating hormone and the long-acting thyroid stimulating hormone. *Clin. Endocrinol. (Oxf)*, 3:363-374.

Brown, J., and Munro, D. S. (1967). A new in vitro assay for thyroid stimulating hormone. *J. Endocrinol.*, 38:439-449.

Buerklin, E. M., Schimmel, M., and Utiger, R. D. (1976). Pituitary-thyroid regulation in euthyroid patients with Graves' disease previously treated with antithyroid drugs. *J. Clin. Endocrinol. Metab.*, 43:419-427.

Chapman, R. S., Malon, P. G., and Ekins, R. P. (1976). The effects of microunit
    doses of thyrotropin on iodothyronine release from mouse thyroid lobes in
    vitro. In *Thyroid Research*, J. Robbins and L. E. Braverman (eds.). Excerpta
    Medica, Amsterdam, pp. 217-220.

Chayen, J., Daly, J. R., Loveridge, N., and Bitensky, L. (1976). The cytochemi-
    cal bioassay of hormones. *Recent Prog. Horm. Res.*, 32:33-79.

Chayen, J., Gilbert, D. M., Robertson, W. R., Bitensky, L., and Besser, G. M.
    (1980). A cytochemical section bioassay for thyrotrophin. *J. Immunoassay*,
    1:1-13.

Chopra, I. J., Chopra, U., and Orgiazzi, J. (1973). Abnormalities of hypothalamo-
    hypophyseal-thyroid axis in patients with Graves' ophthalmopathy. *J. Clin.
    Endocrinol. Metab.*, 37:955-967.

Chopra, I. J., Chopra, U., Vanderlaan, W. P., and Solomon, D. H. (1974). Com-
    parison of serum prolactin and thyrotropin responses to thyrotropin-releasing
    hormone in patients with Graves' ophthalmopathy. *J. Clin. Endocrinol.
    Metab.*, 38:683-687.

Clifton-Bligh, P., Silverstein, G. E., and Burke, G. (1974). Unresponsiveness to
    thyrotropin-releasing hormone (TRH) in treated Graves' hyperthyroidism and
    in euthyroid Graves' disease. *J. Clin. Endocrinol. Metab.*, 38:531-538.

Döhler, K.-D., Hashimoto, T., and von zur Mühlen, A. (1978). Use of a cyto-
    chemical bioassay for determination of thyroid stimulating hormone in clini-
    cal investigation. In *Radioimmunoassay and Related Procedures in Medicine*,
    Vol. I. International Atomic Energy Agency, Vienna, pp. 297-307.

Ealey, P. A., Bidey, S. P., Marshall, N. J., and Ekins, R. P. (1980). Use of the sec-
    tion cytochemical bioassay for the measurement of thyroid stimulators. *J.
    Endocrinol.*, 87:35P-36P.

Elte, J. W. F., Haak, A., Fröhlich, M., Wiarda, K. S., and van Wermeskerken,
    R. K. A. (1976). Autonomously functioning euthyroid multinodular goitre.
    Neth. J. Med., 20:1-4.

Emrich, D., and Bähre, M. (1978). Autonomy in thyroid goitre. Maladaptation
    to iodine deficiency. *Clin. Endocrinol. (Oxf)*, 8:257-265.

Emrich, D., Bähre, M., von zur Mühlen, A., Hesch, R.-D., and Köbberling, J.
    (1976). Insufficient TSH-stimulation after successful treatment for hyperthy-
    roidism. *Horm. Metab. Res.*, 8:408.

Faglia, G., Bitensky, L., Pinchera, A., Ferrari, C., Paracchi, A., Beck-Peccoz, P.,
    Ambrosi, B., and Spada, A. (1979). Thyrotropin secretion in patients with
    central hypothyroidism: Evidence for reduced biological activity of immuno-
    reactive thyrotropin. *J. Clin. Endocrinol. Metab.*, 48:989-998.

Fisher, D. A., and Odell, W. D. (1969). Acute release of thyrotropin in the new-
    born. *J. Clin. Invest.*, 48:1670-1677.

Gemsenjäger, E., Staub, J. J., Girard, J., and Heitz, P. (1976). Preclinical hyper-
    thyroidism in multinodular goitre. *J. Clin. Endocrinol. Metab.* 43:810-816.

Gershengorn, M. C., and Weintraub, B. D. (1975). Thyrotropin induced hyperthyroidism caused by selective pituitary resistance to thyroid hormone. A new syndrome of "inappropriate secretion of TSH." *J. Clin. Invest.*, 56:633-642.

Gilbert, D. M., Besser, G. M., Bitensky, L., and Chayen, J. (1977). Development of a cytochemical section-assay for thyroid stimulators. *J. Endocrinol.*, 75:40P.

Harada, A., Kojima, A., Tsukui, T., Onaya, T., Yamada, T., Ikejiri, K., and Yukimara, Y. (1975). Pituitary unresponsiveness to thyrotropin-releasing hormone in thyrotoxic patients during chronic antithyroid drug therapy and in rats previously treated with excess thyroid hormone. *J. Clin. Endocrinol. Metab.*, 40:942-948.

Hashimoto, T., Döhler, K.-D., Emrich, D., and von zur Mühlen, A. (1979). Clinical application of a cytochemical bioassay for the determination of thyroid stimulating hormone. *J. Endocrinol. Invest.*, 2:395-400.

Hershman, J. M., and Pittman, J. A. (1971). Utility of the radioimmunoassay of serum thyrotropin in man. *Ann. Intern. Med.*, 74:481-490.

Hesch, R. D., Emrich, D., von zur Mühlen, A., and Breuel, H.-P. Der Aussagewert der radioimmunchemischen Bestimmung von Trijodthyronin and thyreotropem Hormon für die Schilddrüsendiagnostik in der Praxis. *Dtsch. Med. Wochenschr.*, 100:805-812.

Hesch, R. D., Hüfner, M., and von zur Mühlen, A. (1972). Erste klinische Ergebnisse mit einer radioimmunchemischen Bestimmung von Trijodthyronin im Plasma. *Dtsch. Med. Wochensch.*, 97:351-353.

Hesch, R. D., Hüfner, M., von zur Mühlen, A., and Emrich, D. (1974). Triiodothyronine levels in patients with euthyroid endocrine exophthalmos and during treatment of thyrotoxicosis. *Acta Endocrinol. (Kbh)*, 75:514-522.

Kirkegaard, C., Faber, J., Friis, T., Lauridsen, U. B., Rogowski, P., and Siersbaek-Nielsen, K. (1977). Intravenous and peroral TRH stimulation in sporadic atoxic goitre. *Acta Endocrinol. (Kbh)*, 85:508-514.

Kumahara, Y., Miyai, Y., and Azukizuwa, M. (1971). Clinical application of synthetic thyrotropin releasing hormone to TSH secretion test (TRH-test). *Med. J. Osaka Univ.*, 22:97-107.

Lemarchand-Beraud, T. (1976). In *Methods of Hormone Analysis,* H. Breuer, D. Hamel, and H. L. Krüskemper (eds.). Georg Thieme Verlag, Stuttgart, pp. 22-35.

Loveridge, N., Zakarija, M., Bitensky, L., and McKenzie, J. M. (1979). The cytochemical bioassay for thyroid-stimulating antibody of Graves' disease: Further experience. *J. Clin. Endocrinol. Metab.*, 49:610-615.

Lucke, C., Hehrmann, R., von Mayersbach, K., and von zur Mühlen, A. (1977). Studies on circadian variations of plasma TSH, thyroxine and triiodothyronine in man. *Acta Endocrinol. (Kbh)*, 86:81-88.

Martino, E., Pinchera, A., Capiferri, R., Macchia, E., Sardano, G., Bartalena, L., Mazzanti, R., and Baschieri, L. (1976). Dissociation of responsiveness to thyrotropin-releasing hormone and thyroid suppressibility following anti-thyroid drug therapy of hyperthyroidism. *J. Clin. Endocrinol. Metab.*, 43: 543-549.

McKenzie, J. M. (1958). The bioassay of thyrotropin in serum. *Endocrinology*, 63:372-382.

Miller, J. M., and Block, M. A. (1970). Functional autonomy in multinodular goiter. *JAMA*, 214:535.

Miyai, K., Takai, S., Kuma, K., and Kumahara, Y. (1974). Serum thyrotropin response to thyrotropin releasing hormone and the concentration of thyroid hormones in patients with hyperfunctioning thyroid nodule. *Endocrinol. Jpn.*, 21:393-397.

Odell, W. D., Wilbur, J. F., Utiger, R. D. (1967). Studies of thyrotropin physiology by means of radioimmunoassay. *Recent Prog. Horm. Res.*, 23:47-85.

Ogihara, T., Yamamoto, T., Azukizawa, M., Miyai, K., Kumahara, Y. (1973). Serum thyrotropin and thyroid hormones in the course of subacute thyroid-itis. *J. Clin. Endocrinol. Metab.*, 37:602-606.

Patel, Y. C., Alford, F. P., and Burger, H. G. (1972). The 24-hour plasma thyro-tropin profile. *Clin. Sci.*, 43:71-77.

Petersen, V. B., McGregor, A. M., Belchetz, P. E., Elkeles, R. S., and Hall, R. (1978). The secretion of thyrotrophin with impaired biological activity in patients with hypothalamic-pituitary disease. *Clin. Endocrinol. (Oxf)*, 8: 397-402.

Petersen, V., Smith, B. R., and Hall, R. (1975). A study of thyroid stimulating activity in human serum with the highly sensitive cytochemical bioassay. *J. Clin. Endocrinol. Metab.*, 41:199-202.

Petersen, V., Smith, B. R., and Hall, R. (1976). Measurement of thyrotropin and thyroid stimulating immunoglobulins with the cytochemical bioassay. In *Thyroid Research*, J. Robbins and L. E. Braverman (eds.). Excerpta Medica, Amsterdam, p. 610.

Sanchez-Franco, F., Garcia, M. D., Cacicedo, L., Martin-Zurro, A., Escobar del Rey, F., and Moreale de Escobar, G. (1974). Transient lack of thyrotropin (TSH) response to thyrotropin-releasing hormone (TRH) in treated hyper-thyroid patients with normal or low serum thyroxine (T4) and triiodothyro-nine (T3). *J. Clin. Endocrinol. Metab.*, 38:1098-1102.

Trowell, O. A. (1959). The culture of mature organs in a synthetic medium. *Exp. Cell Res.*, 16:118-147.

Utiger, R. D. (1965). Radioimmunoassay of human plasma thyrotropin. *J. Clin. Invest.*, 44:1277-1286.

van Cauter, E., Lequercq, R., Vanhaelst, L., and Goldstein, J. (1974). Simultaneous study of cortisol and TSH daily variations in normal subjects and patients with hyper-adrenalcorticism. *J. Clin. Endocrinol.*, 39:645-652.

Vanhaelst, L., van Cauter, E., Degante, J. P., and Goldstein, J. (1972). Circadian variations of serum thyrotropin in man. *J. Clin. Endocrinol.*, 35:479-482.

von zur Mühlen, A., Döhler, K.-D., and Poernomo, J. (1978). Further advancement of the cytochemical bioassay for human TSH. *Ann. Endocrinol. (Paris)*, 39:4.

von zur Mühlen, A., Emrich, D., Hesch, R. D., and Köbberling, J. (1971). Untersuchungen über die Beeinflussung der Thyreotrophin-Inkretion beim Menschen. *Acta Endocrinol. (Kbh)*, 68:669-685.

von zur Mühlen, A., Hesch, R. D., and Köbberling, J. (1975). The TRH-test in the course of treatment for hyperthyroidism. *Clin. Endocrinol. (Oxf)*, 4:165-172.

Webster, B. R., Guansing, A. R., and Paice, J. C. (1972). Absence of diurnal variation of serum TSH. *J. Clin. Endocrinol.*, 34:899-901.

Wecke, J. (1973). Circadian variation of the serum thyrotropin level in normal subjects. *Scand. J. Clin. Lab. Invest.*, 31:337-342.

6

The Thyroid-Stimulating Antibody of Graves' Disease

J. Maxwell McKenzie and Margita Zakarija / University of Miami School of
Medicine, Miami, Florida

Lucille Bitensky and J. Chayen / Kennedy Institute of Rheumatology
London, England

PART I: GENERAL CONCEPTS AND ASSAY MATERIALS

J. Maxwell McKenzie and Margita Zakarija

INTRODUCTION

The hyperkinetic features of Graves' disease were not considered to involve the
thyroid gland until near the end of the last century (Rehn, 1884). Then, with
the recognition of the existence of thyrotropin as a pituitary product (Smith and
Smith, 1922), the theory of hyperthyroidism being due to the action of thyrotropin was conceived (Loeb and Bassett, 1929) and held the limelight for about
25 years. However, in 1956 the modern era in this topic dawned with an observation by Adams and Purves (1956). They had developed a more sensitive bioassay for thyrotropin, using guinea pigs, and recognized that the injection of
serum from some patients with Graves' disease produced an "abnormal response." With the identification of a similar pattern of response when the mouse
was used as the bioassay animal (McKenzie, 1958), and the adoption of the

The work from the authors' laboratory quoted in this chapter was financially supported by
funds from the Medical Research Council of Canada (MT 884 and MT 5190) and the U.S.
Public Health Service (AM04121) when both authors were members of McGill University
Clinic, Royal Victoria Hospital, Montreal, Quebec, Canada.

135

procedure by several laboratories, agreement to name the entity long-acting thyroid stimulator (LATS) was reached by a number of interested investigators on the steps of Church House, London, at the Fourth International Thyroid Congress in 1960.

The next development was the evidence that LATS was not an abnormal thyrotropin, but an immunoglobulin, class IgG (Kriss et al., 1964); thus, an undoubted thyroid stimulator (McKenzie, 1960) being in the blood of some patients with hyperthyroidism, it was accepted that, at least when present, this may well be the cause of the thyroid overactivity (McKenzie, 1972). However, the low incidence of occurrence of LATS in hyperthyroidism (McKenzie, 1972) and its presence in some patients who were not hyperthyroid (Chopra and Solomon, 1970; Wong and Doe, 1972) led to a feeling that this was an "epiphenomenon" (Chopra et al., 1970) and not the cause of the syndrome.

The beginning of rationalization of the controversy occurred with the development of other procedures for the assay of the thyroid-stimulating IgG. With the inevitable clarity of hindsight, it is obvious that a human IgG, presumably an autoantibody against a human antigen, is unlikely to interact with the equivalent molecule in the mouse thyroid; if such heterologous interaction were to occur, it should happen only in instances of very high titer, or affinity of the antibody for the homologous antigen. In other words, one view may be that the mouse bioassay, identifying LATS, constitutes a cross-reaction of the human antibody with a foreign, even distant, but to some degree similar, antigen in the mouse thyroid (Zakarija and McKenzie, 1978a). As known from studies of the experimental production of antibodies, the higher the affinity of an antibody for its homologous antigen, the more likely will there be cross-reaction with a different, but chemically related, antigen (Kabat, 1968). Thus a system using the human thyroid ought to be much more efficient in detecting a thyroid-stimulating antibody in patients' sera, and as detailed below, this is indeed the case.

Regarding the theory that the LATS assay is merely an inefficient way of measuring the antibody, we have accumulated information supporting this contention: comparing the LATS assay and a human system (Zakarija et al., 1980), the thyroid-stimulating antibody has the same isoelectric point for each (Zakarija and McKenzie, 1978a); adsorbing the activities to either human or bovine thyroid, there is the equivalent recovery of each on desorption of the IgG (Zakarija and McKenzie, 1978b); it is overwhelmingly the antibody of high titer, as measured in the human assay, that is positive in the mouse, and preparations of IgG, positive in both systems, show a linear relationship when the dual sets of responses are plotted, one against the other (Zakarija and McKenzie, 1978a). In brief, we consider that LATS is the same IgG that is now measured more efficiently in other systems.

Table 1 Assays for Thyroid-Stimulating Antibody of Graves' Disease

Assays	Name Given to Antibody	Acronym	Method	Reference
Stimulation assays	Long-acting thyroid stimulator	LATS	1. Discharge in mice of radioiodine-labeled thyroid components: increase in blood radioactivity	McKenzie (1958)
			2. Cytochemical change in guinea pig thyroid in vitro	Bitensky et al. (1974)
	Human thyroid stimulator	HTS	Colloid droplet formation or increase in cyclic AMP concentration in human thyroid slices incubated for 10 min	Onaya et al. (1973)
	Human thyroid adenyl cyclase stimulator	H-TACS	Stimulation of adenyl cyclase in human thyroid membranes	Orgiazzi et al. (1976)
	Human-specific thyroid stimulator	HSTS	Generic term for assays using human thyroid end point	Kendal-Taylor (1975)
	Thyroid-stimulating immuno-globulin Thyroid-stimulating antibody	TSAb	Increase in cyclic AMP concentration in human thyroid slices incubated for 2 hr	McKenzie and Zakarija (1977) Zakarija et al. (1980)
Receptor modulation assays	Long-acting thyroid stimulator-protector	LATS-P	Binding to human thyroid preventing ("protecting") subsequent binding of LATS	Adams and Kennedy (1967)
	Thyroid-stimulating immuno-globulins	TSI		Smith and Hall (1974)
	Thyroid-stimulating antibodies	TSAb	Inhibition of ^{125}I-thyrotropin binding to human thyroid membranes	Davies et al. (1977)
	Thyrotropin displacement activity	TDA		Smith (1976) O'Donnell et al. (1978)
	TSH-binding inhibiting immuno-globulins	TBII		Endo et al. (1978)

CURRENT ASSAY SYSTEMS

Table 1 shows a partial list of methods that have been described for what we
shall now refer to only as thyroid-stimulating antibody, or TSAb. They are easily
divisible into two categories: one in which there is a more or less direct measure
of thyroid stimulation and the other in which TSAb interferes with the binding
of, in all but one instance (namely, LATS-protector, LATS-P), radioactive thyrotropin to its receptor in a preparation of human thyroid membranes. The
LATS-P assay is entirely comparable with the other "receptor-modulation"
assays; instead of ^{125}I-labeled thyrotropin being the ligand, a standard preparation of LATS is used. The LATS activity is shown to be bound to the human
thyroid, or it is not, having been "protected" from being bound by the action of
the unknown IgG under test.

Most of the remainder of this part will deal with the thyroid stimulation type
of assay, but first a comment on the receptor-modulation systems may be ger-
maine. Of the various names and acronyms listed in Table 1 for these assays it
seems to us that thyrotropin-binding inhibition (TBI) is the most appropriate
for the system and thyrotropin-binding inhibiting immunoglobulin (TBII) for
the moiety being studied.

Although it has not been directly proven beyond doubt, it should be accepted
that TSAb inhibits the binding of thyrotropin to its receptor and thus should
be positive in a TBI assay. However, it is quite clear that not all TBII are thyroid stimulating, and therefore a TBI-positive preparation is not necessarily
positive in a TSAb assay. Support for these latter statements comes from a
number of reports. Positive TBI assays occurred with IgG in clinical situations,
such as thyroid cancer and Hashimoto's disease, where TSAb was not expected
(Smith, 1976). When direct comparisons were made, i.e., aliquots of individual
preparations of IgG were tested in both the TBI and TSAb systems, there was
incomplete correlation of data (McKenzie et al., 1978; Sugenoya et al., 1979);
a positive TBI might be negative for TSAb, TSAb might be negative in the TBI
system, and the grading of responses when several IgGs were tested in the two
systems did not match (McKenzie et al., 1978). Direct proof that an antibody
may inhibit thyrotropin binding yet not be thyroid stimulating was provided
in the study of monoclonal antibodies derived from hybridomas. The data
shown in Table 2 were obtained in collaboration with Dr. L. D. Kohn and his
colleagues, who have recently published a detailed report of their investigations
(Yavin et al., 1981). Review of these findings leads to the conclusion that it is
now established that an antibody raised against a component of the thyroid
plasma membrane may inhibit thyrotropin binding without stimulating the
gland. In this work (Yavin et al., 1981), a soluble preparation of receptor for
thyrotropin was used as immunogen. In our own studies we have found that

Table 2 Monoclonal Antibodies to TSH Receptor

Test Sample	TBI (cpm)	TSAb (pmol cAMP per mg wet wt.)
Control	18,000	0.49 ± 0.02[a]
Ab 1	18,000	0.48 ± 0.07
Ab 2	12,000	0.44 ± 0.08
Ab 3	1000	0.41 ± 0.01
TSH 50 mU	—	5.00 ± 0.96

[a]Mean ± SD; n = 4. Monoclonal antibodies (Ab 1-3) were provided by Dr.
L. D. Kohn. They had been produced by hybridomas developed from
spleen cells of mice immunized with a soluble preparation of bovine
thyroid receptor for thyrotropin. The TSAb assay was with slices of
bovine thyroid (Zakarija and McKenzie, 1978a).
Note: Ab 3 was a potent inhibitor of binding of thyrotropin; no antibody was active in the TSAb assay.

soluble receptor, extensively purified by gel and affinity chromatography in
which thyrotropin is the ligand, remains heterogeneous with multiple protein
bands on polyacrylamide gel electrophoresis (Koizumi et al., 1982). So, extrapolation of these findings toward understanding of molecular events requires caution. It is known that binding of cholera toxin (Mullin et al., 1976) and tetanus
toxin (Ledley et al., 1977) to the thyroid plasma membrane will inhibit or enhance the binding of thyrotropin to its receptor, and this may be associated with
stimulation of the gland (Zakarija and McKenzie, 1980). The action of thyrotropin itself results in major perturbation of the total surface of the thyroid cell,
as established by scanning electron microscopy (Zamora et al., 1979).

Consequently, it seems likely that the combination of an antibody with an
antigen on the plasma membrane, at a site remote from the receptor for thyrotropin, may perturb the entire membrane so that the binding of thyrotropin is
affected. These considerations readily explain the comparative nonspecificity of
the TBI technique as an assay for TSAb.

Despite this polemic against the specificity of the TBI assay, it is most likely
that a component of the receptor for thyrotropin is in fact the antigen with
which TSAb interacts. The most convincing data in support are that the guinea
pig fat-cell membrane, long known to contain a receptor for thyrotropin, also
specifically binds TSAb (Endo et al., 1981); as in thyroid systems, the binding of
TSAb causes an inhibition of thyrotropin binding. That antibodies other than
TSAb might bind to the thyroid membrane and influence the binding of thyro-

tropin, as discussed above, is not difficult to understand. That a human thyroid antibody might interact with a guinea pig fat-cell component other than the receptor for thyrotropin seems highly unlikely.

CLINICAL CORRELATIONS OF THE ASSAY
OF TSAb BY THE cAMP SLICE ASSAY

The procedure we have developed for the assay of TSAb is outlined in Table 3. Applying this technique to studies of the clinical significance of the antibody, we have obtained the following results.

1. TSAb may be measured in over 90% of patients who develop hyperthyroidism of Graves' disease (Zakarija et al., 1980). This is an incidence only slightly greater than that reported by Bech and her colleagues, who used an adenylate cyclase assay with thyroid homogenate as the source of enzyme (Bech and Nistrup Madsen, 1979). Others who applied the adenylate cyclase technique have obtained a lower incidence of positive results (Orgiazzi et al., 1976; Kendal-Taylor, 1975), but this is probably attributable to a lesser sensitivity of the various techniques when compared with the thyroid slice system. Certainly the range of responses is much lower for data reported with homogenate (Bech and Nistrup Madsen, 1979) than with those obtained using thyroid slices (Zakarija et al., 1980).

2. Persistence of TSAb in the blood at the end of a course of anti-thyroid drug therapy is associated with relapse of hyperthyroidism (Zakarija et al., 1980). Similar data were reported by others with different assays for TSAb (Davies et al., 1977; O'Donnell et al., 1978; Schleusener et al., 1978). By retrospective review, it may be further inferred that persistence of TSAb occurs most commonly with a high concentration of the antibody, and this results in the need for eventual ablation of the thyroid gland to avoid relapse; i.e., the clinical course is characterized by recurrent relapse and further therapy, until hypothyroidism eventuates (Zakarija et al., 1980).

3. In the syndrome of euthyroid ophthalmopathy of Graves' disease, TSAb may occur but some abnormality of thyroid function appears always to be associated. Thus, in our admittedly limited experience, these patients showed either no suppression of thyroid function on administration of thyroid hormone, or no response of the pituitary thyrotrope to the injection of thyrotropin-releasing hormone; these abnormalities were associated with the presence of serum antibodies to the thyroid microsomal antigen and to thyroglobulin (Zakarija et al., 1980). Conversely, failure to find TSAb in the blood of these patients coincided with none of these other findings; i.e., thyroid function was normal in all respects. These observations are similar to those of Solomon et al. (1977), who used a LATS-P assay in a comparable analysis, except that two of

Table 3 Method of Assay of Thyroid-Stimulating Antibody

Basis	Increase in $3',5'$-adenosine monophosphate (cyclic AMP) concentration in human thyroid in vitro
Tissue	Slices of fresh "normal" thyroid from operating room: usually paranodular tissue at lobectomy for "cold" nodule
Test material	IgG from patients' sera; $(NH_4)_2SO_4$ precipitate or IgG purified by diethylaminoethylcellulose chromatography
Control	IgG from sera of normal human subjects
Incubation	2 hr at $37°C$ in Krebs-Ringer-bicarbonate buffer with 0.1% glucose, 0.1% human serum albumin, and 10 mM theophylline
End point	Cyclic AMP extracted from the tissue, measured by radioimmunoassay, and expressed as pmol per mg wet weight

their patients had LATS-P in the blood and no abnormality of thyroid function whatsoever. In view of the factors that may influence the receptor-modulation type of assay when radioactive thyrotropin is the ligand (see above), it seems to us that false-positive assays for TSAb ought to occur with the LATS-P technique as they do with the TBI procedure.

4. The monitoring of TSAb in pregnancy may enable forecasting of neonatal Graves' disease (McKenzie and Zakarija, 1978). The neonatal syndrome, long recognized to be associated with Graves' disease in the mother, is now known to be due to the transplacental passage of TSAb. Our recent data indicate that the antibody has to be of high enough concentration in the maternal blood, at least 500% response in the thyroid slice assay (Zakarija and McKenzie, unpublished data), and this value has to be with blood taken in the third trimester. As with other immune phenomena (Froelich et al., 1980), it appears that the titer of TSAb tends to fall throughout pregnancy (with a rebound postpartum), so that a value of 500% early in pregnancy may fall to a concentration low enough to have no major detrimental effect on the fetus or neonate (Zakarija and McKenzie, unpublished data).

The importance of the concentration of TSAb to the development of neonatal Graves' disease is a conclusion substantiated by another study in which the LATS-P assay was used (Dirmikis and Munro, 1975) and in some isolated reports with LATS-P or TBI assays (Nutt et al., 1974; Thomson et al., 1975); however, a high concentration of TBII in a pregnancy not associated with neonatal Graves' disease has been described (Hales et al., 1980).

The implication of this synoptic review of clinical correlations of TSAb assays is that it should be a useful laboratory adjunct in the management of patients with Graves' disease. However, in view of the insensitivity of the LATS assay, the nonspecificity of the current TBI assays, and the cumbersome nature of a true TSAb procedure, general application of the information to clinical care is going to have to await the development of a better procedure combining the essential elements of specificity, precision, sensitivity, and reproducibility by clinical laboratories.

THE IMMUNOCHEMISTRY OF TSAb: THE POSSIBILITY OF DEVELOPING AN IMMUNOASSAY

Shortly after it was shown that TSAb was unequivocally an IgG, there appeared reports that it was polyclonal (Kriss, 1968; Ochi and DeGroot, 1968; Maisey, 1972). These investigations relied upon the LATS mouse assay and apparent partial inhibition of this activity by the addition of antibody to either the K or λ light chain of human IgG. Seemingly in support of a polyclonal identity for the antibody was the claim that nonspecific κ chain, i.e., Bence Jones protein, could be used to combine with the heavy chain of TSAb-IgG, after the heavy and light chains of that preparation had been separated, to result in 100% recovery of biological activity of the antibody (Mehdi and Kriss, 1978). The latter finding is, however, incompatible with current understanding of requirements for antibody activity. Although in some studies the heavy chain alone was shown to combine with the homologous antigen (Fougereau et al., 1964; Bridges and Little, 1971), any intrinsic activity is minor. Indeed, in earlier experiments with LATS, the heavy chain of the IgG had minimal (Dorrington and Munro, 1966) or no effect (McKenzie, 1965). Rather it is accepted that for full antibody potency both the original, specific heavy and light chains, in combination, are required (Braun et al., 1976; Huser et al., 1975). Even the product of recombined heavy and light chains that were originally separated from a single serum total IgG has minimal, if any, immunological activity (Fougereau et al., 1964), presumable because recombination is largely at random and any antibody is a small component of the total IgG.

A reassessment of the older data for TSAb by more recently developed techniques brought out some important differences from the original findings (Zakarija, 1980). Heavy chains of pure preparations of TSAb-IgG, when combined with the original total light-chain fraction or with the κ or λ light chain (Bence Jones proteins) showed no TSAb activity. Isolation of TSAb-IgG that contained only κ or λ light chain was effected by affinity chromatography that removed the IgG containing the alternate light chain; thyroid-stimulating activity was associated with IgG containing only one or the other light chain. Of the eight preparations tested, TSAb was IgGλ in seven and IgGκ in one. Since there are

four antigenically distinct subclasses of IgG (Kabat, 1968), the relation of TSAb
to subclass identity was examined. With three preparations of TSAb-IgG, isolation of IgG1 by sequential removal of IgG3 using staphylococcal protein A linked
to Sepharose (Skvaril, 1976) and IgG4 and IgG2 by affinity chromatography
using monospecific antisera to those subclasses showed TSAb activity to remain
with IgG1. (In two instances this was with $\sim$ 100% recovery of activity and, in
one, incomplete recovery.)

These findings may be aligned with earlier data showing that TSAb-IgG, fractionated by preparative isoelectric focusing, has a relatively constant isoelectric
point of 8.5-9.0, although obtained from different patients (Zakarija and Mc-
Kenzie, 1978c). Altogether they suggest that TSAb, at least regarding the potent
preparations used (of necessity) in these studies, indeed acts in the manner of a
conventional antibody, requiring the combined properties of specific heavy and
light chains, and has restricted heterogeneity. The latter term is used to indicate
that an antibody is not polyclonal in origin but is the product of only a few
clones of lymphocytes. One implication of this conclusion is that it may be possible to develop an anti-idiotypic antibody to TSAb since the antigenic determinants to an antibody are related to the hypervariable regions that are also specific for antigen-binding properties. That such an anti-idiotypic antibody, raised
against the idiotype of a single TSAb-IgG, would "recognize" the idiotype of
TSAb-IgG in another patient is a vital point that cannot be decided in advance
from current data. However, if such were the case, the development of an immunoassay for TSAb would be feasible. Then the general availability of an assay
for TSAb, with resultant widespread application of clinically related knowledge
of this autoantibody, will become a viable project.

PART II: THE CYTOCHEMICAL BIOASSAY

Lucille Bitensky and J. Chayen

PROCEDURE

The procedure for the cytochemical segment bioassay (Bitensky et al., 1974) is
essentially the same as that for thyroid-stimulating hormone (TSH) (Chapter 5)
except that, at dilutions appropriate for the potency, the time of response to
thyroid-stimulating antibody (TSAb) takes considerably longer: e.g., 20 min
(Bitensky et al., 1974) or 25 min (Petersen et al., 1975) as against 7 min for TSH.

The procedure for the cytochemical section bioassay of TSAb is identical
with that for TSH (Chayen et al., 1980), with appropriate dilutions of TSH having their maximal effect at 90 sec whereas appropriate dilutions of TSAb produce their maximum effect at 3 or 3.5 min (Ealey and Smyth, 1980; Chayen
et al., 1980).

Because of its greater practicability, allowing many more samples to be assayed each week, most recent workers have used the cytochemical section bioassay (Gilbert et al., 1977; Chayen et al., 1980).

VALIDATION

The section bioassay was validated in a detailed study by Ealey and Smyth (1980). The coefficient of variation among 24 follicles within a section was 9.2%; the mean intersegment coefficient of variation was 8.4%. Measurement of every cell from four follicles showed that the coefficient of variation was 7.9%.

The specificity of the assay was demonstrated by the fact that an antibody to human TSH had no effect whereas anti-human IgG almost completely abolished the activity. Anti-human IgG on its own produced no response. Neither bovine nor human thyroglobulin (at concentrations of 47 ng/ml to 2 mg/ml), that are active in the McKenzie assay for LATS (Florsheim et al., 1970; Burke and Szabo, 1972) and which interact with thyroid membranes (Consiglio et al., 1979), had no significant stimulatory activity when tested alone, and did not alter the response when added to a plasma sample that had been shown to have TSAb activity (Smyth et al., 1982).

PREPARATION OF THE SAMPLE FOR ASSAY

It has been emphasized by Chayen (1980), and reemphasized in Chapter 9, that the material used for cytochemical bioassays should be plasma, not serum. For example, it is well known that white cells contain considerable concentrations of polyamines; if these are liberated during the clotting process they can seriously interfere with the assay of thyroid stimulators. Moreover, some polypeptide hormones are inactivated when serum is prepared (Chapter 9). In spite of this, there is a tendency not only to use serum, but to prepare a relatively crude IgG fraction (by precipitation with 40% ammonium sulfate and then resolubilizing this fraction), and to use this material for assaying TSAb. This seems to be a somewhat blinkered procedure, since it ensures that any other thyroid stimulators, that might possibly be present in the circulation, are excluded from the assay. Furthermore, it assumes that the precipitation and resolubilization will have no effect on the biological activity of the IgGs; this may be valid for immunoassay but remains to be proven for assay by biological activity. In fact, Ealey and Smyth (1980) found that whereas a plasma gave responses that were parallel to those obtained with the MRC Research Standard LATS-B reference preparation, they obtained nonparallel responses to the same sample that had been left at room temperature for 2.5 hr before separation.

TIME COURSE OF THE RESPONSE

As is discussed in Chapter 3, samples that contain very high activity may produce a maximum response more quickly than samples that contain lower activities. With polypeptide hormones there appears to be a range of concentrations, extending over perhaps four or sometimes five decades of concentration (e.g., 0.5 or 5 fg to 5 pg/ml) at which the response time is virtually constant, the different concentrations being expressed by the quantitative extent of the response produced at that time. At higher concentrations, the time course is accelerated. Moreover, as discussed in Chapter 2, all bioassays are dilution assays: that is, they depend on diluting the sample sufficiently to bring the concentration of the hormone within the assayable levels (i.e., the four or five decades of concentration in which the response time is constant). Normally this is readily achieved in the cytochemical bioassays by diluting plasma 1:100 and 1:1000. The proofs that the dilution is indeed adequate are the findings that the former gives a larger response than the latter, and that the two responses are parallel to the responses obtained with a standard reference preparation of that hormone. If these results do not pertain, it may be necessary to use greater dilutions of the sample, until the assayable concentrations are achieved, as emphasized by Loveridge et al. (1979). When they are achieved, the higher dilutions will yield lesser responses than the lower dilutions. For these reasons it is suggested in Chapter 2 that three dilutions of plasma (e.g., $1:10^2$, $1:10^3$, and $1:10^4$) should always be used, so that there is a better chance that at least two will fall within the assayable concentrations.

It is regrettable that these basic concepts of bioassay either have been ignored, or were not appreciated, by some workers who have endeavored to use the cytochemical bioassay for TSAb, e.g., Figure 7 in Ealey et al. (1981) and their misleading advocacy of using either "the ascending or descending section of the response curve": they should have diluted their more potent sample until it reached the potency concentrations at which it could be validly assayed.

The question of "appropriate dilution" may be more difficult in assaying TSAb than when assaying polypeptide hormones, possibly because of the greater range of potencies found in the IgG fractions from different thyrotoxic patients. Moreover, when assaying a polypeptide hormone, one has a calibration graph of the response elicited by suitable concentrations of a standard preparation of the hormone to act as a guide: the unknown sample must then be diluted until it produces responses that parallel those of the standard preparation. This procedure may be more difficult when assaying TSAb, but merits attention; it is further complicated by the fact that the TSAb may comprise only a small fraction of the IgG fraction, and that the relative potencies of the TSAbs may vary widely. As a guide, the concentrations of the MRC Research Standard LATS-B

that were assayable in the cytochemical section bioassay were 1.5×10^{-8}-
1.5×10^{-5} μU/ml (Ealey and Smyth, 1980).

DISCRIMINATION BETWEEN TSH AND TSAb

It has been suggested by some workers (Hall et al., 1975) that it is possible to
separate thyroid-stimulating activity due to TSAb or to TSH *solely* on the basis
of the time course of their activities. This suggestion remains valid provided that
the dose responses, of at least two concentrations, at the selected time are parallel to the responses elicited by standard reference preparations of either TSH or
of TSAb. Because there seems to have been some confusion in the minds of
some workers, it may be helpful to amplify this statement. Suppose a sample
of plasma, at a certain dilution, produces a peak of activity at 90 sec in the section bioassay. If this is due to TSH, then a dose response done at this time will
be parallel to a dose response to a standard preparation of TSH. Without this
parallelism of dose responses, the mere fact of obtaining a response at 90 sec
does not prove that the sample contains TSH.

If a nonparallel dose response, or a negative dose response, is obtained, the
plasma must be diluted until a positive dose response is achieved. If it does contain TSH, then, at suitable dilutions, a dose response will be obtained that parallels the response to a standard reference preparation of TSH (10^{-4}-10^{-1} μU/ml
of the MRC Research Standard A, of human TSH). However, in view of the fact
that highly potent samples of TSAb can elicit responses as rapid as those normally associated with TSH (Loveridge et al., 1979; Ealey et al., 1981), it may be
that much greater dilutions of our hypothetical plasma will show no response at
90 sec, but will show a positive dose response, parallel to that induced by the
MRC Research Standard LATS-B preparation (at dilutions of 1.5×10^{-8}-
1.5×10^{-5} μU/ml), when assayed at 3 min.

Thus while, in principle, it may be possible to discriminate between TSH and
TSAb by judicious use of the time of response, the procedure would be cumbersome and unlikely to be used in practice. It is far more reasonable to discriminate between these activators by using specific antibodies, either to the TSH or
to the immunoglobulin, to annul the response, as suggested by Loveridge et al.
(1979). This is particularly pertinent to the question of whether there may be
measurable levels of TSH in the plasma of thyrotoxic patients where the concentrations of TSH may be low and the potency of the TSAbs may vary considerably. Döhler et al. (see Chapter 5) have been able to resolve this question by the
use of a specific antibody to TSH and so discriminate between these two types
of thyroid stimulators even when they occurred in the same sample.

COMPARISON WITH OTHER ASSAYS

The cytochemical bioassay is done on segments or sections of guinea pig thyroid, and it has been suggested that, while LATS acts on the mouse thyroid gland, other thyroid-stimulating immunoglobulins of Graves' disease may be specific for the human gland, as discussed by McKenzie and Zakarija (1977; Zakarija and McKenzie, 1978), (see also Chapter 6). Thus, LATS-negative plasma from a thyrotoxic patient might contain only the human-specific immunoglobulins. This was disproved when McKenzie and Zakarija (1977) obtained positive results in the cytochemical bioassay with a LATS-negative as well as a LATS-positive serum. It was subsequently confirmed by Loveridge et al. (1979) in the IgG fractions from six LATS-negative sera. In that study, nine preparations (three LATS positive, six LATS negative) with a very wide range of potency of TSAb, as assessed by the human thyroid-slice cAMP assay, were tested. In general the ranking order of potency, assessed by the cytochemical bioassay, agreed with that obtained by the cAMP assay. Such agreement is in contrast with the results obtained comparing the thyrotropin-binding inhibition assay, the cyclic AMP assay, and the colloid-droplet assay (Biro, 1982) on the same samples.

RESULTS WITH THE CYTOCHEMICAL BIOASSAY

Ealey and Smyth (1980) studied 13 patients with untreated Graves' disease and 5 euthyroid subjects. They expressed their results as the greatest dilution at which an activity, significantly greater than that induced by a control plasma pool, was obtained. All 13 were positive, with titers ranging from $1:10^2$ to $1:10^5$, 8 having a titer of $1:10^4$. In contrast, none of the plasma from the euthyroid subjects was positive. In a more extensive study (Smyth et al., 1982), 56 hyperthyroid patients were investigated. All 26 patients with classic Graves' disease, and all 24 with Graves' disease superimposed on a multinodular goiter, were positive. This was also true of the 6 patients with Plummer's disease (toxic nodular goiter) at titers of between $1:10^3$ and $1:10^4$. The plasmas of 4 of the 9 patients with diffuse nontoxic goiter, and in 10 of the 18 with idiopathic nontoxic nodular goiter, were also positive. None of the plasmas from the 17 control subjects, who had no history of thyroid disease, was positive even at high concentration. The finding of TSAb in these classes of patients, and the possible significance of the findings, is discussed by Smyth et al. (1982).

TSAb PRODUCED IN VITRO

Following their earlier studies with the cytochemical bioassay (Petersen et al., 1975), McLachlan et al. (1977) used the assay to detect the production of TSAb by human lymphocytes in vitro.

REFERENCES TO PART I

Adams, D. D., and Kennedy, T. H. (1967). Occurrence in thyrotoxicosis of a gammaglobulin which protects LATS from neutralization by an extract of thyroid gland. *J. Clin. Endocrinol. Metab.*, 27:173-177.

Adams, D. D., and Purves, H. D. (1956). Abnormal responses in the assays of thyrotropin. *Proc. Univ. Otago. Med. Sch.*, 34:11-12.

Bech, K., and Nistrup Madsen, S. (1979). Thyroid adenylate cyclase stimulating immunoglobulin in thyroid disease. *Clin. Endocrinol. (Oxf)*, 11:47-58.

Bitensky, L., Alaghband-Zadeh, J., and Chayen, J. (1974). Studies on the thyroid stimulating hormone and the long-acting thyroid stimulating hormone. *Clin. Endocrinol. (Oxf)*, 3:363-374.

Braun, D. G., Riesen, W. F., and Huser, H. (1976). Active heterologous chain recombinants of monoclonal antibodies raised in related rabbits. *Eur. J. Immunol.*, 6:819-822.

Bridges, S. H., and Little, J. R. (1971). Recovery of binding activity in reconstituted mouse myeloma proteins. *Biochemistry*, 10:2525-2526.

Chopra, I. J., and Solomon, D. H. (1970). Graves' disease with delayed hyperthyroidism. *Ann. Intern. Med.*, 73:985-990.

Chopra, I. J., Solomon, D. H., Johnson, D. E., and Chopra, U. (1970). Thyroid gland in Graves' disease. Victim or culprit? *Metabolism*, 19:760-772.

Davies, T. F., Yeo, P. B., Evered, D. E., Clark, F., Smith, B. R., and Hall, R. (1977). Value of thyroid stimulating antibody determinations in predicting short-term thyrotoxic relapse in Graves' disease. *Lancet*, i:1181-1182.

Dirmikis, S. M., and Munro, D. S. (1975). Placental transmission of thyroid-stimulating immunoglobulins. *Br. Med. J.*, 2:665-666.

Dorrington, K. J., and Munro, D. S. (1966). The long-acting thyroid stimulator. *Clin. Pharmacol. Ther.* 7:788-806.

Endo, K., Amir, S. M., and Ingbar, S. H. (1981). Development and evaluation of a method for the partial purification of immunoglobulins specific for Graves' disease. *J. Clin. Endocrinol. Metab.*, 52:1113-1123.

Endo, K., Kasagi, K., Koniski, J., Ikekubo, K., Okuno, T., Takeda, Y., Mori, T., and Torizuka, K. (1978). Detection and properties of TSH-binding inhibitor immunoglobulins in patients with Graves' disease and Hashimoto's thyroiditis. *J. Clin. Endocrinol. Metab.*, 46:734-739.

Fougereau, M., Olins, D. E., and Edelman, G. M. (1964). Reconstitution of anti-phage antibodies from L and H polypeptide chains and the formation of interspecies molecular hybrids. *J. Exp. Med.*, 120:349-358.

Froelich, C. J., Goodwin, J. S., Bankhurst, A. D., and Williams, R. C. (1980). Pregnancy, a temporary fetal graft of suppressor cells in autoimmune disease? *Am. J. Med.*, 69:329-331.

Hales, I. B., Luttrell, B. M., and Saunders, D. M. (1980). Placental transmission of thyrotropin-binding inhibitor immunoglobulins without neonatal thyrotoxicosis. In *Thyroid Research VIII, Proceedings Eighth International Thyroid Congress,* Sydney, Australia, J. R. Stockigt and S. Nagataki (eds.). Australian Acad. Sci., Canberra, pp. 591-593.

Huser, H. J., Haimovich, J., and Jaton, J. C. (1975). Antigen binding and idiotypic properties of reconstituted immunoglobulins G derived from homogeneous rabbit anti-penumococcal antibodies. *Eur. J. Immunol.,* 5:205-210.

Kabat, E. A. (1968). *Structural concepts.* In *Immunology and Immunochemistry.* J. D. Ebert, A. G. Loewy, and H. A. Schneiderman (eds.). Holt, Rinehart and Winston, New York, pp. 112-132.

Kendall-Taylor, P. (1975). LATS and human specific thyroid stimulator, their relation to Graves' disease. *Clin. Endocrinol. Metab.,* 4:319-339.

Koizumi, Y., Zakarija, M., and McKenzie, J. M. (1982). Solubilization, purification, and partial characterization of thyrotropin receptor from bovine and human thyroid gland. *Endocrinology,* 110:1381-1391.

Kriss, J. P. (1968). Inactivation of long-acting thyroid stimulator (LATS) by anti-kappa and anti-lamda antisera. *J. Clin. Endocrinol. Metab.,* 28:1440-1444.

Kriss, J. P., Pleshakov, V., and Chien, J. (1964). Isolation and identification of the long-acting thyroid stimulator and its relation to hyperthyroidism and circumscribed pretibial myxedema. *J. Clin. Endocrinol. Metab.,* 24:1005-1028.

Ledley, F. D., Lee, G., Kohn, L. D., Habig, W. H., and Hardegree, M. C. (1977). Tetanus toxin interactions with thyroid plasma membranes. *J. Biol. Chem.,* 252:4049-4055.

Loeb, L., and Bassett, R. B. (1929). Effect of hormones of anterior pituitary on thyroid gland in the guinea pig. *Proc. Soc. Exp. Biol. Med.,* 26:860-862.

Maisey, M. N. (1972). The Ig class and light chain type of the long-acting thyroid stimulator. *Clin. Endocrinol. (Oxf),* 1:189-198.

McKenzie, J. M. (1958). Delayed thyroid response to serum from thyrotoxic patients. *Endocrinology,* 62:865-868.

McKenzie, J. M. (1960). Further evidence for a thyroid activator in hyperthyroidism. *J. Clin. Endocrinol. Metab.,* 20:380-388.

McKenzie, J. M. (1965). The gammaglobulin of Graves' disease: Thyroid stimulation by fraction and fragment. *Trans. Assoc. Am. Physicians,* 78:174-186.

McKenzie, J. M. (1972). Does LATS cause hyperthyroidism in Graves' disease? (A review based towards the affirmative). *Metabolism,* 21:883-894.

McKenzie, J. M., and Zakarija, M. (1977). LATS in Graves' disease. *Recent Prog. Hormone Res.,* 33:29-57.

McKenzie, J. M., and Zakarija, M. (1978). Pathogenesis of neonatal Graves' disease. *J. Endocrinol. Invest.,* 2:183-189.

McKenzie, J. M., Zakarija, M., and Sato, A. (1978). Humoral immunity in Graves' disease. *Clin. Endocrinol. Metab.*, 7:31-45.

Mehdi, S. Q., and Kriss, J. P. (1978). Preparation of radiolabeled thyroid-stimulating immunoglobulins (TSI) by recombining TSI heavy chains with ^{125}I-labeled light chains: Direct evidence that the product binds to the membrane thyrotropin receptor and stimulates adenylate cyclase. *Endocrinology*, 103:296-301.

Mullin, B. R., Aloj, S. M., Fishman, P. H., Lee, G., Kohn, L. D., and Brady, R. O. (1976). Cholera toxin interactions with thyrotropin receptors on thyroid plasma membranes. *Proc. Natl. Acad. Sci. USA*, 73:1679-1683.

Nutt, J., Clark, F., Welch, R. G., and Hall, R. (1974). Neonatal hyperthyroidism and long-acting thyroid stimulator protector. *Br. Med. J.*, 4:695-696.

Ochi, Y., and DeGroot, L. J. (1968). Studies on the immunological properties of LATS. *Endocrinology*, 83:845-854.

O'Donnell, J., Trokoudes, K., Silverberg, J., Row, V. V., and Volpe, R. (1978). Thyrotropin displacement activity of serum immunoglobulins from patients with Graves' disease. *J. Clin. Endocrinol. Metab.*, 46:770-777.

Onaya, T., Kotani, M., Yamada, T., and Ochi, Y. (1973). New in vitro tests to detect the thyroid stimulator in sera from hyperthyroid patients by measuring colloid droplet formation and cyclic AMP in human thyroid slices. *J. Clin. Endocrinol. Metab.*, 36:859-866.

Orgiazzi, J., Williams, D. E., Chopra, I. J., and Solomon, D. H. (1976). Human thyroid adenyl cyclase-stimulating activity in immunoglobulin G of patients with Graves' disease. *J. Clin. Endocrinol. Metab.*, 42:341-354.

Rehn, L., (1884). Uber die extirpation des Kropfs bei Morbus Basedowii. *Klin. Wochenschr.*, 21:163.

Schleusener, H., Finke, R., Kotula, P., Wentzel, K. W., Meinhold, H., and Roedler, H. D. (1978). Determination of the thyroid stimulating immuno-globulins (TSI) during the cource of Graves' disease. A reliable indicator for remission and persistence of this disease? *J. Endocrinol. Invest.*, 2:155-161.

Skvaril, F. (1976). The question of specificity in binding human IgG subclasses to protein A-sepharose. *Immunochemistry*, 13:871-872.

Smith, B. R. (1976). Immunology of the thyrotropin receptor. *Immunol. Commun.*, 5:345-360.

Smith, B. R., and Hall, R. (1974). Thyroid stimulating immunoglobulins in Graves' disease. *Lancet*, ii:427-430.

Smith, P. E., and Smith, I. P. (1922). The repair and activation of the thyroid in the hypophysectomized tadpole by the parenteral administration of fresh anterior lobe of the bovine hypophysis. *J. Med. Res.*, 43:267-285.

Solomon, D. H., Chopra, I. J., Chopra, U., and Smith, F. J. (1977). Identification of subgroups of euthyroid Graves' ophthalmopathy. *N. Engl. J. Med.*, 296:181-186.

Sugenoya, A., Kidd, A., Row, V. V., and Volpe, R. (1979). Correlation between thyroid-displacing activity and human thyroid-stimulating activity by immunoglobulins from patients with Graves' disease and other thyroid disorders. *J. Clin. Endocrinol. Metab.*, 48:398-402.

Thomson, J. A., Dirmikis, S. M., Munro, D. S., Smith, B. R., Hall, R., and Mukhtar, E. D. (1975). Neonatal hyperthyroidism and long-acting thyroid stimulator protector. *Br. Med. J.*, 2:36.

Wong, E. T., and Doe, R. P. (1972). Suppressibility of thyroid function despite high levels of long-acting thyroid stimulator. *Ann. Intern. Med.*, 76:77-84.

Yavin, E., Yavin, Z., Schneider, M. D., and Kohn, L. D. (1981). Monoclonal antibodies to the thyrotropin receptor: Implications for receptor structure and the action of antibodies in Graves' disease. *Proc. Natl. Acad. Sci. USA*, 78: 3180-3184.

Zakarija, M. (1980). Thyroid-stimulating antibody (TSAb) of Graves' disease: Evidence for restricted heterogeneity. *Horm. Res.*, 13:1-15.

Zakarija, M., and McKenzie, J. M. (1978a). Zoological specificity of human thyroid stimulating antibody of Graves' disease. *J. Clin. Endocrinol. Metab.*, 47: 249-254.

Zakarija, M., and McKenzie, J. M. (1978b). Adsorption of thyroid stimulating antibody (TSAb) of Graves' disease by homologous and heterologous thyroid tissue. *J. Clin. Endocrinol. Metab.*, 47:906-908.

Zakarija, M., and McKenzie, J. M. (1978c). Isoelectric focusing of thyroid-stimulating antibody of Graves' disease. *Endocrinology*, 103:1469-1475.

Zakarija, M., and McKenzie, J. M. (1980). Influence of cholera toxin on in vitro refractoriness to thyrotropin of thyroids from rats fed propylthiouracil. *Endocrinology*, 107:2051-2054.

Zakarija, M., McKenzie, J. M., and Banovac, K. (1980). Clinical significance of the assay of thyroid-stimulating antibody of Graves' disease. *Ann. Intern. Med.*, 93:28-32.

Zamora, P. O., Waterman, R. E., and Kerkof, P. R. (1979). Early effects of thyrotropin on the surface morphology of thyroid cells in culture. *J. Ultrastruct. Res.*, 69:196-210.

REFERENCES TO PART II

Biro, J. (1982). Thyroid-stimulating antibodies in Graves' disease and the effect of thyrotrophin-binding globulins on their determination. *J. Endocrinol.*, 92: 175-184.

Bitensky, L., Alaghband-Zadeh, J., and Chayen, J. (1974). Studies on thyroid stimulating hormone and the long-acting thyroid stimulating hormone. *Clin. Endocrinol. (Oxf)*, 3:363-374.

Burke, G., and Szabo, M. (1972). Effects of thyroglobulin on thyroid function. *J. Clin. Endocrinol. Metab.*, 35:552-560.

Chayen, J. (1980). *The Cytochemical Bioassay of Polypeptide Hormones.* Monographs on Endocrinology, vol. 17. Springer, Berlin.

Chayen, J., Gilbert, D. M., Robertson, W. R., Bitensky, L., and Besser, G. M. (1980). A cytochemical section bioassay for thyrotrophin. *J. Immunoassay,* 1:1-13.

Consiglio, E., Salvatore, G., Rall, J. E., and Kohn, L. D. (1979). Thyroglobulin interaction with thyroid plasma membranes. *J. Biol. Chem.*, 254:5065-5076.

Ealey, P. A., Marshall, N. J., and Ekins, R. P. (1981). Time-related thyroid stimulation by thyrotropin and thyroid-stimulating antibodies, as measured by the cytochemical section bioassay. *J. Clin. Endocrinol. Metab.*, 52:483-487.

Ealey, P. A., and Smyth, P. P. A. (1980). Validation of the cytochemical section bioassay for thyroid stimulating antibodies. *J. Immunoassay,* 1:175-194.

Florsheim, W. H., Williams, A. D., and Schonbaum, E. (1970). On the mechanism of the McKenzie bioassay. *Endocrinology,* 87:881-888.

Gilbert, D. M., Besser, G. M., Bitensky, L., and Chayen, J. (1977). Development of a cytochemical section-bioassay for thyroid stimulators. *J. Endocrinol.* 75: 40P.

Hall, R., Smith, B. R., and Mukhtar, E. D. (1975). Thyroid stimulators in health and disease. *Clin. Endocrinol. (Oxf),* 4:213-230.

Loveridge, N., Zakarija, M., Bitensky, L., and McKenzie, J. M. (1979). The cytochemical bioassay for thyroid-stimulating antibody of Graves' disease: Further experience. *J. Clin. Endocrinol. Metab.*, 49:610-615.

McKenzie, J. M., and Zakarija, M. (1977). LATS in Graves' disease. *Recent Prog. Horm. Res.*, 33:29-57.

McLachlan, S. M., Smith, B. R., Petersen, V. B., Davies, T. F., and Hall, R. (1977). Thyroid-stimulating autoantibody production in vitro. *Nature,* 270: 447-449.

Petersen, V., Smith, B. R., and Hall, R. (1975). A study of thyroid stimulating activity in human serum with the highly sensitive cytochemical bioassay *J. Clin. Endocrinol. Metab.*, 41:199-202.

Smyth, P. P. A., Neylan, D., and O'Donovan, D. K. (1982). The prevalence of thyroid-stimulating antibodies in goitrous disease assessed by cytochemical section bioassay. *J. Clin. Endocrinol. Metab.*, 54:357-361.

Zakarija, M., and McKenzie, J. M. (1978). Zoological specificity of human thyroid-stimulating antibody. *J. Clin. Endocrinol. Metab.*, 47:249-254.

7

Thyroid Growth Stimulating and Blocking Immunoglobulins

H. A. Drexhage,* Gian Franco Bottazzo, and Deborah Doniach / Middlesex Hospital Medical School, London, England

INTRODUCTION

Since the discovery by Adams and Purves (1956) of a long-acting thyroid stimulator (LATS) in the serum of patients with Graves' disease, many workers have confirmed the presence of these thyroid-stimulating immunoglobulins (TSI) in these patients (Chapter 6). TSI have been clearly shown to be IgG molecules (McKenzie, 1958), whose thyroid-stimulating activity is formed by the combination of the heavy and light chains in the Fab part of the molecule, suggesting that they are in fact antibodies (Smith et al., 1969). There is evidence (discussed in Chapter 6) that the antibodies stimulate the thyroid gland because they have specificity for the thyroid cell's receptor to thyrotropin (TSH-R), which normally regulates the gland (Smith and Hall, 1974). The main and best known circuit of regulation by TSH involves the stimulation of plasma membrane adenylate cyclase, resulting in increased cytoplasmic levels of cyclic AMP, which will enhance the synthesis and secretion of thyroid hormones (Orgiazzi et al., 1976).

However, evidence is accumulating that some TSH effects are not mediated by cAMP; these include the stimulation of phosphatidylinositol turnover and the activation of the pentose shunt, both of which systems are involved in thyroid growth (Dumont et al., 1981). This has led to the suggestion that two

*Present affiliation: Free University Hospital, Amsterdam, The Netherlands.

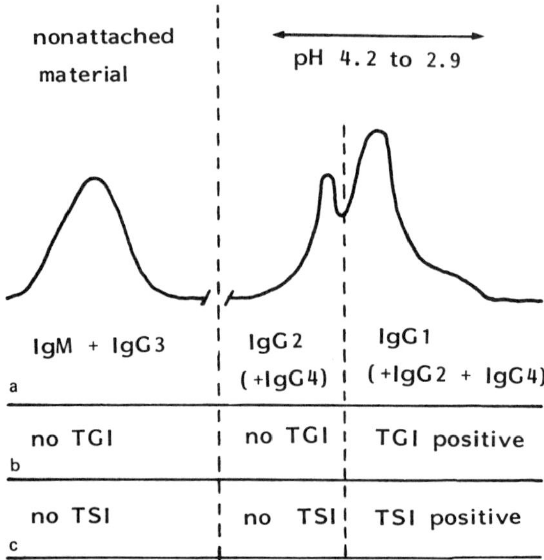

Figure 1   A pH gradient elution profile from protein A-Sepharose of an immunoglobulin from a patient. Different classes and subclasses were obtained as
has been described in detail by Duhamel et al. (1979). These fractions have been
tested for TGI and TSI activity (a: measured by immunoelectrophoresis; b:
measured by Feulgen densitometry and glucose-6-phosphate dehydrogenase
activity; c: measured by NADPH-oxidation activity). From these experiments it
became clear that the TSI and TGI activity resided in the IgG1 subclass.
(Column separations were done by Dr. Russell Buchanan, Immunogenetics Laboratory, Kennedy Institute of Rheumatology, London.)

different TSH-receptor sites, the one with cAMP-dependent effects (the β-
receptor site) and the other with effects not involving cAMP, the α receptor
site (van Herle et al., 1979).

  Antibodies to the TSH-receptor might likewise be dissociated, and some of
them might trigger metabolic pathways separate from those which result in hormone synthesis and secretion, but which will stimulate thyroid growth. Such
immunoglobulins, primarily stimulating follicle-cell growth, the TGI, have now
been identified in an experimental bioassay system of guinea pig thyroid segments cultured in the presence of the immunoglobulins (Drexhage et al., 1980).
Two metabolic effects have been used to measure growth in these explants,
namely, DNA synthesis measured by nucleic acid cytophotometry and stimulation of pentose shunt activity. It was found that TGI was thyroid specific and

of the IgG1 subclass (Figure 1). Clinically, TGI are present in selected cases of simple sporadic goiter and in patients with Graves' disease showing a large goiter. They are absent in nongoitrous patients with Graves' hyperthyroidism.

Apart from antibodies which bind to the receptor and stimulate the cell's metabolism, another variety of receptor antibodies has been identified which block TSH-induced activities of the cell. Orgiazzi et al. (1976) found that some immunoglobulins from thyrotoxic patients inhibited the rise in cAMP normally evoked by the addition of TSH. These "blocking" antibodies (TSI-block) have also been detected in patients with primary myxedema (Matsuura et al., 1980). Immunoglobulins not blocking thyroid hormone synthesis, but blocking follicle cell growth (TGI-block), have now been identified as well, by means of a similar bioassay system as that described for the detection of TGI (Drexhage et al., 1981): immunoglobulins from myxedema patients with atrophic thyroiditis had no effect on DNA synthesis in guinea pig thyroid explants by themselves, but when both TSH and the immunoglobulins were incubated with the thyroid segments, there was blocking of the growth effect usually seen with TSH alone.

TWO QUANTITATIVE CYTOCHEMICAL TECHNIQUES FOR THE DETECTION OF TGI AND TGI-BLOCK

Two quantitative cytochemical techniques have been applied to the detection of thyroid growth-stimulating or thyroid growth-blocking immunoglobulins; one is based on nucleic acid cytophotometry and the other on the measurement of glucose-6-phosphate dehydrogenase (G6PD) activity.

Nucleic Acid Cytophotometry

This method is based on the measurement of the amount of DNA in individual follicle-cell nuclei of a guinea pig thyroid segment. The DNA is visualized by the Feulgen reaction and is quantified in individual nuclei (in relatively arbitrary units) by means of a microdensitometer or cytophotometer (Sandritter, 1979). Many nuclei are measured, and the results are plotted as a population histogram, with the DNA content along the horizontal axis and the number of nuclei showing a particular content of DNA on the vertical axis (Figure 2).

The cell cycle forms the basis of the assessments. In differentiated tissues, we will find cells with only the diploid (2c) content of DNA; but in proliferating tissues, some nuclei will be in S-phase and these nuclei will have DNA content values intermediate between the 2c and 4c amounts. From the population histograms, the percentage of cells in S-phase can easily be seen, and these high values together with 4c values are taken as evidence for proliferation.

For the detection of TGI the assay system is as follows. Thyroids were removed from guinea pigs of the Hartley strain, weighing 250-350 g and fed on a

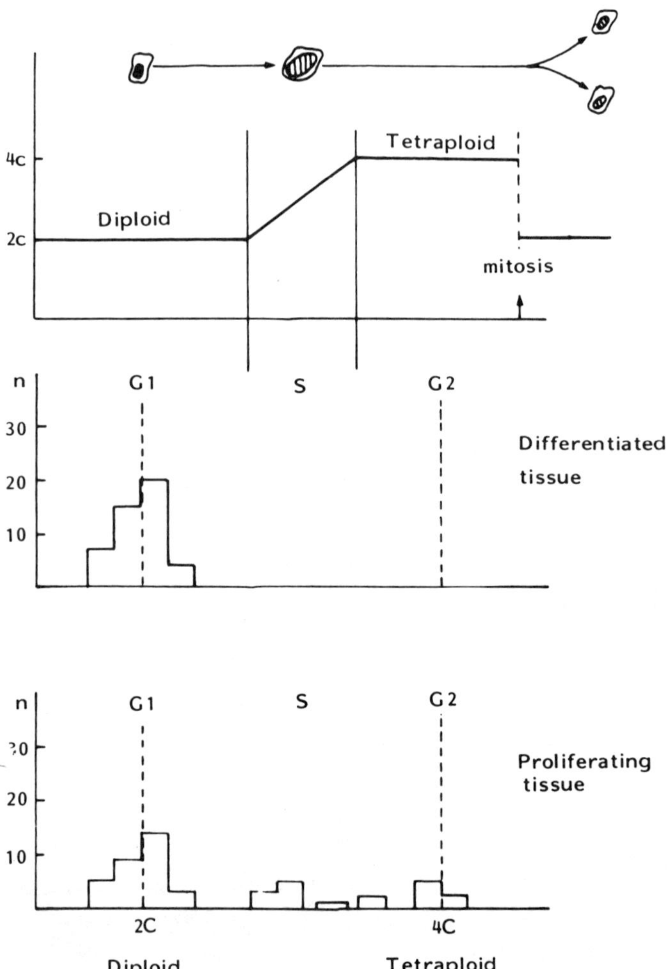

Figure 2   The top graph shows the DNA cycle in a cell from interphase to mitosis. The population histogram of the differentiated tissue shows only the 2c values in the diploid cells. The population histogram of the proliferating tissue additionally shows 4c values corresponding to cells in G2 and values intermediate to 2c and 4c, corresponding to cells in S-phase.

diet of pellets supplemented with cabbage. The animals were killed by asphyxiation in nitrogen.

The thyroids were divided into six segments, and each segment was maintained for 5, 24, or 48 hr at 37°C in nonproliferative organ culture using

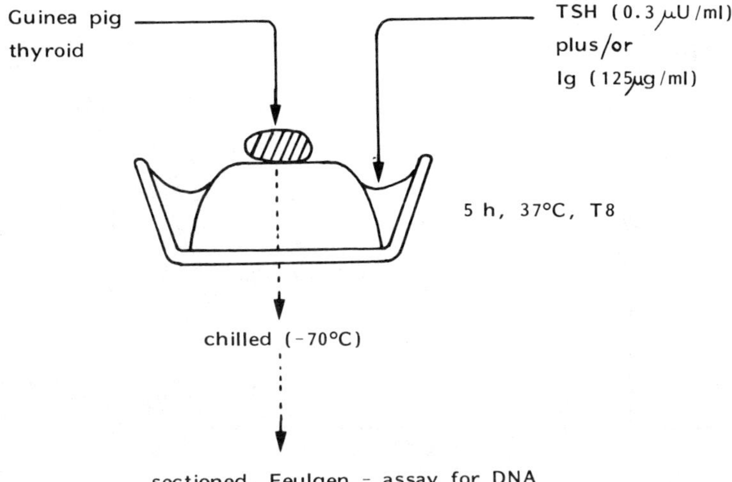

Figure 3 This figure shows the culture system for the bioassay for thyroid growth. One-third of a guinea pig thyroid lobe was cultured on a metal grid for 5 hr at 37°C in T8 culture fluid. The medium had been enriched with either TSH or an immunoglobulin. After culture the thyroid segment was chilled in n-hexane and sectioned in an automatically driven cryostat. Sections were reacted with the Feulgen reaction for DNA.

Trowell's T8 medium (Gibco). The medium was enriched with an Ig preparation obtained by ammonium sulfate precipitation (Hudson and Hay, 1976) of the patient's serum. In each study a control segment was maintained without Ig (the control culture) and in some cases 0.3 μU/ml TSH (MRC Research Standard A) was added as a positive control (Figure 3).

After culture the segments were chilled in n-hexane at –70°C, and then sectioned at 16 μm in an automatically driven Bright's cryostat with a cooled knife (solid carbon dioxide).

After a 15 min fixation in acetic acid-ethanol (1:3 v/v), the sections were immersed in 5 N HCl at room temperature for 20 min; they were then transferred to Schiff's reagent (Lamb Ltd.) for 1 hr in the dark and then washed in three changes of SO_2 water to remove unbound reagent (Chayen et al., 1973). The amount of Feulgen stain in 50-100 individual follicle-cell nuclei from three to six separate sections was measured in a randomized way with a Vickers M85 scanning and integrating microdensitometer (λ = 550 nm, $\times$100 oil-immersion objective, 0.2 μm scanning spot). The relative absorption values (relative DNA content values per nucleus) were plotted as population histograms, from which the diploid (2c) content could be defined (Figure 4).

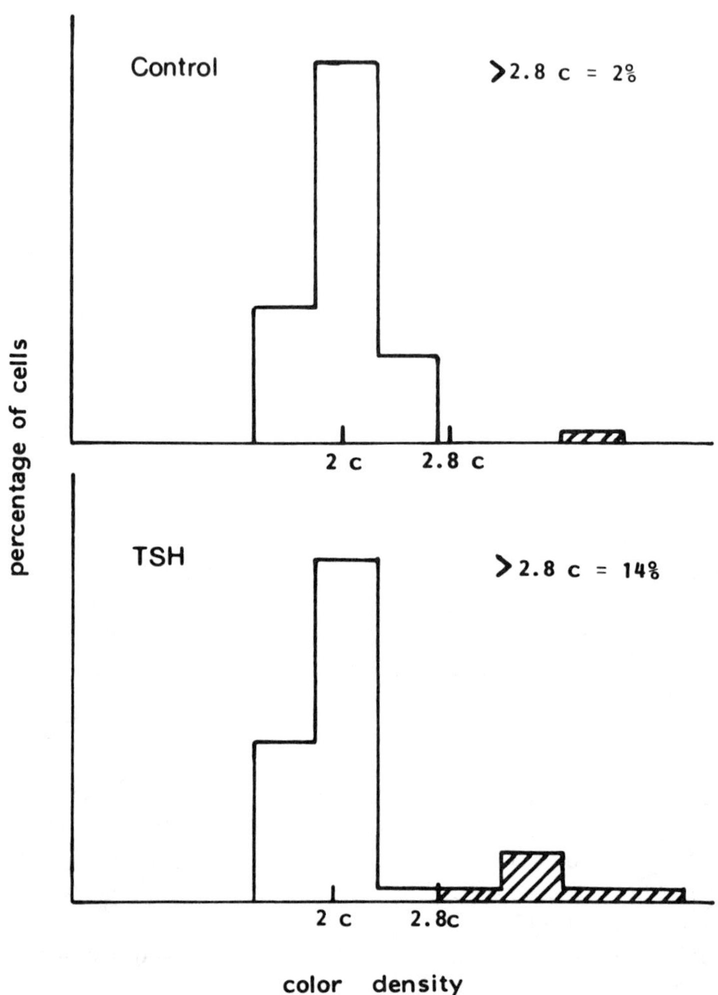

Figure 4 Effect of TSH on DNA synthesis in guinea pig thyroid. The color density of the Feulgen-reacted nuclei was measured and plotted as a population histogram. From these, the diploid (2c) was defined. A higher percentage of nuclei in S-phase (over 2.8c : hatched) was taken as evidence for thyroid growth.

Cells with a content of more than 2.8c were taken as being in, or having completed, the S-phase. This value was chosen as it included 3 standard deviations from the mean of 2c in histograms of normal thyroid segments. Results were expressed as the percentage of cells in S-phase.

Other new ways of assessing DNA-synthetic activity from nucleic acid cyto-photometry have been developed (Coulton et al., 1981). In these methods an index is calculated which gives a numerical representation of both the numbers of cells in S-phase and the amount of DNA present in these cells. This way of assessment might also prove to be reliable and useful for these assay systems.

Measurements of Glucose-6-Phosphate Dehydrogenase (G6PD) Activity

Glucose-6-phosphate dehydrogenase is the first and rate-controlling enzymatic step of the pentose shunt, which is the major source of ribose sugars needed for the synthesis of both DNA and RNA, the latter being important in protein synthesis (Figure 5). It also provides much of the NADPH of the cytosol, which is essential for many biosynthetic steps required for cellular growth. NADPH is also used by the microsomal respiratory pathway in many hydroxylation and mixed-function oxidation processes, as well as being involved in the generation of H_2O_2, important for iodination in the synthesis of thyroid hormones (Figure 5).

The total generation of NADPH by G6PD in a particular cell can be measured by reacting tissue sections with G6P and NADP in the presence of the intermediate hydrogen carrier, phenazine methosulfate (PMS), which transfers reducing equivalents quantitatively from NADPH to the final acceptor, neotetrazolium chloride, which on reduction precipitates as a formazan (Figure 5). The amount of formazan per follicle cell is measured by microdensitometry.

The proportion of NADPH-hydrogen reoxidized by the microsomal respiratory pathway can be determined by excluding PMS from the reaction medium; under these conditions the reducing equivalents are transferred by the microsomal respiratory pathway from the relatively negative electrode potential of the NADP: NADPH couple (E'_0 of -320 mV) to a more positive electrode potential at which the neotetrazolium chloride can be reduced in the absence of the intermediate carrier PMS (E'_0 of +170 mV) (Figure 5). These reducing equivalents have been defined as type 1 hydrogen (Chayen et al., 1974) and have been shown to be associated with thyroid hormone synthesis (Drexhage et al., 1982). In our system, type 1 hydrogen comprised only 5-10% of the total NADPH generated by G6PD activity. Most of the reducing equivalents were kept at the high electronegative potential of –320 mV and thus could be used for biosynthetic processes; these reducing equivalents are referred to as type 2 hydrogen.

The advantage of this type of quantitative cytochemical study of the pentose shunt is that it permits an analysis of how reducing equivalents from NADPH are apportioned between growth activities (type 2 pathway) and hormone-production activities involving the microsomal pathway (type 1 pathway).

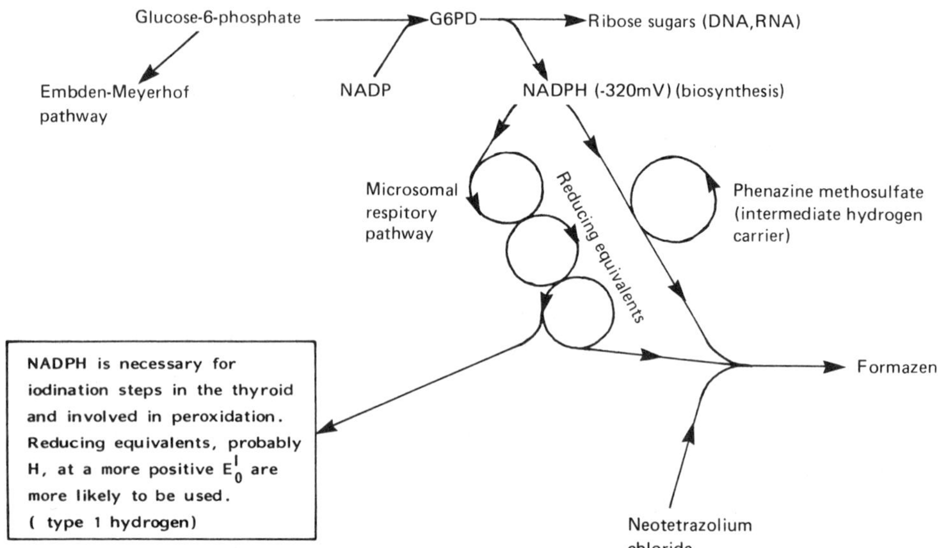

Figure 5 Diagrammatic representation of the utilization of reducing equivalents from the pentose shunt. G6PD = glucose-6-phosphate dehydrogenase. NADP = nicotinamide adenine dinucleotide phosphate. (Adapted from Drexhage, Hammond, et al., 1982.)

The ratio of the type 2/type 1 pathway in a particular thyroid follicle cell will indicate its tendency to be involved either in growth or in hormone synthesis: high ratios will suggest mainly growth activity, whereas low ratios will suggest processes of hormone synthesis.

The measurements were made on 10 μm sections cut from the guinea pig thyroid segments which were also used for the nucleic acid cytophotometry assay and had been exposed to Igs. The sections were made with an automatically driven Bright's cryostat with the cabinet temperature at $-25°C$, and the knife cooled with solid carbon dioxide. The sections were taken off the knife onto glass slides as recommended by Chayen et al. (1973). Six serial sections were cut from each specimen. Three of these sections were used to measure G6PD activity, with neotetrazolium chloride as the only hydrogen acceptor present, and the other three were used to measure this activity in the presence of the intermediate carrier PMS. Thus, the first measures the activity of the reoxidation of NADPH (type 1 pathway), whereas the second measures the total dehydrogenase activity (type 2 + type 1 pathway). The sections were incubated as described by Chayen et al. (1973) in a reaction medium consisting of 0.2 M glycylglycine buffer, containing 30% (w/v) polyvinyl alcohol (grade GO4/140); 4.5 mM (3 mg/ml) neotetrazolium chloride, 4.6 mM (1.5 mg/ml) glucose-6-phosphate; 3.2 mM (2.5 mg/ml) NADP; and, in some cases, as described, 0.65 mM (0.2 mg/ml) PMS. The final pH of the medium was 8.0. The reaction was allowed to proceed till the formazan precipitate was just visible, and the time at which this occurred was recorded. There is a linear relationship between the amount of formazan produced by G6PD and the time of incubation.

The precise amount of formazan present in 50-100 follicle cells was quantified using a Vickers M 85 scanning and integrating microdensitometer. Measurements were made at 585 nm, the isobestic point of the formazan produced (Chayen, 1980), with a ×100 oil-immersion objective and a scanning spot of 0.2 μm diameter. In each segment exposed to a particular stimulus (Ig or TSH) the activity of the type 1 and the total NADPH pathway of 100 follicle cells was measured and the activity per minute was calculated. Because there was a considerable difference between G6PD activities of thyroids of different animals, this value was expressed as a percentage of the value found in the control segment, i.e., the segment of the same thyroid kept without any addition of Ig or TSH. All results were analyzed by Student's t-test.

RESULTS WITH THE NUCLEIC ACID
CYTOPHOTOMETRY ASSAY SYSTEM

The amount of Ig and the time of exposure in this assay system were defined by a time course (1, 2, 3, 4, 5, 24, and 48 hr of culture) and dose-response curves (30, 60, 120, 240, and 480 μg/ml), using Ig of a healthy individual (normal Ig) or an Ig obtained from a thyrotoxic, LATS-positive patient with a large goiter (Figure 6). Normal Ig inhibited DNA synthesis somewhat compared to the control cultures. This was already evident with dosages as low as 60 μg/ml. With the LATS-Ig, the percentage of cells in S-phase rose rapidly with 60 μg Ig/ml culture fluid and reached a plateau at about 120 μg/ml. The DNA synthesis could already be measured at 2 hr but was clearly visible at 3, 5, 24, and 48 hr, the last showing the greatest stimulation. The 5 hr culture was chosen for convenience.

The effect of TSH and the Ig obtained from different groups of patients and controls is shown in Figure 7. In thyroid cultures kept without any addition (the control cultures) 3.9% ± 1.7 (mean ± standard deviation, SD) of cells were found to be in S-phase. A slightly lower percentage (3.3% ± 1.4) was seen after exposure to Ig obtained from a group of healthy individuals or from patients with thyroid disorders of nonautoimmune origin, such as dyshormonogenetic goiter, medullary carcinoma, cyst, and single autonomous adenoma. TSH, and more markedly TSH in combination with normal Ig, increased DNA synthesis in the explants (9.8% ± 1.8, and 12.0% ± 1.6, respectively). A similar enhancing

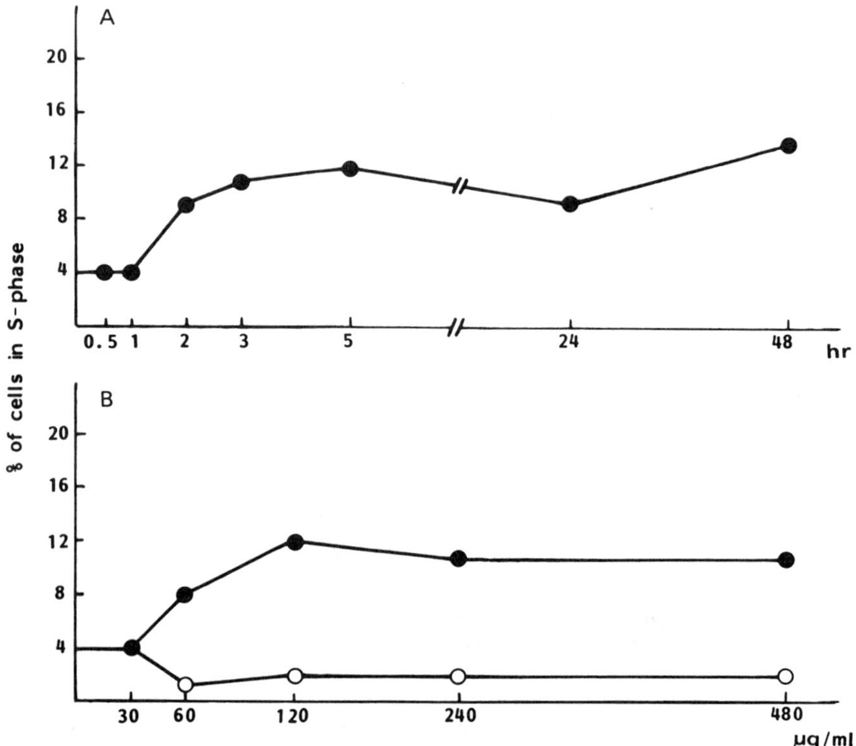

Figure 6 Time-response curve (A) and dose-response curve (B) of a LATS-positive Ig (closed circle) and of a normal Ig (open circle). The percentage of cells in
S-phase is plotted on the vertical axis and was determined by Feulgen densitometry. (Adapted from Drexhage, Hammond, et al., 1982.)

effect was obtained with Ig from patients with Graves' disease with goiter
(10.7% ± 2.3). In contrast, Ig from patients with Graves' disease without goiter
formation showed minimal stimulating influence (4.3% ± 1.5). In Graves' disease
the stimulation of growth by Ig in vitro correlated well with the size of the
goiter in vivo ($r = 0.76$, $p < 0.01$), rather than with the degree of hyperthyroidism, i.e., the T_3 level ($r = 0.1$, $p > 0.05$), or the presence of thyrotropin-displac-
ing immunoglobulin (TBII) ($r = 0.40$, $p > 0.01$). The Ig of 17 cases of sporadic
simple goiter were tested in this assay system as well. A proportion of these
patients had some features resembling thyrotoxicosis, including raised iodine
uptake values and flat or absent TRH responses. Half of them had first-degree
relatives with thyrotoxicosis or Hashimoto's disease, and in 6 of 15, the cyto-

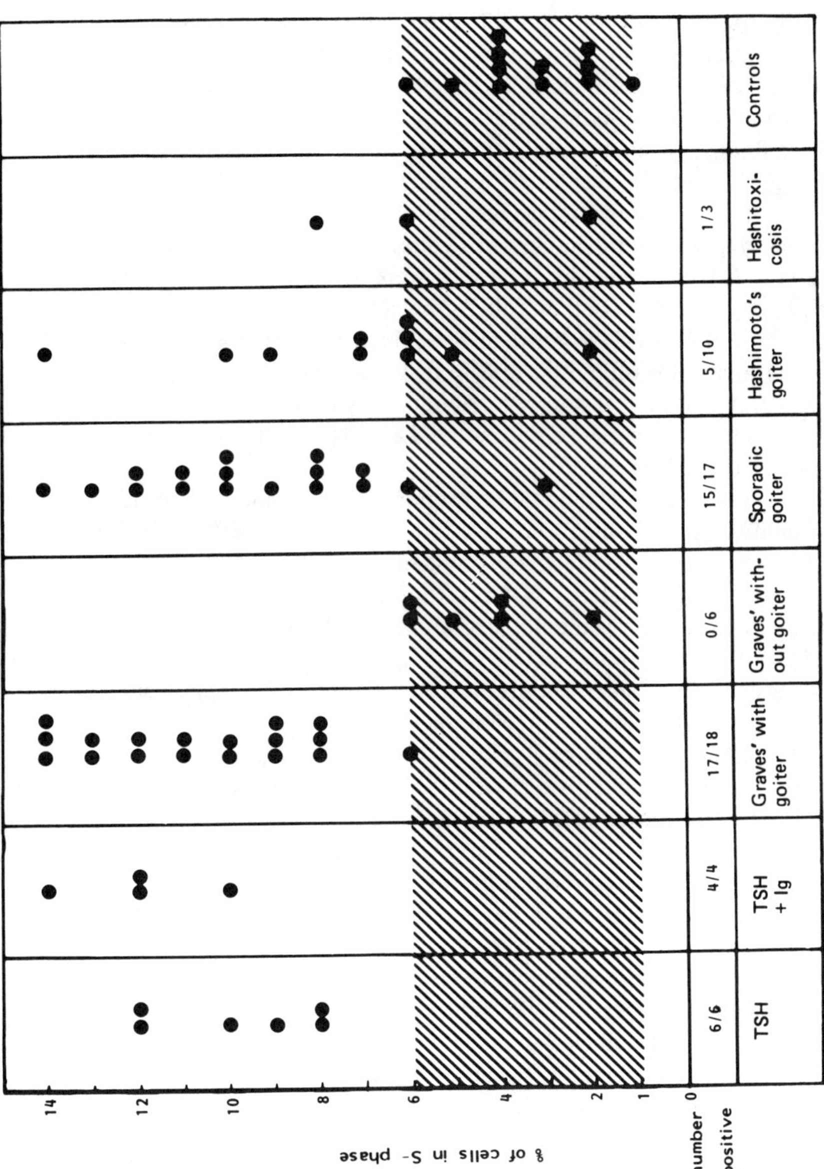

Figure 7   Thyroid growth Ig (TGI) in sera from patients with and without goiter and the effect of TSH alone and in combination with normal Ig. Results are expressed as percentage of nuclei in S-phase. The hatched area represents the range found with control immunoglobulins, including Ig of healthy individuals, of dyshormonogenic goiter, single autonomous adenoma, medullary carcinoma, and cysts. (Adapted from Drexhage, Hammond, et al., 1982.)

chemical bioassay for TSI activity (Chayen et al., 1980) had given positive results (Ealey, personal communication). Several other authors have also reported TSI levels in approximately 40% of sporadic simple goiter cases (Brown et al., 1978); however, negative results have been published as well (Bolk et al., 1979). Practically all our selected patients had detectable TGI; only one long-standing case aged 72 years did not show any growth-stimulating activity (2% of cells in S-phase). These findings suggest that at least some of the simple sporadic euthyroid goiters might be considered as a new form of the thyroid autoimmunity, in that the growth of the gland might have been caused by the circulating growth-stimulating immunoglobulin. In Hashimoto's goiter and hashitoxicosis, TGI was detected in 6 of 13 patients; the antibodies were found especially in those cases whose goiters failed to shrink on full T4 replacement therapy. There was no correlation with the size of the goitre in vivo. However, it is known that up to 70% of the goiter mass in Hashimoto's disease may consist of infiltrating lymphocytes, and in addition, the growth stimulus may derive mostly from pituitary TSH, as clinical experience shows that T4 replacement is most effective in patients having a raised serum TSH.

In primary myxedema, no TGI could be found; in this disease most of the patients appeared to have Ig which were able to block the tropic effects of TSH in the culture system. A detailed report of these blocking studies is given elsewhere (Drexhage et al., 1981). Such TGI-block immunoglobulins might be important in the pathogenesis of glandular atrophy in this disease (Editorial, *Lancet*, 1981).

Until now, tests have been carried out only on groups of patients by adding fixed aliquots of ammonium sulfate-precipitated Ig (125 μg/ml) to a 5 hr culture system. The kinetics of this assay system (the dose-response and time-response curves), using different dosages of different Ig with possibly different avidities, have not been studied yet. It is possible that Ig with strong TGI activity due to high-avidity antibodies will show optimal responses at other times and other dosages. Such a study is a prerequisite for the use of this assay system in clinical practice, where one wants to know the exact TGI activity of a certain Ig preparation.

The nucleic acid cytophotometric method is time consuming, and only a few Ig preparations can be tested each week. Especially the microdensitometric measurement of the 50-100 follicle-cell nuclei takes time and requires experience. We have tested two different microdensitometers for the measurement of the DNA content of large numbers of individual follicle-cell nuclei, the Vickers M85 (Vickers Instrum. Ltd., YO 3 7 SD York, U.K.) and the Cytoscan SMP (Zeiss, Oberkochen, Federal Republic of Germany) interfaced to a PDP 11/10 computer (Digital Equipment Corporation, Maryland, Mass., U.S.A.) with the HIDACSYS Computer program (van der Ploeg et al., 1977). Although the

Cytoscan SMP avoids the use of a mask of a fixed diameter, the Vickers M85 is more suitable for measuring large series of nuclei since the actual measurements can be made more quickly.

At present, experiments are being undertaken to simplify the methods by using more conventional biochemical techniques, such as the incorporation of ^{3}H-thymidine and the estimation of thymidine kinase activity. Both are known to be increased in proliferating lymphocytes (Olsen and Harris, 1975). However, in the majority of cultured or normal guinea pig thyroids, we were not able to detect the presence of any noteworthy amount of thymidine kinase activity, and neither was the incorporation of ^{3}H-thymidine increased.

The incorporation of ^{3}H-orotic acid and ^{3}H-aspartic acid have also been tried with little success: in long-term cultures (48 and 72 hr) TSH generally induces an increased uptake of ^{3}H-orotic acid or ^{3}H-aspartic acid, but the results vary considerably, and some guinea pig thyroids do not respond to TSH with an increased uptake.

RESULTS WITH CYTOCHEMICAL MEASUREMENTS OF GLUCOSE-6-PHOSPHATE DEHYDROGENASE ACTIVITY

The amount of Ig for studying the stimulation of G6PD activity was defined from the dose-response curve derived from the same normal Ig and the Ig of the LATS-positive patient which had been used for the Feulgen dose-response curve.

Figure 8 shows the results. Although the actual differences between the LATS and the normal Ig were small, the most discrimination was found with a concentration of 120 μg/ml culture fluid. We therefore used this concentration in our other studies, but the actual values found in this dose-response curve indicate that a more extensive study on the kinetics of the response is required before definite conclusions can be drawn on potencies of particular Ig preparations. Only results obtained with groups of patients and individuals will be presented.

The stimulation of G6PD activity (expressed as NAPDH) in segments exposed to TSH (0.3 μU/ml), to TSH in combination with normal Ig (125 μg/ml), and to Ig of different groups of patients and controls is shown in Figure 9. With TSH, TSH in combination with Ig, and with 12 of 14 of the Ig of goitrous Graves' patients, the generation of NADPH was stimulated to above control values. The Ig of the five nongoitrous Graves' patients did not produce any stimulation of G6PD activity. Thus, in general, a good correlation was obtained with the results of this assay system and those of nucleic acid cytophotometry (r = 0.71; n = 41; p < 0.01).

With the Ig of the sporadic simple goiter cases, a more complex picture arose. In only 13 of 22 of these cases was the generation of NADPH above the control

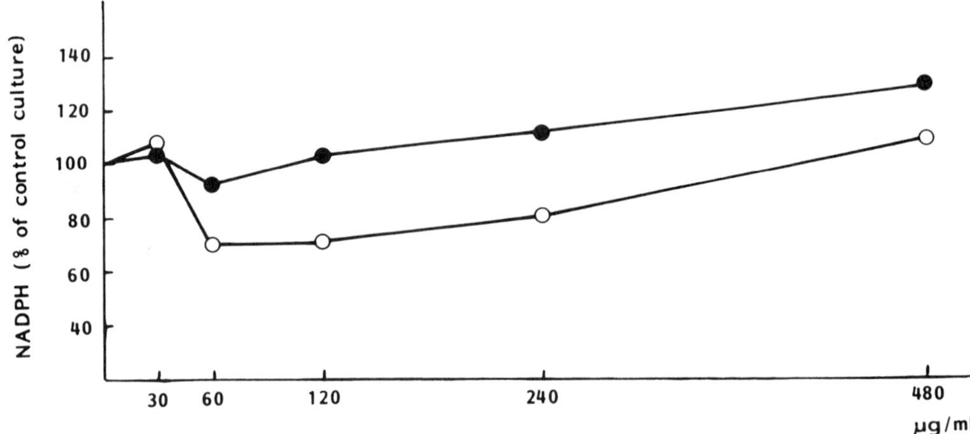

Figure 8   Dose-response curve of normal Ig (open circle) and of Ig of a LATS-positive patient (closed circle). The generation of NADPH due to G6PD activity is plotted on the vertical axis; these values are expressed as a percentage of the values found in the control culture. On the horizontal axis the concentration of Ig is given in micrograms protein per milliliter culture fluid.

limits, whereas practically all had shown TGI activity as tested by nucleic acid cytophotometry. However, when the total NADPH was divided into type 2 hydrogen—namely, that type used for growth—and type 1 hydrogen—that may be reoxidized for H_2O_2 formation—practically all (14 of 17) of the Ig of sporadic simple goiter induced a marked shift toward type 2 hydrogen (Figure 10). Thus, whereas the total generation of NADPH was only affected to a minor extent by these Ig, they did affect the apportioning.

In TGI-positive Hashimoto's goiter, a similar picture was apparent: these Ig also mainly affected the apportioning of NADPH in favor of the type 2 pathway (Figure 10). Interestingly, TSH did not affect the type 2/type 1 ratio (Figure 10), indicating that all excessively generated NADPH (see Figure 9) was used in equal amounts for biosynthesis and hormone production. The extra addition of 125 μg/ml of normal Ig to TSH in the culture had marked effects: the excessively generated NADPH was now mainly apportioned to the type 2 hydrogen pathway, indicating a shift toward growth. The nucleic acid cytophotometry is in accordance with these findings and showed an increased DNA synthesis (see Figure 7). This might indicate that receptors for Ig, maybe Fc receptors, play an important part in the function of the TSH-receptors.

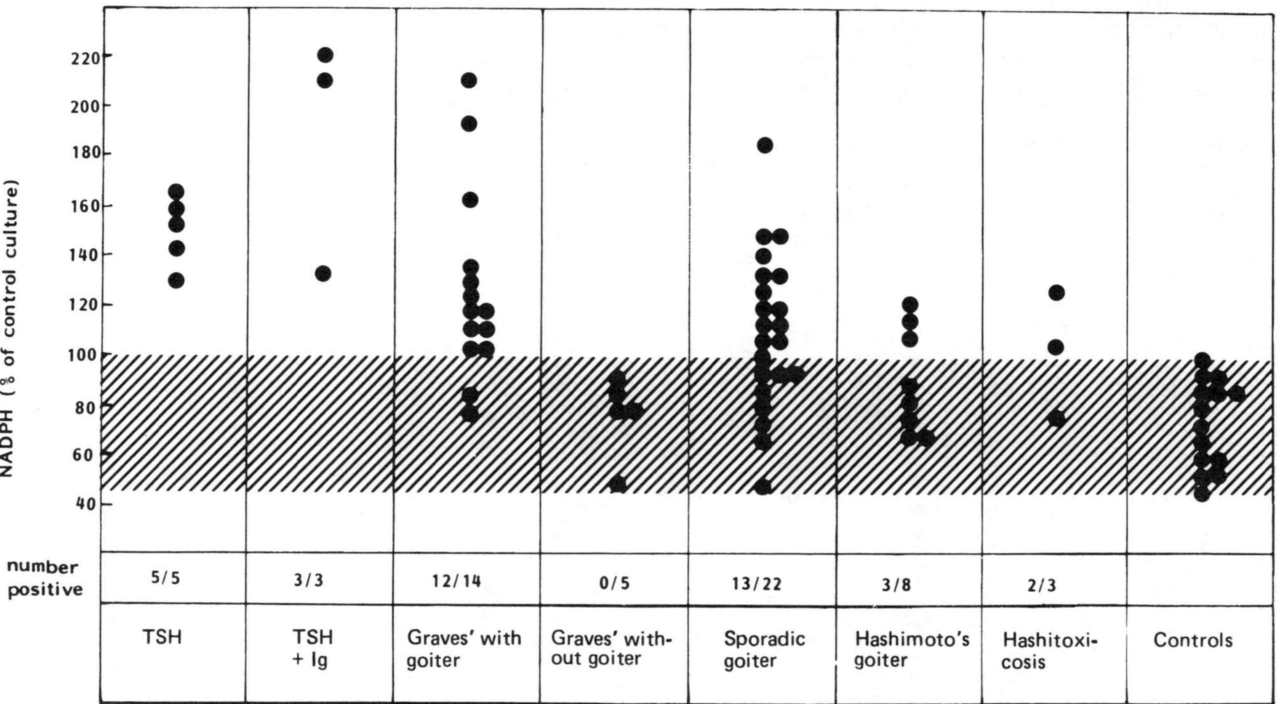

Figure 9   The generation of NADPH in guinea pig thyroid segments cultured for 5 hr with TSH (0.3 μU/ml); TSH in combination with Ig (125 μg/ml); or with Ig (125 μg/ml) from patients with or without goiter. The hatched area represents the range found with the control immunoglobulins. See also Figure 7, where the Feulgen densitometry results of the same patients are included.

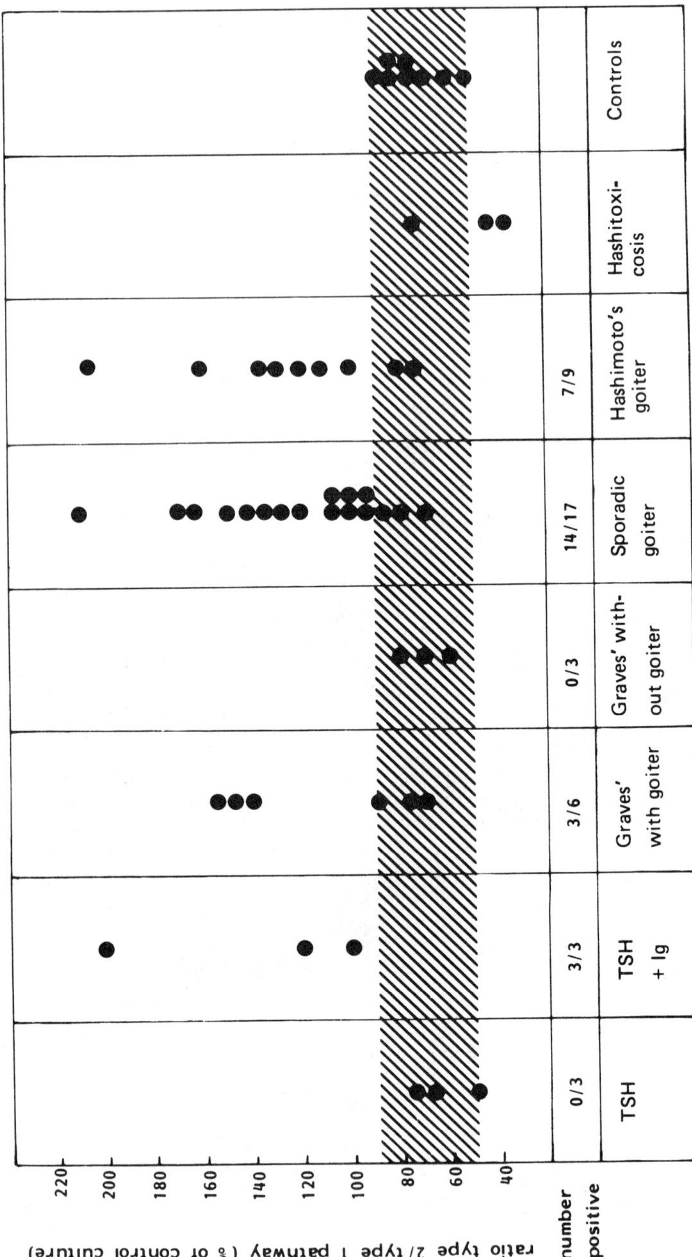

Figure 10   The ratio of the type 2/type 1 NADPH-hydrogen found in guinea pig thyroid segments exposed to TSH (0.3 μU/ml); TSH in combination with Ig (125 μg/ml); or to Ig (125 μg/ml) of patients with or without goiter. The hatched area represents the range found with the control immunoglobulins. See also Figure 9, where the results of these immunoglobulins on the generation of NADPH are included.

In goitrous Graves' disease, a few cases showed high type 2/type 1 ratios, whereas others did not. A clear correlation with preferential goiter growth was not evident in these Graves' cases, but probably a larger series of patients needs to be studied to establish such a correlation.

The results on the type 1 hydrogen generation in segments exposed to TSH or to the various Ig are given in detail in Drexhage, Hammond, et al. (1982). These studies show that Ig stimulating the production of type 1 hydrogen were found in practically all thyrotoxic patients whether or not presenting with goiter, and also in patients with hashitoxicosis. Type 1 pathway-stimulating Ig were absent in the majority of patients with simple sporadic goiter and Hashimoto's goiter.

In two of three Ig from patients with hashitoxicosis, an enhanced total NADPH production was found (Figure 9). These two samples also stimulated DNA synthesis in the nucleic acid cytophotometry assay. The study of the apportioning indicated that in hashitoxicosis, NADPH was used predominantly for reoxidation in the microsomal respiratory pathway (low ratios in Figure 10) and hence is employed for the stimulation of hormone synthesis. This gives further support to the concept that TSI acts additionally to the ordinary microsomal and thyroglobulin antibodies and contributes to the clinical picture in this disease (Doniach et al., 1960).

FUTURE DEVELOPMENTS

From our results summarized here, and also from earlier reports (Drexhage, Bottazzo, et al., 1982), it is clear that TSI, TGI, TSI-block and TGI-block play an important role in functional and morphological disturbances of the thyroid in a whole spectrum of thyroid disorders, including primary myxedema, Graves' disease, simple sporadic goiter, and Hashimoto's disease. The determination, in the serum of patients, of immunoglobulins which affect the biological activity of the thyroid, will probably become increasingly more important for clinical practice. A study of the kinetics of these effects, as demonstrated cytochemically, is necessary to provide a prognosis in individual cases. In addition, the development of simplified methods, to avoid the laborious densitometric measurements, would be advantageous because more sera can then be tested.

Our preliminary data urge a more extensive study on the incidence of TGI and TGI-block in Graves' disease and simple sporadic diffuse or nodular goiter and the degree of correlation between the size of the gland and the presence of these particular antibodies. Moreover, it is known that a continuous glandular TSH-stimulation in vivo results, first, in diffuse but later in nodular goiter development in animals (Williams, 1982). In these nodular goiters, areas of malignancy may finally occur. It is therefore also important to study TGI levels in patients with thyroid carcinoma to see whether a continuous antibody-induced growth stimulus may play a part in the pathogenesis.

Relapses after drug treatment might be predicted more precisely taking into consideration the simultaneous presence and the titer of the different varieties of antibodies before and during antithyroid drug treatment. Studies have been published which indicate that HLA-B8 and HLA-DR3 negative Graves' patients, showing a rapid decline in thyroid-displacing immunoglobulins during drug treatment, are not likely to relapse (McGregor et al., 1980).

It could well be that a HLA-DR3 haplotype together with the simultaneous presence of a few thyroid-reactive antibodies in high titer suggest the expression of a genetic tendency toward immunologic "dysregulation" and might be a sign of poor prognosis. Immunoglobulins capable of blocking cAMP stimulation by TSH were first detected in a small number of thyrotoxic sera by Orgiazzi et al. (1976). A Japanese team demonstrated these antibodies in the serum of a myxedema patient (Endo et al., 1978) and published further evidence which showed that TSI-blocking antibodies caused a transient neonatal myxedema by transplacental passage from a myxedematous mother to her offspring (Matsuura et al., 1980). Using our assay systems, we were also able to detect these antibodies and in addition were able to show the existence of another variety of thyroid-blocking antibodies, namely, TGI-glock. We have found that some myxedema patients have both growth-blocking and hormone synthesis-blocking Igs; however, preliminary results also suggest that TGI-block is more common in this disease than TSI-block. We even have some evidence that some sera may block growth effects of TSH and yet stimulate the reoxidation of NADPH, which is suggestive of a coexistent TSI-like effect.

The study of blocking antibodies therefore requires extension and might especially be relevant in pregnant women to prevent some cases of transient hypothyroidism of the newborn. Furthermore, it is tempting to speculate that employing cytochemical techniques on cultured segments of other organs, such as adrenal, parathyroid, and kidney, other biologically active Ig might be detected in several clinical entities of unknown pathogenesis.

ACKNOWLEDGMENTS

The authors thank Linda J. Hammond for excellent technical assistence. They thank the National Institute for Biological Standards and Control for providing the MRC Research Standard A. We are grateful to Drs. N. J. Marshall, S. Bidey, and P. Ealey for the naphthylamidase assays. We also thank Mrs. Ylva C. Abraham-Dolman and Mr. W. R. Dolman for careful typing of the manuscript. H. A. Drexhage was supported by a Royal Society–Z. W. O. Exchange Fellowship.

REFERENCES

Adams, D. D., and Purves, H. D. (1956). Abnormal responses in the assay for thyrotrophin. *Proc. Univ. Otago School,* 34:11-12.

Bolk, J. H., Bussemaker, J. K., Elte, J. W. F., Haak, A., and von der Heide, D. (1979). Thyroid stimulating immunoglobulins do not cause non-autonomous, autonomous, or toxic multinodular goiters. *Lancet,* i:61-63.

Brown, R. S., Pohl, S. L., Jackson, I. M. D., and Reichlin, S. (1978). Do thyroid stimulating immunoglobulins cause nontoxic and toxic nodular goitre? *Lancet,* i:904-906.

Chayen, J. (1980). *The Cytochemical Bioassay of Polypeptide Hormones.* Springer, Berlin.

Chayen, J., Bitensky, L., and Butcher, R. G. (1973). *Practical Histochemistry.* John Wiley, London.

Chayen, J., Bitensky, L., Butcher, R. G., and Altman, F. P. (1974). Cellular biochemical assessment of steroid activity. *Adv. Steroid Biochem. Pharmacol.,* 4:1-60.

Chayen, J., Gilbert, D. M., Robertson, W. R., Bitensky, L., and Besser, G. M. (1980). A cytochemical section bioassay for thyrotrophin. *J. Immunoassay,* 1:1-13.

Coulton, L. A., Henderson, B., and Chayen, J. (1981). The assessment of DNA-synthetic activity. Histochemistry 72:91-99.

Doniach, D., Hudson, R. V., and Roitt, I. M. (1960). Human auto-immune thyroiditis: clinical studies. *Br. Med. J.,* 1:365-373.

Drexhage, H. A., Bottazzo, G. F., Bitensky, L., Chayen, J., and Doniach, D. (1981). Thyroid growth-blocking antibodies in primary myxoedema. *Nature,* 289:594-596.

Drexhage, H. A., Bottazzo, G. F., Doniach, D., Bitensky, L., and Chayen, J. (1980). Evidence for thyroid-growth-stimulating immunoglobulins in some goitrous thyroid disease. *Lancet,* ii:287-292.

Drexhage, H. A., Donaich, D., Bottazzo, G. F., and Bitensky, L. (1982). Detection and clinical significance of antibodies stimulating or blocking thyroid growth in disorders of the thyroid gland. In *Current Endocrine Concepts,* E. D. Williams (ed.), London, Praeger, pp. 37-55.

Drexhage, H. A., Hammond, L. J., Bitensky, L., Chayen, J., Bottazzo, G. F., and Doniach, D. (1982). The involvement of the pentose shunt in thyroid metabolism after stimulation with TSH or with immunoglobulins from patients with thyroid disease. II. The reoxidation of NADHP and stimulation of hor-

Duhamel, R. C., Schur, P. H., Brendel, K., and Meezan, E. (1979). pH Gradient elution of human IgG1, IgG2, and IgG4 from protein A-sepharose. *J. Immunol. Meth.,* 31:211-217.

Dumont, J. E., Takeuchi, A., Lamy, F., Gerwy-Decoster, C., Cochaux, P., Roger, P., van Sande, J., Lecock, R., and Mockel, J. (1981). Thyroid control: an example of complex cell regulation network. *Adv. Cyclic Nucleotide Res.,* 14:479-489.

Editorial (1981). Thyroid autoimmune disease: A broad spectrum. *Lancet* 1: 874-875.

Endo, K., Kasagi, K., and Konishi, J. (1978). Detection and properties of TSH-binding inhibitor immunoglobulins in patients with Graves' disease and Hashimoto thyroiditis. *J. Clin. Endorcinol. Metab.,* 46:734-739.

Hudson, L., and Hay, F. C. (1976). *Practical Immunology.* Blackwell Scientific, Oxford.

Matsura, N., Yamada, U., Nohara, Y., Konishi, J., Endo, K., Konjuna, H., and Wataya, K. (1980). Familial transient hypothyroidism dur to maternal TSH-binding inhibitor immunoglobulin. *N. Engl. J. Med.,* 303:738-741.

McGregor, A. M., Smith, B. R., Hall, R., Petersen, M. M., Miller, M., and Dewar, P. J. (1980). Prediction of relapse in hyperthyroid Graves' disease. *Lancet,* i: 1101-1103.

McKenzie, J. M. (1958). Fractionation of plasma containing the long-acting thyroid stimulator. *J. Biol. Chem.,* 237:3571-3572.

Olsen, I., and Harris, G. (1975). Thymidine kinase of mouse spleen cells in vivo and in vitro. *Biochem. J.,* 146:489-496.

Orgiazzi, J., Williams, D. E., Chopra, I. J., and Solomon, D. H. (1976). Human thyroid adenyl cyclase-stimulating activity in immunoglobulin G of patients with Graves' disease. *J. Clin. Endocrinol. Metab.,* 42:341-354.

Sandritter, W. (1979). A review of nucleic acid cytophotometry in general pathology. In *Quantitative Cytochemistry and Its Applications,* J. R. Pattison, L. Bitensky, and J. Chayen (eds.). Academic, London, pp. 1-8.

Smith, B. R., Dorrington, K. J., and Munro, D. S. (1969). The thyroid-stimulating properties of long-acting thyroid stimulator gamma G subunits. *Biochim. Biophys. Acta,* 142:277-285.

Smith Rees, B., and Hall, R. (1974). Thyroid-stimulating immunoglobulins in Graves' disease. *Lancet,* ii:427-431.

van der Ploeg, M., van der Broek, Smeulders, A. W. M., Vossepoel, A. M., and van Duyn, P. (1977). HIDACSYS: Computer programs for interactive scanning cytophotometry. *Histochemistry,* 54:273-288.

van Herle, A. J., Vassart, G. J., and Dumont, J. E. (1979). Control of thyroglobulin synthesis and secretion. *N. Engl. J. Med.,* 301:239-294; 307-314.

Williams, E. D. (ed.) (1982). *Current Endocrine Concepts.* Praeger, London.

8

Luteinizing Hormone: A New Generation of Bioassays

Judith Weisz / The Milton S. Hershey Medical Center, The Pennsylvania State University, Hershey, Pennsylvania

INTRODUCTION

The story of the bioassays for luteinizing hormone (LH) largely parallels that of many other protein hormones. The development of radioligand displacement assays for LH in the 1960s (Midgley, 1966; Odell et al., 1967) led to a temporary loss of interest in bioassays for this hormone. The sensitivity, convenience, and reproducibility of these new assays, in particular of the radioimmunoassay, were far beyond what could be achieved with the bioassays then available. However, in recent years a new generation of in vitro assays for evaluating the biological potency of the hormone has emerged. An outstanding characteristic of these new bioassays is their sensitivity. In this respect they equal and even surpass the radioimmunoassay and radioreceptor assays.

The interest in developing and applying these new bioassays stems largely from the realization that the structure of what we term LH is not fixed and invariant but that the hormone can exist as a class of related molecules. Only the core of the molecule appears to be fixed for each species. Not only may LH undergo a sequence of modifications during the process of its biosynthesis, but the form in which it is released may vary and be influenced by the reproductive status and hormonal milieu of the organism as well as by the rate of LH secretion (Azhar et al., 1978; Liu et al., 1979; Mukhopadhyay et al., 1979; Qazi et al., 1974; Sharpe et al., 1975; Dufau et al., 1976; Wakabayashi, 1977; Robertson et al., 1979; Solano et al., 1980). Furthermore, LH may undergo additional

Supported by NICHD Grant HD-09542

modifications in the circulation (Ascoli and Puett, 1976; Campbell et al., 1978). This heterogeneity or pleomorphism of the hormone is reflected in differences in such structural features as its molecular size and degree of glycosylation, its biological half-life in vivo, and last, but not least, biological potency in vivo and in vitro. In the case of the gonadotropins, this heterogeneity made itself evident in differences in the ratio of biological to immunological potency identified first for FSH and then for LH, differences that could be related to the hormonal status of the donor. These findings led Bogdanove and coworkers (1974) to formulate the concept that gonadal hormones may serve not only to modulate the quantity but also the quality of the gonadotropin released. A basis for this pleomorphism of LH is to be found in the structure of the hormone, in particular its glycoprotein moiety, discussed below. The new generation of in vitro bioassays for LH is providing an essential tool for probing the question of the significance of these differences in molecular form of the hormone for normal and abnormal regulation of gonadal function.

Our present understanding of the nature and function of LH has evolved over the last five decades (Rowlands and Parkes, 1966; Greep, 1974). The dependence of gonadal function in mammals on the adenohypophysis was demonstrated conclusively only a little over half a century ago (Smith and Engel, 1927). The two complementary gonadotropic hormones of the adenohypophysis, responsible for maintaining gonadal function in the two sexes, were identified within the next decade (Greep et al., 1936). Their designation as follicle-stimulating hormone (FSH) and luteinizing hormone was based on their most conspicuous effects on the female gonad in mammals. The essentially identical nature of the gonadotropins in males and females, as well as similarities between the biological effects of LH and human chorionic gonadotropin, (HCG), the gonadotropin of placental origin in the human, were also recognized by early investigators (Rowlands and Parkes, 1966). The alternative term for LH, interstitial cell stimulating hormone (ICSH), describing the hormone's effect on males, i.e., maintenance of interstitial or Leydig-cell function, did not gain general acceptance. However, a few investigators have remained staunchly faithful to the masculine designation (Papkoff et al., 1973).

The term *interstitial cell stimulating hormone* does, indeed, define adequately the hormone's action in the male. Leydig cells are the sole direct targets of its action in the mammalian testes. The effects of LH or ICSH on other components of the testis, the seminiferous tubules and spermatogenesis and on the supporting structural components of the organ, appear all to be mediated by androgen secreted by the Leydig cells in response to LH. In the absence of LH, the Leydig cell and its steroidogenic machinery regress. Besides providing trophic support for Leydig cells, LH can stimulate acutely the synthesis and release of androgen, an effect thought to result from stimulation of 20α-hydroxylation of cholesterol, the presumed rate-limiting step in steroidogenesis (Hall and Young, 1968). With respect to ovary, however, the term *LH* describes only limited aspects of the hormone's role in maintaining normal ovarian function. The transformation of a mature follicle into a corpus luteum follows inevitably the expulsion of the ovum that occurs following a surge of LH. Underlying the process is luteinization. This involves a rapid morphological transformation, primarily of the granulosa cells lining the *mature* follicle, and a striking increase in their ability to form and secrete progesterone (Rothschild, 1981). The potential to luteinize appears to be an intrinsic feature of mature granulosa cells, a potential that is held in check by the follicular environment. According to this concept, the primary role of LH in initiating luteinization is to change this environment irreversibly by causing ovulation. The corpus luteum itself is intrinsically an ephemeral structure. The prolongation of its life span depends on hormonal support. It is LH either by itself or in combination with other hormones (e.g., prolactin or estrogen) that is used by many species to achieve this end (Rothschild, 1981). But the term LH gives no hint of the hormone's essential role in bringing the follicle to an appropriate stage of maturity so that it can respond to LH by ovulation. LH acts synergistically with FSH, and probably also with prolactin, to produce a complex endocrine structure able to synthesize steroids as well as a variety of proteins, that act both as hormones and local modulators of ovarian function (Richards, 1980). A corpus luteum and mechanisms to delay its demise are specialized requirements of viviparous species. In contrast, follicular development is a requirement for all oviparous and oviviviparous species, as well as viviparous. The close structural similarity between LH and FSH, described below, has led to the hypothesis that these two hormones may have evolved from a common single ancestral hormone. However, in their systematic comparative studies, Licht and coworkers (1977) have found evidence for the coexistence of two hormones, analogous to LH and FSH, in representatives of several phylogenetic classes of vertebrates representing presumptive "intermediate" stages in mammalian evolution.

STRUCTURE OF LH

The view that the gonadotropins in male and female are basically identical has not been challenged by the information on the structure of LH and FSH that has emerged during the 1970s (Papkoff et al., 1973; Ward et al., 1973; Reichert et al., 1973; Shome and Parlow, 1974; Bishop et al., 1976). The elucidation of the structure of LH concurrently with that of FSH and TSH led, furthermore, to a recognition of the fundamental similarities among these three pituitary glycoprotein hormones, as well as their resemblance to the placental gonadotropin, HCG (Pierce et al., 1971; Ward et al., 1973; Papkoff et al., 1973; Canfield et al., 1976; Vaitukaitis et al., 1976). All four glycoprotein hormones are composed of a protein core with branched carbohydrate side chains. This protein core in each is formed of the two dissimilar subunits, designated α and β, that are linked noncovalently. In addition to these overall similarities in quaternary structure, the α subunit of the different members of the glycoprotein family of hormones in each species is essentially identical. The hormonal specificity is imparted by the amino acid sequence of the β subunit. The β chain of HCG is essentially homologous with the β chain of human LH, except that it contains an additional sequence of 30 amino acids at the carboxyl terminus (Birkens and Canfield, 1977).

Pleomorphism of structure and related differences in immunological and biological characteristics of LH and HCG appear to be achieved through modifications of the carbohydrate side chains that embellish the protein core of the hormone. The branched carbohydrate side chains usually terminate in sialic acid, which appears to protect the glycoprotein hormone from degradation by the liver (Canfield et al., 1976). Consequently, the biological half-life of the hormone can be related to the sialic acid content of the molecule. Thus, the prolonged survival of HCG in the circulation, and its long half-life, relative to LH, can be explained by the much higher sialic acid content (10% by weight for HCG as compared to only about 1% for LH) (Vaitukaitis et al., 1976). The longer half-life of HCG in the organism makes this hormone more potent than LH when the two are compared in bioassays in vivo, a difference that largely disappears when the hormones are assayed in vitro (Dufau et al., 1971; Vaitukaitis et al., 1976).

The influence of experimental modifications of the carbohydrate side chain on the biological and immunological characteristics of LH and HCG has been examined in a systematic manner by several groups of investigators (Moyle et al., 1975; Vaitukaitis et al., 1976; Channing and Bahl, 1978a,b). As shown in these studies, modifications of the carbohydrate side chain can clearly affect, differentially, the biological functions of the hormone (i.e., half-life in vivo, and its ability to bind to the receptor and elicit a biochemical response), on the one hand, and its immunological characteristics on the other. These studies have also provided a glimpse of how, in the organism, modulation of hormone action could be achieved by specific modifications of the carbohydrate side chain. The development of sensitive in vitro bioassays for LH have provided one of the essential tools for exploring this interesting and potentially important aspect of the actions of gonadotropins.

BIOASSAYS FOR LH AND HCG: OLD AND NEW GENERATIONS

Until the introduction of radioimmunoassays, quantification of LH, like that of other protein hormones, was largely restricted to in vivo assays. Among the first of these was the ventral prostate assay proposed by Greep (Greep et al., 1941).

It proved dependable and gained wide acceptance. The end point used was the increase in prostatic weight of immature hypophysectomized rats secondary to stimulation of testosterone secretion from Leydig cells by LH. Like all the first generation of bioassays, the sensitivity of the ventral prostate assay was poor. The ovarian ascorbic acid depletion (OAAD) assay, introduced by Parlow in 1958, offered much greater sensitivity, though its precision was often less than ideal. The assay was based on the decrease in ascorbic acid content of corpora lutea caused by LH. In corpora lutea, as in the adrenal cortex, depletion of ascorbic acid appears to be an essential component of the acute steroidogenic response (Chayen et al., 1976). For this assay immature rats were superovulated by pretreatment with gonadotropin so that large ovaries packed with corpora lutea were produced. The sensitivity of the assay was greater than that of the ventral prostate assay, and it filled an obvious need. This assay was rapidly adapted for use in various laboratories following only the briefest description of the method by Parlow (1958). In spite of the frustrations attendant on making this assay work (Bogdanove and Gay, 1967), it proved to be the source of a great deal of valuable information during the next decade. However, the assay was still not sensitive enough to measure LH in plasma precisely, even in castrated animals. The sensitivity as well as the precision of Parlow's OAAD assay was greatly improved by Bogdanove and Gay (1967), who added estrogen to the pretreatment regimen. This increased the sensitivity of the assay so that a response could be obtained with about 30 ng LH (NIH-LH-S11). However, because radioimmunoassays for LH became available about this time, there was little further interest in using such an exacting in vivo bioassay.

The biological responses used as end points in the two classic in vivo bioassays for LH, namely, testosterone secretion by Leydig cells or depletion of ascorbic acid from corpora lutea, form the basis also of two bioassays of the new generation. In each case, enhanced sensitivity was made possible by the development of new techniques for quantifying the end point of the hormone's action: radioimmunoassay to measure the amount of testosterone secreted by the Leydig cells, on the one hand, and microspectrophotometry to detect changes in ascorbic acid content of luteal cells on the other.

In the updated, in vitro counterpart of the ventral prostrate assay, testosterone secreted by the Leydig cells during a short-term incubation is measured by radioimmunoassay (Dufau et al., 1972; Van Damme et al., 1974). To date, this assay has been the one used most extensively. It has been adopted by several groups of investigators; problems associated with its application have been identified and various modifications proposed (Dufau et al., 1974; Garfink et al., 1976).

The standard technique involves incubation of Leydig cell-enriched fractions isolated from enzymatically dispersed rat or mouse testes. Various spurious factors that may influence the assay, the presence of serum or steroid precursors, such as progesterone, have been identified and approaches to control for them developed. The degree of sensitivity achieved in different laboratories appears to vary, though exact comparisons are difficult because of differences in LH standards used. As is the case with a number of bioassays, there can be seasonal variations in the sensitivity of the rodent Leydig-cell LH assay (Vaitukitis, personal communication), a point generally not commented on but of interest in itself. However, it is clear that it is possible to achieve with this in vitro bioassay a sensitivity that is at least one order of magnitude greater than that obtained with standard radioimmunoassays for LH.

Some interesting variations in the basic assay technique have been proposed. They include the use of whole testes of 18-day-old rat fetuses to measure LH activity in blood of fetal rats (Picon and Habert, 1981), so avoiding the need to use enzymatic treatment that has been reported to decrease the responsiveness of Leydig cells to LH (Schumacher et al., 1979). The assay also takes advantage of the exquisite sensitivity of the testes of rat fetuses of this age to LH.

The cytochemical assays proposed for LH represent the updated, in vitro version of Parlow's original in vivo OAAD assay and a logical extension to LH and the ovary of the approach used for cytochemical assay for ACTH, described elsewhere in this volume. Two versions of the cytochemical OAAD assay have been described. In the first, a segment assay, the ovaries used are from immature superovulated rats as in the original Parlow assay (Kramer et al., 1974). Portions (segments) of such ovaries are first exposed either to different concentrations of standard LH or to dilutions of the unknown sample. The segments are then chilled, sectioned, and reacted for reducing potential. Microspectrophotometric measurements can then be made of individual luteal cells identified under the microscope. The principle of the cytochemical OAAD assay was subsequently adapted by Buckingham and Hodges (1981) to cryostat sections. For this version of the assay, ovaries are obtained from adult rats on the second day of diestrus when progesterone secretion from corpora lutea is maximal. The ovaries are chilled after a brief (4-min) exposure to low concentrations of LH (10^{-7} IU/ml) in Trowell's medium, and sectioned on a cryostat. It is these cryostat sections that are then exposed to various concentrations of standard hormone or unknown before being stained for reducing potential. The initial brief "priming" with LH is stated to prevent loss of ascorbate (reducing potential) that may occur spontaneously during the subsequent incubation.

The sensitivity of the section assay is stated to be 10^{-7} IU/ml (WHO 68/40). It was shown to have the requisite hormonal specificity. No significant loss of ascorbate was observed following exposure of the sections to low concentrations of other pituitary tropic hormones, to angiotensin II, to vasopressin, or to a variety of steroid hormones. The α and β subunits of LH caused some depletion of reducing potential only when used in high concentrations. An interesting and unexplained feature was the markedly smaller response obtained with HCG than LH. Whether this is due to a difference in the time course of the response caused by LH and HCG, respectively, remains to be determined. As in other cytochemical section assays, the response to the hormone occurs with remarkable rapidity; the time responses in these assays tend also to be highly reproducible. In the case of LH, the maximum depletion of reducing potential occurs in 80 sec. The time course of the response to HCG remains to be determined.

The cytochemical OAAD segment assay has been adapted by Schallenberger and Schams (1980) to measure LH in bovine serum. The serum was subjected to chromatography on CM-Sephadex-C-50 to remove proteins in blood, including FSH, thought by the investigators to potentiate the ascorbic acid depletion by LH. The sensitivity of the assay is stated to be 10 fg/ml, in terms of a bovine LH standard with a biological potency comparable to that of NIH-LH-S1. A good correspondence was obtained between estimates of biological and immunological potency for bovine serum.

The cytochemical OAAD assay has been taken little beyond its initial descriptive stage. It has not been applied in a systematic manner to any one problem or tried by different investigators and within different experimental contexts. Thus, the cytochemical OAAD assay has not yet undergone the process of honing through use essential for establishing any sensitive assay system.

For a number of protein hormones, quantitative cytochemistry has provided the first, and in most cases has remained, the sole means for measuring biological potency with a sensitivity and precision that meets the standards set by the radioligand assays. This is clearly not the case for LH: the Leydig-cell assay is providing an effective, practical, specific, and sensitive bioassay. The question may then be raised; What need is there for alternative bioassays, and what specific advantages might a quantitative cytochemical assay offer?

Development of alternative assays would first of all permit a comparison of the potency of any given hormonal preparation with respect to more than one of its target cells. Specifically, the cytochemical OAAD bioassay for LH should make it possible to determine whether a given modification of the basic LH or HCG molecule affects to the same degree the interaction of that molecule with Leydig cells, on the one hand, and luteal cells on the other. If different biological potencies of a given hormone preparation were found, depending on the nature of the target cell, this would point to interesting possibilities concerning specialization at the level of the LH receptor. Thus the new generation of bioassays, in particular the cytochemical assays, that permit precise quantification of a biochemical response of defined populations of target cells, do not simply represent a means for measuring biological potency of a hormone. They do, in fact, provide systems in which the mechanism of hormone action could be studied and defined in intact cells under controlled conditions.

APPLICATIONS OF THE NEW GENERATION OF BIOASSAY

The relationship of values obtained by bioassay and radioimmunoassay has been the central issue in most of the studies reported to date in which the new generation of bioassays has been used. As discussed above, so far only the Leydig-cell assay has been used extensively for this purpose. The question addressed in these studies has been the one already raised by Midgley in 1966, in his first publication of an immunoassay for LH: this is the question of the degree of correspondence between the immunological and biological determinants of the hormone molecule.

Studies involving comparison of values registered by bioassays and immunoassays could be viewed simply as serving the rather pedestrian task of policing of immunoassay by bioassay. However, as shown in the studies of FSH in rats and monkeys, changes in ratio of biological to immunological activity may well reflect physiologically significant aspects of the regulation of gonadotropin secretion (Diebel et al., 1973; Peckham et al., 1973; Bogdanove et al., 1974). That the same principle may apply to LH is suggested by the findings from many studies of humans and experimental animals, during the last decade, in which circulating LH levels were measured in parallel by immunoassay and bioassay. While the *profile* of values obtained in the two assay systems has proved to be quite similar under a wide range of physiological and experimental conditions, a number of situations have been identified in which there is a change in ratio of biological to immunological activity (B/I ratio). For example, in the human and the rhesus monkey, puberty has been found to be associated with a more marked increase in biological than immunological activity of circulating LH (Lucky, Rich, Rosenfield, Victor, and Roche-Bender, 1980; Williams and Hodgen, 1982; Reiter et al., 1982). In contrast, during the midcycle surge of LH in normal women, the B/I ratio decreases (Dufau et al., 1976; Robertson et al., 1979). The nature of LH released in women in response to an injection of gonadotropin-releasing hormone (GnRH), appears to be influenced also by the stage of the cycle. During the early follicular phase, the values obtained by bioassay and immunoassay remain parallel throughout the LH surge initiated by GnRH. However, during the late follicular and luteal phases, there is a marked increase in B/I ratio in the LH released (Dufau et al., 1976). Both the rate of secretion of LH and the gonadal hormonal milieu have been implicated as factors that may cause changes in the quantity of LH released and, thereby, in the B/I ratio of circulating LH. Evidence for the importance of both the rate of hormone secretion and hormonal milieu in determining the nature of the LH released has been obtained, primarily from studies of isolated rat pituitaries in vitro (Azhar et al., 1978; Liu et al., 1979; Mukhopadhyay et al., 1979). In the presence of low concentrations of GnRH, the LH secreted has a high molecular weight, is immunoreactive but possesses little bioactivity. The carbohydrate content of this big form of LH is low. On stimulation with higher concentrations of LHRH, there is a shift to secretion of LH that is smaller but has higher biological potency (Azhar et al., 1978; Liu et al., 1979).

The cytochemical OAAD assay, as discussed earlier, has been used only to a very limited extent. It has been used to compare bioassay and radioimmunoassay values for LH in blood samples obtained from normal women throughout the menstrual cycle. With the segment assay, two peaks of bioactivity were found in midcycle, the first before and the second after the peak obtained with radioimmunoassay (Kramer et al., 1976). On the other hand, Buckingham and Hodges (1981) found good correspondence between the timing and magnitude of the LH peak defined by radioimmunoassay or by the cytochemical section assay. The basis for this discrepancy has not yet been explored.

In interpreting the data from studies in which immunoassays and bioassays are compared, there are certain experimental variables that need to be considered. The absolute *magnitude* of any discrepancy between the two assays is greatly influenced by the nature of the reagents used, i.e., the purity of the standard, the iodination procedure, and the characteristics of the antiserum (Robertson et al., 1979; Suginami et al., 1981). Certain hormone standards clearly contain more immunologically active but biologically inactive molecules than others. This has been demonstrated by Robertson et al. (1979) and suggested by Reiter et al. (1982). Therefore, the significant information obtained from B/I ratios is the differences in this value obtained under different physiological or experimental conditions, or between individuals when using the same reagents in the assay. A second consideration is that when values obtained with one of the assay systems are near the limits of sensitivity for that assay, the ratios obtained tend to be unreliable.

The potential for structural variations exhibited by LH could, theoretically, lead to the formation and release of LH with inadequate biological potency that need not be reflected in radioimmunoassay. This possibility has been examined in a limited number of patients and disease states, specifically, precocious and delayed puberty (Lucky, Rich, Rosenfield, Fang, and Roche-Bender, 1980; Reiter et al., 1982), gonadal dysgenesis (Beitins et al., 1980), and primary Leydig-cell hyperplasia (Schedwie et al., 1981). To date only one patient has been identified whose abnormality of a hypogonadal state could be related unequivocally to an abnormal, biologically inactive form of circulating LH (Beitins et al., 1981). The patient had higher than normal immunologically active plasma LH which, however, did not stimulate either testosterone or cAMP production by isolated rat Leydig cells.

Finally, in one recent study, the Leydig-cell bioassay has been used to follow the fate of HCG internalized in luteal cells (Ziminski et al., 1982). The hormone was recovered from rat luteal cells at various times, up to 24 hr, following administration of HCG in vivo, and characterized in terms of its biological as well as immunological potency. The internalized HCG was shown to retain its biological activity for a surprisingly long period of time.

With this last study, then, the changes that LH may undergo in terms of its biological activity have been followed from the time the hormone is synthesized and processed in the anterior pituitary to when it has interacted and has become internalized within the target cell.

PROJECTIONS FOR THE FUTURE

In both bioassays of the new generation described in this chapter, a single class of target cells is isolated for quantifying of the biochemical consequences of the hormone's action. This is clearly one of the factors contributing to the high degree of sensitivity and specificity of these assays. In the Leydig-cell assay, the target cells are separated out physically after enzymatic and/or mechanical disruption of the target organ. With quantitative cytochemistry, as in the OAAD bioassay, the separation is achieved visually, using microscopy to identify specific cell types in situ, without the need to disrupt the target organ. This feature of quantitative cytochemistry, coupled with the remarkable sensitivity of microspectrophotometry, should make it possible to develop, in the future, new approaches to quantifying the biochemical consequences of the action of LH on its different target-cell populations within the same organ. This can in turn form the basis to develop specific bioassays on the one hand, and to gain insight into the mechanism of hormone action on the other: two sides of the same coin.

Microspectrophotometry is clearly sensitive enough to provide precise, quantitative data on enzyme activity or cell constituents even in a single cell. It, as well as other measuring devices operating through a microscope, e.g., interferometry or microfluorimetry, should make it possible to relate specific effects of a particular hormone preparation to individual cell types.

The existence of different steroidogenic cell types responsive to LH within the ovary has long been recognized. More recent is the realization that in the testes also there may be more than one type of Leydig cell. There appear to be two generations of Leydig cells. The first differentiates during fetal life, the second in relation to puberty. In the rat, one point of difference between these two generations of cells appears to be the mode of regulation of LH receptors (Hutaniemi et al., 1981). In adult rats, also, there appear to be two subpopulations of Leydig cells (O'Shaughnessy et al., 1981). Population II is characterized by its greater responsiveness to LH-HCG in terms of testosterone secretion. The difference between the two populations appears to lie beyond the HCG-LH receptor. Quantitative cytochemistry should provide an ideal tool for characterizing these cell types and their response to LH, without problems associated with trying to separate them from each other physically.

The ability of microspectrophotometry to define functional differences between cells in close juxtaposition to each other in the gonads has been demonstrated in quantitative cytochemical studies of the membrana granulosa of preovulatory follicles of rats. Cells in the peripheral portion of the membrana granulosa are ones that have the biochemical characteristics associated with steroidogenesis. It is probably only these cells that respond to LH before ovulation, first with increased estrogen formation and then with a surge of progesterone secretion (Weisz and Zoller, 1979).

Bioassays now available for LH represent essentially a research tool. As a research tool, the cytochemical bioassays, in particular, have great potential. A great deal of new information is becoming available on the mode of action of LH at the cellular level. This could serve as a basis for establishing new cytochemical approaches to quantifying LH. Cytochemistry can, however, also be a source of new insights into the interactions between LH and its different target cells, taking into account that both the hormone and its target organs are heterogeneous.

REFERENCES

Ascoli, M., and Puett, D. (1976). Biotransformation of pituitary luteinizing hormone in serum and urine. I. Association with serum components. *Endocrinology*, 99:1229-1236.

Azhar, S., Reel, J. R., Pastushok, C. A., and Menon, K. M. J. (1978). LH biosynthesis and secretion in rat anterior pituitary cell cultures. Stimulation of LH glycosylation and secretion by GnRH and an agonistic analogue and blockade by an antagonistic analogue. *Biochem. Biophys. Res. Commun.*, 80:659-666.

Beitins, I. Z., Axelrod, L., Ostrea, T., Little, R., and Badger, T. M. (1981). Hypogonadism in a male with an immunologically active, biologically inactive Luteinizing hormone: Characterization of the abnormal hormone. *J. Clin. Endocrinol. Metab.*, 52:1143-1149.

Beitins, I. Z., Dufau, M. L., O'Loughlin, K., Bercu, B. B., McArthur, J. W., Crawford, J. D., and Catt, K. J.(1980). Biological and immunological activities of serum LH in normal women during LHRH infusion and in Turner's Syndrome during estrogen treatment. In *The LH-Releasing Hormone,* C. G. Beling and C. Wertz (eds.). Masson, New York, pp. 135-153.

Birkins, D., and Canfield, R. E. (1977). Isolation and amino acid sequence of COOH-terminal fragments from beta subunits of human chorionic gonadotrophin. *J. Biol. Chem.*, 252:5386-5392.

Bishop, W. H., Nureddin, A., and Ryan, R. J. (1976). Pituitary luteinizing and follicle-stimulating hormones. In *Peptide Hormones,* J. Parsons (ed.). Macmillan, New York, pp. 273-298.

Bogdanove, E. M., Campbell, G. T., Blair, E. D., Mula, M. E., Miller, A. E., and Grossman, G. H. (1974). Gonad-pituitary feedback involves quantitative changes: Androgens alter the type of FSH secreted by the rat pituitary. *Endocrinology*, 95:219-228.

Bogdanove, E. M., and Gay, V. L. (1967). Enhancement of ovarian ascorbic acid depletion response during estrogen-prolonged pseudopregnancy: An improved bioassay for LH. *Endocrinology*, 81:1104-1116.

Buckingham, J. D., and Hodges, J. R. (1981). A cytochemical bioassay method for the determination of luteinizing hormone in biological fluids and tissues. *Br. J. Pharmacol.*, 73:111-118.

Campbell, G. T., Nansel, D. D., Meinzer, III, W. M., Aiyer, M. S., and Bogdanove, E. M. (1978). Prolonged infusion of rat luteinizing hormone alters its metabolic clearance pattern. Indirect evidence for post-secretory modulation of luteinizing hormone. *Endocrinology*, 103:683-693.

Canfield, R. E., Birkens, S., Morse, J. H., and Morgan, I. J. (1976). Human chorionic gonadotrophin. In *Peptide Hormones*, J. A. Parsons (ed.). Macmillan, New York, pp. 299-315.

Channing, C. P., and Bahl, O. P. (1978a). Role of carbohydrate residues of human chorionic gonadotrophin in binding and stimulation of adenosine $3'5'$ monophosphate accumulation by porcine granulosa cells. *Endocrinology*, 103:341-348.

Channing, C. P., and Bahl, O. P. (1978b). Role of carbohydrate residues of human gonadotrophin in stimulation of progesterone secretion by cultures of monkey granulosa cells. *Biol. Reprod.*, 18:707-711.

Chayen, J., Daly, J. R., Loveridge, N., and Bitensky, L. (1976). The cytochemical bioassay of hormones. *Recent Prog. Horm. Res.*, 32:33-79.

Diebel, N. D., Yamamoto, M., and Bogdanove, E. M. (1973). Discrepancies between radioimmunoassay and bioassay for rat FSH: Evidence that androgen treatment and withdrawal can alter bioassay-immunoassay ratios. *Endocrinology*, 92:1065-1078.

Dufau, M. L., Beitins, I. Z., McArthur, J. W., and Catt, K. J. (1976). Effects of luteinizing hormone releasing hormone (LHRH) upon bioactive and immunoactive serum LH levels in normal subjects. *J. Clin. Endocrinol. Metab.*, 43:658-667.

Dufau, M. L., Catt, K. J., and Tsurubara, T. (1971). Retention of in vitro biological activities of desialylated human luteinizing hormone and chorionic gonadotropin. *Biochem. Biophys. Res. Commun.*, 44:1022-1029.

Dufau, M. L., Catt, K. J., and Tsurubara, T. (1972). A sensitive gonadotrophin responsive system: Radioimmunoassay of testosterone production by the rat testis in vitro. *Endocrinology*, 90:1032-1040.

Dufau, M. L., Mendelson, C. R., and Catt, K. J. (1974). A highly sensitive in vitro bioassay for luteinizing hormone and chorionic gonadotrophin: Testosterone production by dispersed Leydig cells. *J. Clin. Endocrinol. Metab.*, 39:610-613.

Garfink, J. E., Moyle, W. R., and Bahl, D. P. (1976). A rapid sensitive bioassay for luteinizing hormone-like gonadotropins. *Gen. Comp. Endocrinol.*, 30:292-300.

Greep, R. O. (1974). History of anterior hypophysial hormones. In *Handbook of Physiology*, Vol. IV, part 2. R. O. Greep, E. B. Astwood, E. Knobil, W. H. Sawyer, and S. R. Geiger (eds.). Williams and Wilkins, Baltimore, pp. 1-27.

Greep, R. O., Fevold, H. L., and Hisaw, I. L. (1936). Effect of two hypophyseal gonadotropic hormones on the reproductive system of the male rat. *Anat. Rec.*, 65:261-269.

Greep, R. O., Van Dyke, H. B., and Chau, B. F. (1941). Use of the anterior lobe of the prostrate gland in the assay of metakentrin. *Proc. Soc. Exp. Biol. Med.*, 46:644-649.

Hall, P. F., and Young, D. (1968). Site of action of trophic hormones upon the biosynthetic pathways to steroid hormones. *Endocrinology*, 82:559-568.

Hutaniemi, I. T., Katikineni, M., and Catt, K. J. (1981). Regulation of luteinizing hormone receptors and steroidogenesis in the neonatal rat testis. *Endocrinology*, 109:588-595.

Kramer, R., Holdaway, I. M., Crighton, D. B., McNeilly, A. S., Rees, L. H., and Chard, T. (1976). Comparison of the redox bioassay with other assays for luteinizing hormone. *J. Endocrinol.*, 69:205-211.

Kramer, R., Holdaway, I. M., Rees, L. H., McNeilly, A. S., and Chard, T. (1974). Technical aspects of the redox bioassay for luteinizing hormone. *Clin. Endocrinol. (Oxf)*, 3:373-381.

Licht, P., Papkoff, H., Farmer, S. W., Muller, C. H., Tsui, H. W., and Crews, D. (1977). Evolution of gonadotropin structure and function. *Recent Prog. Horm. Res.*, 33:169-243.

Liu, T., Ax, R. L., and Jackson, G. L. (1979). Characterization of luteinizing hormone synthesized and released by rat pituitaries in vitro: Dissociation of immunological and biological activities. *Endocrinology*, 105:10-15.

Lucky, A. W., Rich, B. H., Rosenfield, R. L., Fang, V. S., and Roche-Bender, N. (1980). Bioactive LH, a test to discriminate true precocious puberty from premature thelarche and adrenarche. *J. Pediatr.*, 97:214-216.

Lucky, A. W., Rich, B. H., Rosenfield, R. L., Victor, V. S., and Roche-Bender, N. (1980). LH bioactivity increases more than immunoreactivity during puberty. *J. Pediatr.*, 97:205-213.

Midgley, A. R. (1966). Radioimmunoassay: A method for human chorionic gonadotrophin and human luteinizing hormone. *Endocrinology*, 79:10-18.

Moyle, W. R., Bahl, G. P., and Marz, L. (1975). Role of carbohydrate of human chorionic gonadotrophin in the mechanism of hormone action. *J. Biol. Chem.*, 250:9163-9169.

Mukhopadhyay, A. K., Leidenberger, F. A., and Lichtenberg, V. (1979). A comparison of bioactivity and immunoactivity of luteinizing hormone stored in and released in vitro from pituitary glands of rats under various gonadal states. *Endocrinology*, 104:925-931.

Odell, W. A., Ross, G. T., and Rayford, P. L. (1967). Radioimmunoassay for luteinizing hormone in human plasma and serum. *J. Clin. Invest.*, 46:248-255.

O'Shaughnessy, P. J., Wong, K., and Payne, A. H. (1981). Differential steroidogenic enzyme activities in different populations of rat Leydig cells. *Endocrinology*, 109:1061-1066.

Papkoff, H., Sairam, M. R., Farmer, S. W., and Li, C. H. (1973). Studies on the structure and function of interstitial cell stimulating hormone. *Recent Prog. Horm. Res.*, 29:563-590.

Parlow, A. F. (1958). A rapid bioassay method for LH and factors stimulating LH secretion, Abstract #1587. *Fed. Proc.*, 17:402.

Peckham, W. D., Yamaji, T., Dierschke, D. J., and Knobil, E. (1973). Gonadal function and the biological and physiochemical properties of follicle stimulating hormone. *Endocrinology*, 92:1660-1666.

Picon, R., and Habert, R. (1981). A sensitive bioassay for luteinizing hormone-like activity applied to systemic plasma of foetal rats. *Acta Endocrinol. (Kbh)*, 97:176-180.

Pierce, J. G., Liao, T., Howard, S. M., Shome, B., and Cornell, J. S. (1971). Studies on the structure of thyrotropin: Its relationship to luteinizing hormone. *Recent Prog. Horm. Res.*, 27:165-212.

Qazi, M. H., Romani, R., and Diczfalusy, E. (1974). Discrepancies in plasma LH activities as measured by radioimmunoassay and in vitro bioassay. *Acta Endocrinol. (Kbh)*, 77:675-685.

Reichert, L. O., Leidenberger, F. A., and Trowbridge, F. (1973). Studies on luteinizing hormone and its subunits. *Recent Prog. Horm. Res.*, 29:497-532.

Reiter, E. O., Beitins, I. Z., Ostrea, T., and Gutai, J. P. (1982). Bioassayable luteinizing hormone during childhood and adolescence in patients with delayed pubertal development. *J. Clin. Endocrinol. Metab.*, 54:155-160.

Richards, J. S. (1980). Maturation of ovarian follicles, actions and interactions of pituitary and ovarian hormones on follicular differentiation. *Physiol. Rev.*, 60:51-89.

Robertson, D. M., Puri, V., Lindberg, M., and Diczfalusy, E. (1979). Biologically active luteinizing hormone in plasma. V. A reanalysis of the differences in the ratio of biological to immunological LH activities during the menstrual cycle. *Acta Endocrinol. (Kbh)*, 92:615-626.

Rothschild, I. (1981). The regulation of the mammalian corpus luteum. *Recent Prog. Horm. Res.*, 37:183-284.

Rowlands, I. W., and Parkes, A. S. (1966). Hypophysectomy and the gonadotrophins. In *Marshall's Physiology of Reproduction,* Vol. III, A. S. Parkes
(ed.). pp. 26-146.

Schwedwie, H. K., Reiter, E. O., Beitins, I. Z., Seyed, S., Wooten, V. J., Jiminez,
J. F., Aiman, E. J., DeVane, G. W., Redman, J. F., Elders, M. J. (1981). Testi-
cular Leydig cell hyperplasia as a cause of familial sexual precocity. *J. Clin.
Endocrinol. Metab.,* 52:271-278.

Schumacher, M., Schafer, G., Lichtenberg, V., and Hilz, H. (1979). Maximal
steroidogenic capacity of mouse Leydig cells. *FEBS Lett.,* 107:398-402.

Shallenberger, E., and Schams, D. (1980). Adaptation of the cytochemical luteinizing hormone assay for measurement in bovine blood. *Endocrinology,* 76:
29-35.

Sharpe, R. M., Shahmanesh, M. G., Ellwood, M. G., Hartog, M., and Brown, P. S.
(1975). Discrepancy between radioimmunoassay and radioligand receptor
assay of luteinizing hormone released in vitro by pituitary tissue from male
rats of different ages. *J. Endocrinol.,* 65:265-275.

Shome, B., and Parlow, A. F. (1974). Human follicle stimulating hormone: First
proposal for the amino acid sequence of the hormone-specific β subunit. *J.
Clin. Endocrinol. Metab.,* 39:203-205.

Smith, P. E., and Engel, E. T. (1927). Experimental evidence regarding the role
of the anterior pituitary in the development and regulation of genital system.
Am. J. Anat., 40:159-217.

Solano, A. R., Garcia-Vela, A., Catt, K. J., and Dufau, M. L. (1980). Modulation
of serum and pituitary luteinizing hormone bioactivity by androgen in the
rat. *Endocrinology,* 106:1941-1948.

Suginami, H., Koizumi, Y., and Nakajima, A. (1981). Measurement of human luteinizing hormone in plasma by in vitro bioassay and by conventional and improved radioimmunoassay. *Endocrinol. Jpn,* 28:87-97.

Vaitukaitis, J. L., Ross, G. T., Braunstein, G. D., and Rayford, P. L. (1976).
Gonadotrophins and their subunits: Basic and clinical studies. *Recent Prog.
Horm. Res.,* 32:289-331.

Van Damme, M. P., Robertson, D. M., and Diczfalusy, E. (1974). An improved
in vitro bioassay method for measuring luteinizing hormone (LH) activity
using mouse Leydig cell preparations. *Acta Endocrinol. (Kbh),* 77:655-671.

Wakabayashi, K. (1977). Heterogeneity of rat luteinizing hormone revealed by
radioimmunoassay and electrofocusing studies. *Endocrinol. Jpn.,* 24:473-479.

Ward, D. N., Reichert, L. E., Liu, W., Nahm, H. S., Lamkin, W. M., and Jones,
N. S. (1973). Chemical studies of luteinizing hormone from human and ovine
pituitaries. *Recent Prog. Horm. Res.,* 29:533-562.

Weisz, J., and Zoller, L. C. (1979). Quantitative cytochemistry in the study of
regional specialization in the membrane granulosa of the ovulable type of

follicle. In *Quantitative Cytochemistry and Its Applications,* J. R. Pattison, L. Bitensky, and J. Chayen (eds.). Academic, New York, pp. 269-283.

Williams, R. F., and Hodgen, G. D. (1982). Initiation of the primate ovarian cycle with emphasis on perimenarchial and postpartum events. In *Reproductive Physiology,* Vol. V, R. O. Greep (ed.). University Park Press, Baltimore,

Zimniski, S. J., Rorke, E. A., Sickel, M. A., and Vaitukaitis, J. L. (1982). Fate of HCG bound to rat corpora lutea. *Endocrinology* (in press).

9

Technique and Clinical Relevance of a Cytochemical Bioassay for Gastrin-like Activity

Ronald W. Hoile / St. Stephens Hospital, London, England

INTRODUCTION

In this chapter I will describe the technique and clinical application of a cyto-chemical bioassay for gastrin-like activity. The results obtained from this assay and the discrepancies between gastrin radioimmunoreactivity and biological activity suggest that the ability to assign a level of biological activity to a given sample of plasma or tissue may have more clinical significance than results provided by radioimmunoassay.

THE NATURE OF GASTRIN

Gastrin was first purified by Gregory and Tracy (1964), and it was the first gut hormone with a known structure and clearly defined structure-function relationship (Tracy and Gregory, 1964). Although gastrin is described as a gastrointestinal hormone we now know that its distribution is more widespread and it has been identified in both central and peripheral neurons (Rehfeld, 1978; Uvnäs-Wallensten et al., 1977), and indeed a completely new system of peptidergic nervous transmission is now proposed (Sundler et al., 1978).

A continuous spectrum of endocrine cells extends throughout the alimentary tract from the esophagus to the rectum; the gastrin-containing or elaborating cells are called *G cells.* The greatest number of G cells is found in the antro-pyloric mucosa of the stomach, and production of the hormone then decreases in the duodenum, jejunum, and so on distally to the rectum (Bloom and Polak,

```
1   2   3   4   5   6   7   8   9   10  11  12  13  14  15  16  17
Pyro-Gly-Pro-Trp-Leu-Glu-Glu-Glu-Glu-Glu-Ala-Tyr-Gly-Trp-Met-Asp-Phe-NH₂
                                                    |
                                                    R
```

Figure 1   The amino acid sequence of human G-17 (molecular weight 2098): Pyro-, pyroglutamyl residue; Gastrin I, R = H; Gastrin II, R = SO3H.

1978). Gastrin may also be secreted by cells in hyperplastic pancreatic islets and certain APUD tumors (gastrinomas).

The structural form of the polypeptide hormone which Gregory and Tracy initially isolated was the heptadecapeptide designated little gastrin, or G-17 (Figure 1). Each molecule exists in the sulfated (gastrin II) or nonsulfated (gastrin I) form; the biological activity occurs at the carboxyl-terminal tetrapeptide of the molecule, and there are no differences between the biological activity of the sulfated and nonsulfated forms. We now know that there is an array of gastrins; this molecular heterogeneity has been identified mainly by radioimmunological techniques. The following forms of gastrin have been identified (Gregory, 1976):

1. Little gastrin (G-17), 17 amino acid residues
2. "Big" gastrin (G-34), 34 amino acid residues
3. Minigastrin (G-14), 14 amino acid residues
4. "Big big" gastrin, molecular weight about 20,000
5. Component I, size intermediate between "big" and "big big" gastrin
6. NH_2-terminal fragment of G-17, 13 amino acid residues

Of these, 1, 2, 3, and 6 have been isolated from gastrointestinal tissues and fully characterized chemically. The remaining forms have been identified, using immunological techniques, in the circulation and in tissue extracts but they have not yet been chemically identified and it may be that they represent artifacts produced by nonspecific interference in radioimmunoassay systems by serum proteins. It is likely that G-34 is the biosynthetic precursor of G-17, the molecule being split by tryptic cleavage to produce G-17 and an NH_2-terminal peptide, both of which are subsequently stored in antral G cells (Dockray, 1978).

In terms of biological potency, G-17 is about five times more potent than G-34 (Walsh et al., 1976). In view of these different potencies of the circulating gastrins and the fact that G-34 is the major form of gastrin in the circulation during the basal state, it is clear that a single measurement of total serum immunoreactive gastrin gives little information on circulating biological activity. Using radioimmunological techniques with region-specific antibodies it has been shown that both circulating G-17 and G-34 increase after a meal, and in fact the peak increment in G-34 is approximately 50% higher than that of G-17 (Dockray and Taylor, 1976). However, when account is taken of the biological potencies of the two molecules, it can be inferred that G-17 probably contributes 70% or more of the circulating endogenous biological activity after feeding. The problem with these experiments is that true biological activity is not measured and only indirectly estimated from prior knowledge of the biological dose response to exogenous hormone. This is an example of the difficulties of interpreting radioimmunoreactive gastrin levels in terms of clinical significance, and I will refer to these difficulties again.

Since pure gastrin and its synthetic analogs (e.g., pentagastrin) have become available it has been possible to study the spectrum of actions of this hormone. The problem is to decide which are physiological (in that they occur in response to the amount and molecular types of gastrin normally circulating after a meal) and which are pharmacological. The recorded actions of gastrin are many (Walsh and Grossman, 1975), but the main physiological actions appear to be the promotion of gastric secretion (water, electrolytes, and enzymes) and growth of the gastric mucosa and parietal cells. Physiological alterations of gastric blood flow and motility also occur.

CYTOCHEMICAL SECTION BIOASSAY OF GASTRIN-LIKE ACTIVITY

The original cytochemical bioassay for gastrin-like activity was a segment assay, but it proved to be too slow for routine purposes (Loveridge et al., 1974). The section assay described below allows a larger throughput in a within-animal procedure without loss of sensitivity. The procedure to be described below is as previously published (Loveridge, Hoile et al., 1980); a variation of this technique is to take gastric tissue from a fasted guinea pig and eliminate the nonproliferative culture and priming stage (Hoile, 1979a). Results from the two methods are essentially identical although the latter method occasionally shows very small, but theoretically unacceptable, responses to control solutions.

Method

Female guinea pigs (Hartley strain, approximately 400 g body weight) were killed by asphyxiation in nitrogen and the gastric fundus excised and cleaned of debris. Strips of fundus (3 × 2 mm) were maintained in a nonproliferative organ maintenance culture at $37°C$ for 5 hr (Trowell, 1959). After 5 hr the strips were primed by exposing them to $2.3 × 10^{-16}$ M G-17 gastrin in Trowell's T8 medium (pH 7.6) for 5 min. After this, the tissue was chilled in n-hexane to $-70°C$

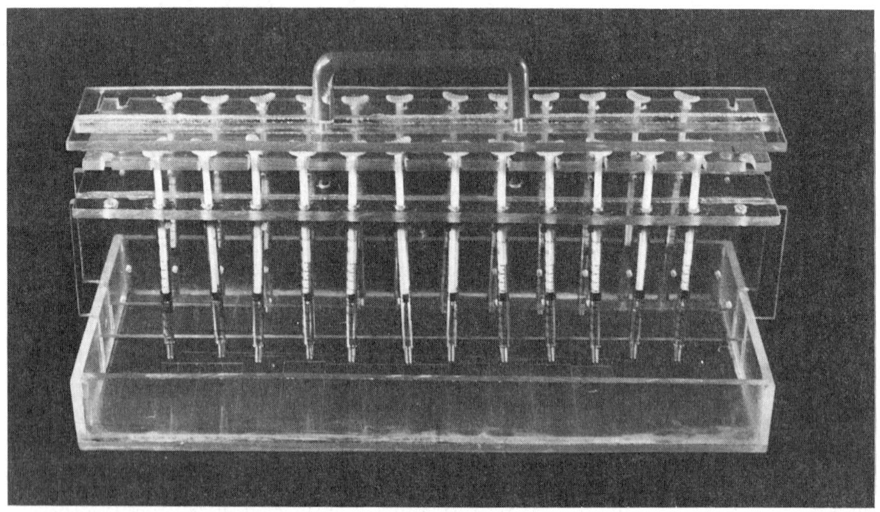

Figure 2 Apparatus for the simultaneous application of the hormone or unknown solutions to duplicate sections.

and stored in dry tubes at this temperature. Within 3 days the tissue was sectioned at 18 μm in a Bright's cryostat with the cabinet temperature at $-25°$C and the knife cooled with solid carbon dioxide to $-70°$C. The sections were then transferred onto glass slides. These methods are described in detal by Loveridge in Chapter 3 and by Chayen et al. (1973).

Serial sections were taken from the cryostat and placed in a trough immediately prior to the assay. Duplicate sections were treated for 75 sec with one of four graded logarithmic doses of gastrin (2.3×10^{-15} to 2.3×10^{-12} M; 0.005-5 pg/ml G-17) or to serial dilutions of the test sample of plasma (usually $1:10^2$ and $1:10^3$). The dilutions of gastrin and test sample were made up in 0.1 M Hepes buffer, pH 7.0, containing gum tragacanth (0.005% w/v). The addition of the gum tragacanth has been shown to produce cleaner sections after the cytochemical reaction, which are easier to measure (Hoile, 1979a). It was necessary to perform the exposure of the sections to hormones and the subsequent cytochemical reaction in a trough in order to allow the escape of evolved CO_2 (see below). The trough and apparatus shown in Figure 2 enabled up to 28 sections to be exposed simultaneously to the hormone or test sample. Thus, four graded concentrations of G-17 and five test samples, each at two dilutions, could be measured each on duplicate sections.

After 75 sec exposure to the hormones the sections were reacted for carbonic anhydrase activity using the method of Loveridge (1978). The reaction media were as follows.

Solution A: 9 ml 0.5 M sulfuric acid (BDH Analar grade)
               9 ml 0.1 M cobalt sulfate ($CoSo_4 \cdot 7H_2O$) (BDH Analar grade)
               3 ml 0.067 M potassium dihydrogen orthophosphate (BDH
                   Analar grade)
               4.5 ml distilled water
Solution B: 1.125 g sodium hydrogen carbonate (BDH Analar grade) in 60 ml
                   0.1 M HEPES (Sigma) buffered to pH 7.4 with sodium
                   hydroxide.

The two solutions were prepared fresh and mixed immediately before use. After mixing, the solution was stirred until the pH had risen to pH 6.8 (1-2 min). The reaction proceeded for 60 sec at $37°C$.

Carbonic anhydrase catalyzes the following reaction (Maren, 1967).

$$HCO_3 \rightleftharpoons OH^- + CO_2$$

The hydroxyl ions are trapped by cobalt as a precipitate of cobalt hydroxide. The accumulation of CO_2 inhibits activity so that this gas must be allowed to diffuse readily from the site of the reaction. It is for this reason that the trough and apparatus described above (Figure 2) were devised. To assist the diffusion of gas, the depth of the reaction medium was kept minimal (0.8 mm) and the trough was gently agitated throughout the reaction time of 60 sec. After the reaction, the sections were washed in running tap water (30 sec), immersed in a saturated solution of hydrogen sulfide, rinsed in distilled water, and mounted in Farrants' medium. The hydrogen sulfide converts the colorless cobalt hydroxide to the intensely colored cobalt sulfide (Figure 3).

Microdensitometric measurements of the sections were made at 550 nm with a $\times 20$ objective, a scanning spot of 1 μm diameter, and a mask of 20 μm diameter, which optically isolated each parietal cell. The carbonic anhydrase activity (expressed as density of the reaction product) was measured in 10 parietal cells in each of two duplicate sections, and the mean of these readings was converted to mean integrated extinction by reference to a standard calibration graph. To correct for nonspecific tissue adsorption of the cobalt and spurious light-scattering by the tissue, 10 similar-sized fields were measured in the muscle, and

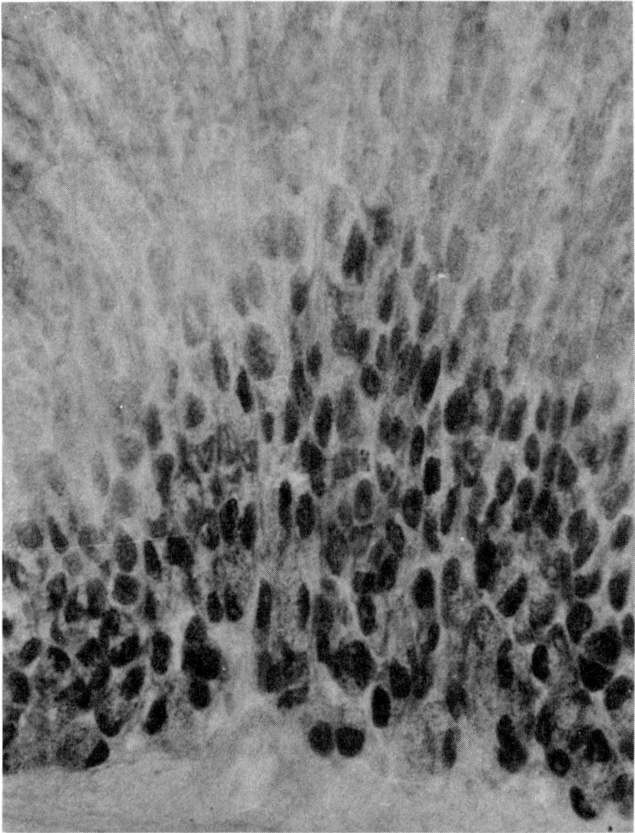

Figure 3 Photomicrograph of a section of guinea pig gastric fundus reacted for carbonic anhydrase activity. The parietal cells are darkly stained and have clear nuclei. (From Hoile, 1982.)

the mean of these readings was subtracted from that obtained from the parietal cells. In general the extinction in the muscle was less than 50% of that found in the nonstimulated parietal cells and did not vary with increasing concentration of the hormone. Thus it became a smaller proportion of the activity measured in the parietal cells, with increasing concentration of the hormone, being only 20-26% at higher concentrations of G-17.

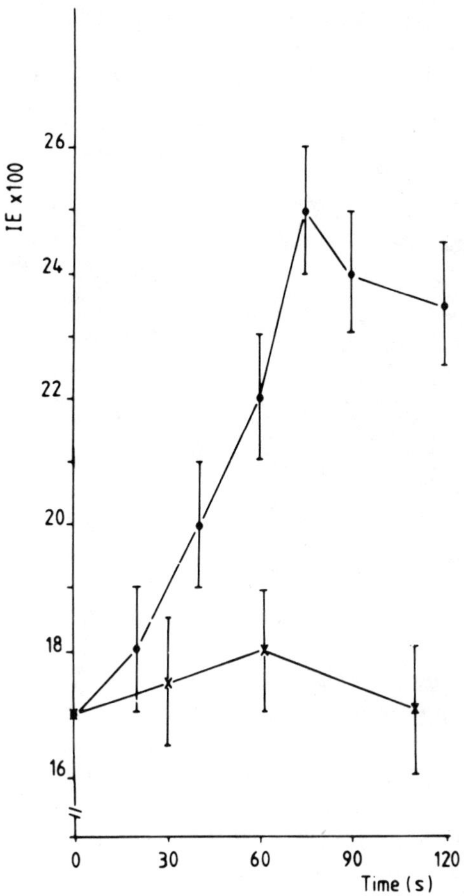

Figure 4   The time course of the carbonic anhydrase activity (IE $\times$ 100) induced in the parietal cells in sections of guinea pig gastric fundus by 2.3 $\times$ 10^{-12} M G-17 (filled circles). The lower graph (crosses) shows the effect of the vehicle alone.

Results

In sections incubated with 2.3 $\times$ 10^{-12} M G-17 there was a detectable increase in carbonic anhydrase activity by 20 sec which was maximal by 75 sec and then decreased slightly from 75 to 120 sec (Figure 4). In contrast, sections incubated with a buffer solution alone did not show any significant increase in enzyme activity.

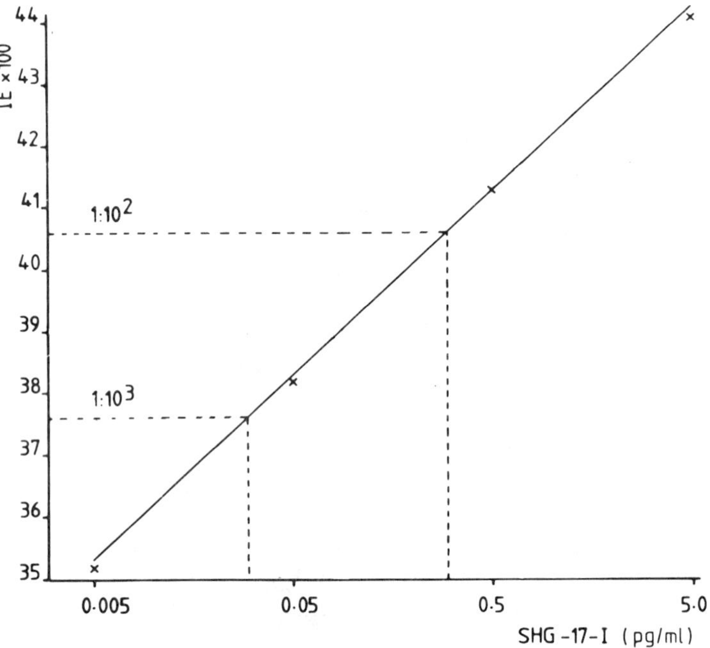

Figure 5   A typical cytochemical section bioassay for gastrin-like activity. The standard graph is constructed from the carbonic anhydrase activity (expressed as integrated extinction, IE × 100) induced in the parietal cells of sections of guinea pig stomach by the application of graded concentrations of SHG-17-I (synthetic human G-17-I). The enzyme activity induced by two dilutions of plasma is shown by the broken lines and the gastrin-like activity is read off the graph. $1:10^2$ = 37.0 pg/ml; $1:10^3$ = 32.0 pg/ml (mean 34.5 pg/ml).

The graded concentrations of hormone produced a linear log dose response. Suitable dilutions of plasma collected in lithium heparin tubes normally gave a response parallel to that obtained with dilutions of the standard preparation, and the gastrin-like bioactivity of these plasma samples could be expressed in terms of activity equivalent to the standard (Figure 5).

VALIDATION OF THE ASSAY

Use of Plasma with Added Aprotinin

The response obtained with plasma collected in lithium heparin tubes was often not shown by samples of serum even from the same blood sample (Figure 6),

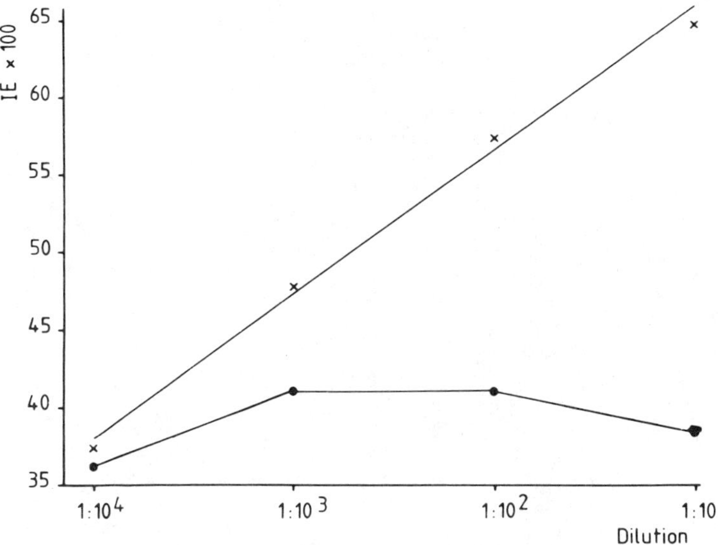

Figure 6   Comparison of the levels of carbonic anhydrase activity (IE × 100) induced by serial dilutions of plasma (crosses) and serum (filled circles). The plasma activity was linearly related to the dilutions of the plasma, but serum produced little activity.

suggesting that gastrin biological activity is lost during the clotting process. It was also found that the levels of activity in plasma increased on storage, even at $-70°C$, possibly due to tryptic activity; such changes could be prevented by the addition of aprotinin to the plasma samples at a concentration of 1000 kIU/ml of whole blood (Figure 7). Therefore all blood for this work is collected into lithium-heparin plastic tubes containing aprotinin (1000 kIU/ml whole blood), rapidly centrifuged and the plasma decanted off and stored at $-70°C$.

Index of Precision (λ) and Fiducial Limits

The mean index of precision for this section procedure in two studies was 0.102 ± 0.05 (n = 8) and 0.113 ± 0.05 (n = 7), compared with 0.14 (n = 36) for the segment assay. The fiducial limits of an assay (P = 0.95) were 75-134%. Inter-assay variation was ±16.3% (n = 3); intra-assay variation was ± –6.4% (n = 4).

Accuracy

The accuracy of the assay was tested by the ability to recover exogenous standard gastrin (G-17) added to plasma. Two samples of plasma from two volunteers

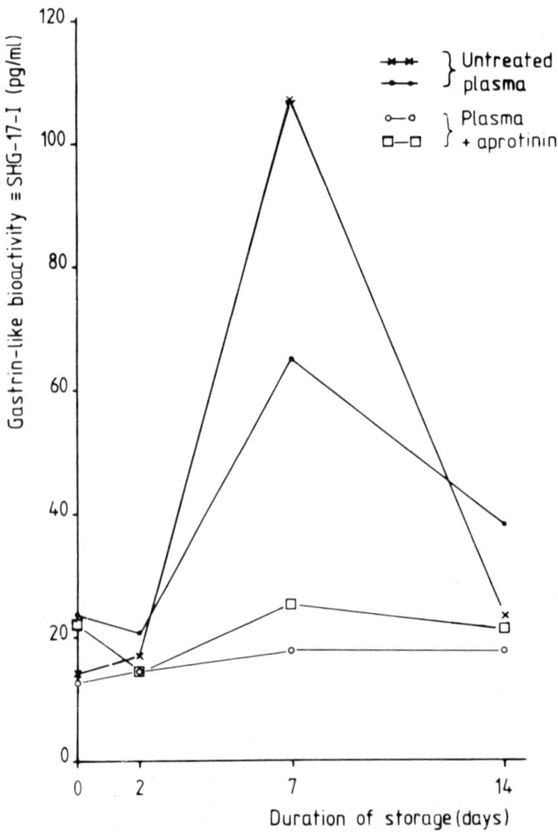

Figure 7   The alterations occurring in fasting plasma gastrin-like bioactivity during storage. The presence of aprotinin is shown to have a marked effect in preventing alteration of the levels during storage.

were first assayed in the usual way to establish the endogenous concentration of the hormonal activity. The gastrin was added to each sample in a concentration of 50 pg/ml and the samples were assayed again. The recovery in the two samples was 96 and 112.8%, respectively (Table 1).

Specificity

A problem common to all biological assay methods for the estimation of gastrointestinal hormones is, inevitably, that other hormones may cross-react in the assay. What is required is a system in which investigators are fairly sure, over a given range of sensitivity, that they are measuring the effect of the hormone

Table 1 Results of Recovery Experiments

Plasma Sample	Mean Gastrin-like Bioactivity Detected (pg/ml)	Mean Gastrin-like Bioactivity in Plasma + 50 pg/ml G-17 (pg/ml)	Recovery (pg/ml)
RWH	30.0	78.0	48.0
PE	6.6	63.0	56.4

under investigation without interference from others. The specificity of this cytochemical section bioassay has been investigated, and the results are essentially as follows:

Pentagastrin and "Big" Gastrin (G-34)
On a molar basis, pentagastrin was equipotent to G-17 (potency ratio of 1.02 ± 0.25, mean ± SEM; n = 5), whereas G-34 was relatively inactive and gave a flat, nonparallel response when tested under these conditions, i.e., at the time at which the activity of G-17 was maximal.

Cholecystokinin
The octapeptide of cholecystokinin (CCK), as might be expected from its structural similarity to the C terminus of G-17, had a molar potency, relative to G-17, of 0.68 ± 0.11 (mean ± SEM; n = 5) when tested alone.

At a concentration of 2.5×10^{-11} M the whole molecule of CCK evoked increased carbonic anhydrase activity that was maximal between 60 and 90 sec and equivalent to 52% of the activity induced by 2.3×10^{-12} M G-17. The effects over a range of concentrations expressed relative to the maximal response in serial sections evoked by 2.3×10^{-12} M G-17 indicated that, when acting alone, CCK was approximately 1000 times less potent than G-17 (Figure 8). To examine the combined effects of G-17 and of CCK, three samples were prepared, each containing 2.3×10^{-11} M (50 pg/ml) G-17. CCK was then added to each sample at concentrations of 3.75×10^{-11} M (150 pg/ml), 1.25×10^{-11} M (50 pg/ml), and 2.5×10^{-12} M (10 pg/ml), respectively. The carbonic anhydrase activity elicited by $1{:}10^2$ and $1{:}10^3$ dilutions of each sample was then measured and compared with that caused by G-17 alone, so allowing an estimate of the gastrin-like activity in each sample to be made. The first two samples, containing 3.75×10^{-11} M and 1.25×10^{-11} M of CCK, gave responses which were not parallel to the standard graph for G-17 alone (Table 2). The response produced to the two dilutions of the third sample, containing 2.5×10^{-12} M CCK, was almost parallel to that of G-17 alone. When corrected for dilution only the third sample assayed at a level equivalent to the concentration of gastrin present in

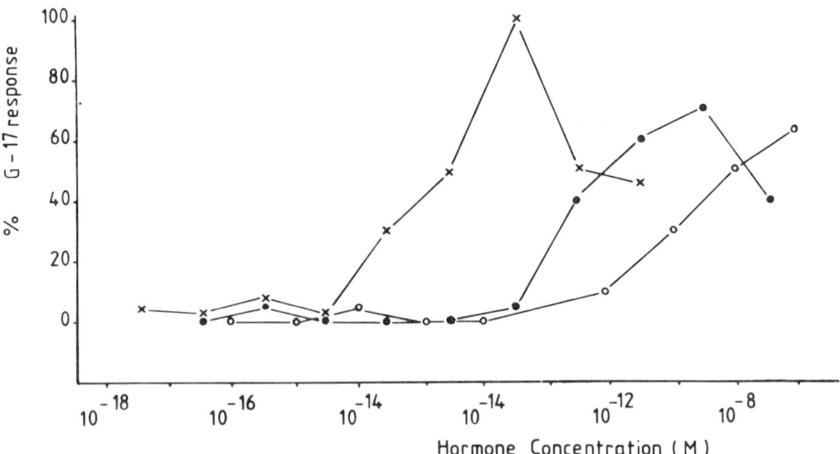

Figure 8   The relative response elicited by various concentrations of secretin
(open circles), CCK (filled circles), and G-17 (crosses), in sections from the same
gastric fundus, taking the maximal response to G-17 (at 2.3×10^{-12} M) as 100%
response.

the sample (i.e., 2.3×10^{-11} M). Thus the higher concentrations of CCK, which
in the absence of gastrin gave only a small elevation of carbonic anhydrase activity (Figure 8), caused some inhibition of the normal response when combined with gastrin, so that the dose-response graph was not parallel to that of
gastrin alone.

Secretin
This hormone was 100 times less potent than CCK and therefore 10^5 times less
potent than G-17 (Figure 8); this finding might reflect contamination of the
sample by CCK. To examine the combined effects of secretin and gastrin the
activity of 10^{-13} M gastrin was tested in the presence of increasing concentrations of secretin (10^{-13}-10^{-10} M). At 10^{-11} and 10^{-12} M the inhibition of
G-17 activity was 21 and 16%, respectively, with no inhibition by 10^{-13} M
secretin; at 10^{-10} M, secretin completely abolished the stimulation of carbonic
anhydrase activity by G-17.

Histamine and Acetylcholine
Using this cytochemical section bioassay we have shown that histamine and
acetylcholine facilitate the action of gastrin on the parietal cells but do not enhance the magnitude of the response (Hoile, 1979a). As regards the possibility
of interference by histamine in the cytochemical bioassay it seems unlikely

Table 2 Effect of CCK on the Measurement of Gastrin-like Activity

Concentrations of Hormones in Sample (M)		Gastrin-like Activity Corrected for Dilution (M)	
G-17	CCK	$1:10^2$ Dilution	$1:10^3$ Dilution
2.3×10^{-11}	3.75×10^{-11}	0.55×10^{-11}	1.7×10^{-11}
2.3×10^{-11}	1.25×10^{-11}	0.83×10^{-11}	1.8×10^{-11}
2.3×10^{-11}	0.25×10^{-11}	1.61×10^{-11}	2.3×10^{-11}

that this will occur over the time course chosen for the assay, given the low concentrations of free histamine in plasma.

Charcoal-Adsorbed Serum

A sample of serum was treated with charcoal in order to adsorb small peptides from it. Serial dilutions of this serum ($1:10$-$1:10^4$) were assayed against graded concentrations of standard hormone. The charcoal-adsorbed serum showed no demonstrable effect on the carbonic anhydrase activity in the cytochemical section assay. Thus it is unlikely that the system reacts to some nonspecific substance in blood to produce an artifactual response.

Effect of a Specific Antibody to Gastrin

A sample of rabbit antigastrin antibody (G/R/26-PF) was added to plasma at $1:10^4$ dilution. This addition produced complete or almost complete (90%) abolition of the detectable gastrin-like bioactivity present in the plasma (Table 3).

Discussion

It seems to be possible to use the cytochemical section assay technique to measure gastrin-like activity in plasma samples, and the assay is certainly accurate, precise, and sensitive. With regard to specificity, complex interactions between the effects of the various gastrointestinal hormones make it unlikely that the response of carbonic anhydrase activity in the parietal cells can be totally specific to any one hormone and some interaction might be expected. It has been shown that secretin, at 10 times the molar concentration of G-17, caused only 16% inhibition of the effect of G-17; even at 100 times the molar concentration of G-17 the inhibitory effect of secretin was only 21%. For most practical purposes, these degrees of inhibition and thus uncertainty as to the true concentration of G-17 may not be serious impediments to the assay of gastrin-like activity. The effect of the intact CCK molecule is more serious, but it should be noted that the marked deviation from parallelism which occurs with appreciable concentrations of CCK is easily detectable; such lack of parallelism indicates that

Table 3   Effect on Plasma Gastrin-like Bioactivity of the Addition of Rabbit
Anti-Gastrin Antibody (G/R/26-PF)

Plasma Sample	Mean Gastrin-like Bioactivity Detected (pg/ml)	Mean Gastrin-like Bioactivity in the Presence of Antibody (pg/ml)
AR	16	0
LB	78	3.9

the biological activity in the sample is not caused predominantly by gastrin and
no value for gastrin-like bioactivity can be assigned to that sample.

The absence of response to control solutions and charcoal-adsorbed serum
and the abolition of bioactivity in plasma by an antigastrin antibody is encour-
aging and suggests that it is the bioactivity of gastrin itself which is being measured rather than any artifact.

APPLICATION OF THE CYTOCHEMICAL SECTION
BIOASSAY FOR GASTRIN-LIKE ACTIVITY

Fasting Normal Subjects

In a group of 24 normal volunteers the fasting level of gastrin-like activity assayed by this procedure was 5.8×10^{-12} M (± 1.7). The range was 1.9-$17.6 \times
10^{-12}$ M and showed a logarithmic normal distribution (Loveridge, Hoile et al.,
1980).

Comparison with Radioimmunoassay in Fasting Normal Subjects

In theory, at least, it is unlikely that any bioassay and immunoassay would produce similar results because of their fundamental conceptual differences, the
former measuring function irrespective of the number of molecules present, and
the latter measuring chemical moieties irrespective of their biological function.

Blood was taken from a group of 15 healthy volunteers and the samples were
assayed, both by the cytochemical section bioassay and by a radioimmunoassay,
kindly performed by Dr. G. Dockray, using antibody 1296 (Dockray and Walsh,
1975). This antibody detects all molecules with a carboxyl-terminal portion
identical to that of little gastrin but is unable to distinguish between the different molecular forms. The mean plasma biological activity from this group was
5.1×10^{-12} M (± 1.8). Results from radioimmunoassay produced levels in three
subjects which were below the threshold of the assay (see Table 4) and are best
ignored for statistical analysis. Comparing only samples 1-12, the levels given by
cytochemical bioassays are significantly lower than those measured by radio-

Table 4 Gastrin Levels Obtained from 15 Normal Fasting Subjects Using Radioimmunoassay and the Cytochemical Section Assay[a]

Sample	Sex	Age	Radioimmunoassay (pmol/liter)	Bioassay (pmol/liter)
1	M	39	22.0	5.8
2	F	24	8.0	6.6
3	M	32	5.0	2.7
4	F	20	6.0	4.2
5	M	24	10.0	6.0
6	M	22	7.0	6.8
7	M	25	9.0	2.8
8	F	25	15.0	6.2
9	F	25	12.0	18.0
10	F	21	6.0	3.2
11	F	35	15.0	12.8
12	M	23	8.0	2.1
13	F	24	< 5.0	7.0
14	M	22	< 5.0	3.0
15	F	28	< 5.0	4.6

[a]Results expressed as activity equivalent to natural human G-17-I.

immunoassay (Hoile, 1979b). This difference is not surprising as the biological assay only measures the smaller molecules with biological activity whereas radioimmunoassay will also detect the larger, less active molecules. The ability of the cytochemical bioassay to measure levels of gastrin-like activity below the threshold of a radioimmunoassay is a tribute to the sensitivity of the biological system.

Normal Response to Food

In our early work with the cytochemical section bioassay we wished to demonstrate the ability of the assay to detect expected changes evoked by a physiological stimulus. We knew that the gastrin levels rose after a meal and, based on the knowledge gleaned from radioimmunoassays, expected a peak in activity at approximately 20 min after the ingestion of food (Blair et al., 1975). The response we actually measured took us completely by surprise.

A healthy female volunteer fasted overnight and then ingested a solution of two Oxo cubes dissolved in 150 ml water. Blood samples were taken at 5 min intervals, beginning during the fasting state, and the gastrin-like response obtained is depicted in Figure 9. It can be seen that the response was rapid and transient. This rapid and transient response to food is at variance with the known data regarding the gastrin response to food based on most radioimmuno-

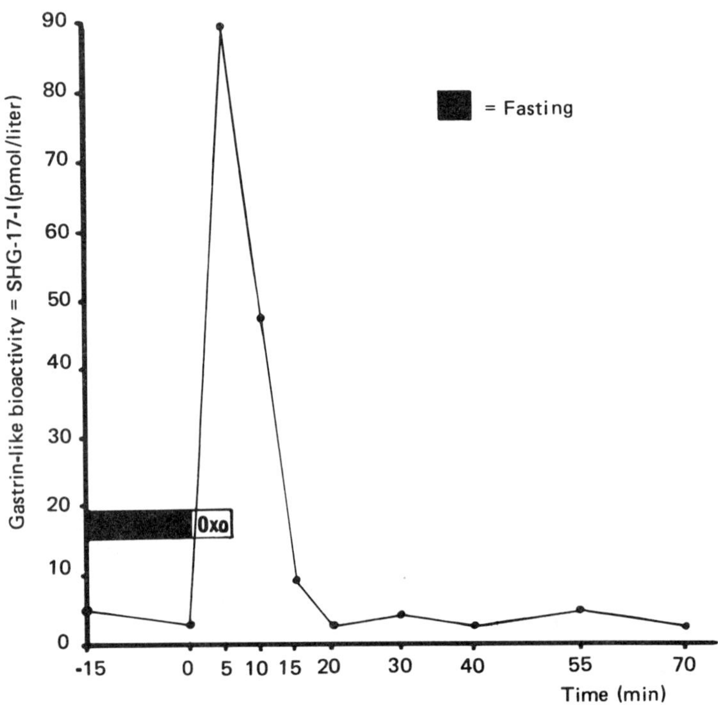

Figure 9   Plasma gastrin-like bioactivity in response to an Oxo "meal" in a healthy female volunteer. The response appears rapidly and is short lived (SHG-17-I = synthetic human G-17-I).

assays, although similar to that produced by gastric distention using a much larger volume of water (Christofides et al., 1978). It is also interesting to note the work of Taylor and his colleagues (1979). They used a double antibody technique to allow the estimation of total gastrin immunoreactivity, G-17 alone, and, by difference, G-34 (G-17 and G-34 being the predominant forms of gastrin in circulation after a meal). They showed, but did not actually state, that whereas the increment in total carboxyl-terminal immunoreactivity after food in normal subjects reached a peak at 30 min, the increment in serum G-17 occurred earlier (appearing to be at 15 min as judged from the published figures). The rapid response detected in our work could reflect a true biological response and the rapid decline of activity may be due to the degradation of the biological activity of the peptide molecules. This is of course conjecture, but at least we had detected a response to a physiological stimulus even though the interpretation of the pattern of this response was difficult.

CLINICAL RELEVANCE OF GASTRIN

The role of gastrin in disease is of clinical interest because of the known physiological actions of gastrin on the gastrointestinal tract and the relationship between the relatively common duodenal ulcer diathesis and gastric hypersecretion. Much of the current knowledge of hypergastrinemia and its consequences in humans is based on radioimmunoassay methods for the measurement of circulating gastrin and observations of the Zollinger-Ellison syndrome. However, comparison between radioimmunoassay results and those produced by earlier, purely qualitative biological assays demonstrated certain discrepancies, particularly in the field of duodenal ulcer (Hansky et al., 1974). The highly sensitive quantitative cytochemical section bioassay for gastrin-like activity (as described above) was initially developed to investigate these discrepancies.

Gastrin in Duodenal Ulceration

Duodenal ulceration is a common condition and is almost certainly multifactorial in origin, being an expression of an imbalance between mucosal protective mechanisms and injury to the mucosa by various aggressive factors. Gastric acid, as one isolated offensive factor, has been studied because of the ease of measurement. On average, patients with duodenal ulceration secrete more gastric acid than normal subjects, both at rest and in response to stimuli, and this hypersecretion may be reduced by gastric or vagal surgery and antacids or antisecretory drugs (Baron, 1973). However, the possible role of gastrin in this gastric acid hypersecretion has proved difficult to interpret. The impression gained from previous studies is that, when all circulating gastrin components are measured, the concentrations of serum gastrin in blood samples after an overnight fast do not differ between duodenal ulcer patients and normal subjects (Korman et al., 1971; McGuigan and Trudeau, 1973). Despite this equivalence some workers have found that within this total gastrin immunoreactivity, levels of G-17 (the most potent molecule biologically) were higher in ulcer patients (Stadil et al., 1975). This "normal" level of fasting gastrin in ulcer patients could in fact be regarded as inappropriately high in view of the strong antral inhibition which should be produced by acid hypersecretion, the known increased sensitivity of the parietal cells of duodenal ulcer patients to gastrin (Isenberg et al., 1975), and the relative predominance of the more potent gastrins.

McGuigan and Trudeau (1973) were among the first to show that patients with duodenal ulcer might have larger responses to food than normal subjects, and this has been confirmed by others using radioimmunoassay (Stadil and

Stage, 1979). Could there be a disproportionate fraction of biologically active G-17 within this total gastrin response? Using G-17-specific antisera and gel filtration, Taylor (1976) found no significant difference in the G-17 response to a meal in normals and duodenal ulcer patients. In fact there was a significantly greater increase in G-34 in the ulcer sufferers; we know that this molecule has a diminished biological potency and if the G-34 accounts for the prolonged post-prandial rise in total gastrin levels then it is unlikely that this response has any biological significance in terms of immediate gastric acid secretion. Taylor et al. (1979) recently confirmed their findings using a double antibody technique rather than gel filtration and again showed that most of the postprandial rise in gastrin radioimmunoreactivity was due to the less biologically active G-34 mole-cules.

In addition to gastric acid hypersecretion it is known that there is an in-creased parietal cell mass in duodenal ulceration, and since gastrin is known to cause hyperplasia of the gastric mucosa the possibility must be considered that gastrin is somehow implicated in the hyperplasia of parietal cells in ulcer dis-ease (Card and Marks, 1960; Helander, 1978).

Could it be that duodenal ulcer patients have an altered ratio of biologically active to inactive types of gastrin within their total immunoreactivity; could there be a circulating stimulant, biologically similar to, but structurally discrete from gastrin, or could duodenal ulcer patients be deficient in some circulating or tissue inhibitor of parietal cell function? Several workers have been able to demonstrate high biological activity in patients, which was not detectable by radioimmunoassay. These discrepancies have been found when studying patients with duodenal ulcer (Hansky et al., 1974), the Zollinger-Ellison syndrome (Bon-fils et al., 1978), and gastric acid hypersecretion (Bugat et al., 1976). In all these instances nonquantitative perfused rat-stomach bioassays were employed. These findings provided some indication of answers to the questions posed above, but the problem could not be resolved without the development of an adequately sensitive and quantitative bioassay.

In conclusion, all we really know about the pathogenesis of duodenal ulcera-tion is that, on average, the acid and pepsin secretion of the afflicted patients is higher than in normals and we assume that this hyperacidity is the fundamental disturbance although in fact not all acid hypersecretors develop ulcers and not all ulcer sufferers are hypersecretors—but of course this is a relative term when considering particular individuals. If one accepts this starting point, then a case can be developed for the role of gastrin. However, radioimmunoassay of this hormone has not provided convincing evidence of a causal relationship between gastrin and the duodenal ulcer diathesis, whereas biological assays (prior to the development of a cytochemical bioassay) had provided interesting glimpses of possible excesses of gastrin-like secretagogues in the sera of duodenal ulcer patients.

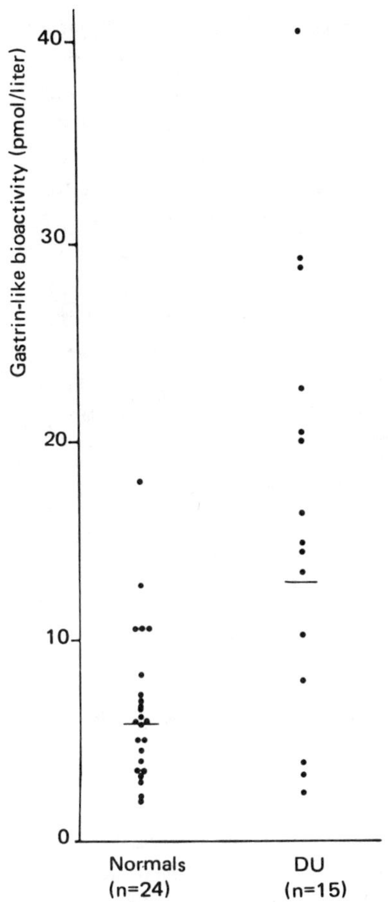

Figure 10   Gastrin-like bioactivity in a group of normal subjects and duodenal ulcer patients. The populations are skewed and the bars represent the logarithmic means of the two groups. The duodenal ulcer patients show an activity significantly in excess of that present in the plasma of normal healthy volunteers (see text). (From Hoile, 1982).

Fasting Gastrin-like Bioactivity in Normals and
Duodenal Ulcer Patients
Blood was taken from fasting duodenal ulcer patients and assayed by the cytochemical bioassay. All these patients had proven duodenal ulcers and were awaiting definitive surgery. The results were compared with those from the normal group detailed above. It should be said that these normal subjects were medical

students, nurses, medical staff, and physiotherapists and all were without gastro-intestinal symptoms or family history of peptic ulcer disease. The two groups were not accurately matched for age and sex, there being a higher age and male predominance in the ulcer group. The fasting levels of gastrin-like bioactivity obtained from the two groups are shown graphically in Figure 10. The mean bioactivity in the ulcer group was 12.8×10^{-12} M (± 2.3) which is significantly higher than the mean gastrin-like bioactivity of 5.8×10^{-12} M (± 1.7) found in the normal group (P <0.0012, Mann-Whitney U test).

Thus the cytochemical section assay for gastrin-like bioactivity has identified an excessive amount of a plasma secretagogue in duodenal ulcer patients which has not previously been quantified either by earlier, less sensitive bioassays or by current radioimmunoassays (Hansky et al.,1974; Bonfils et al., 1978; Bugat et al., 1976). It is likely, from the knowledge of the specificity of the assay, that this activity is due to a form of gastrin, but further studies would be needed to confirm this.

Gastrin-like Response to Food in Normal
Subjects and Duodenal Ulcer Patients
In view of the rapid and transient response to food detected in normal subjects I decided to compare a small group of duodenal ulcer patients with normal sub-jects. The patients were all males with confirmed uncomplicated ulcers awaiting surgery. None were taking cimetidine (see below) at the time of the study. The control group consisted of male volunteers who had no history suggestive of peptic ulceration.

Each subject fasted from midnight. The following morning, blood samples were taken through an indwelling intravenous needle 15 and 5 min before the subject ate a standard meal. The meal consisted of two hard-boiled eggs, two slices of dry toast, and a cup of beef extract (two Oxo cubes in 150 ml of warm water), and was eaten over a period of 10 min. After commencement of the meal, blood samples were taken at 5 min intervals for the first 20 min and then at 10 min intervals for another 40 min. The measurement of gastrin-like activity was performed on plasma containing aprotinin as previously described. Samples were also examined by radioimmunoassay (Dr. Dockray) and for the level of CCK (Dr. C. Marshall).

The patterns of response of the normal subjects and duodenal ulcer patients are given in Figure 11, the radioimmunoassay levels being superimposed on the results from the ulcer patients. From this it is clear that there is no difference between the two groups in terms of gastrin-like response although the ulcer patients tend to have higher fasting levels. One patient in each group showed a peak at 20 min after the meal. Interestingly, the radioimmunoassay results (using antibody 1296 and hence reflecting total immunoreactivity) showed the

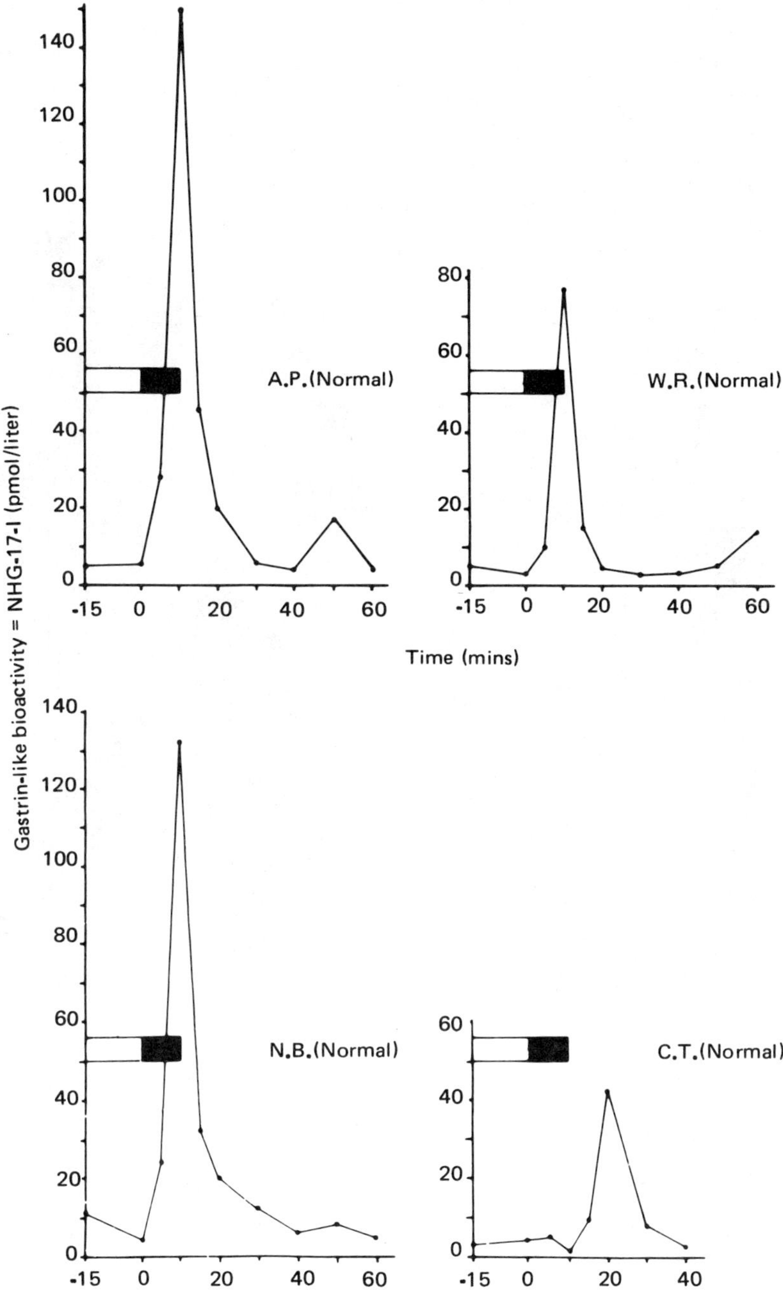

Figure 11   The plasma gastrin-like bioactivity (closed circles) measured in four normal volunteers (above) and four duodenal ulcer patients (p. 210) during the course of a test meal. Radioimmunoreactive gastrin levels are shown for the ulcer patients (open circles). After a fasting period (open box) the patients ate the test meal (closed box).

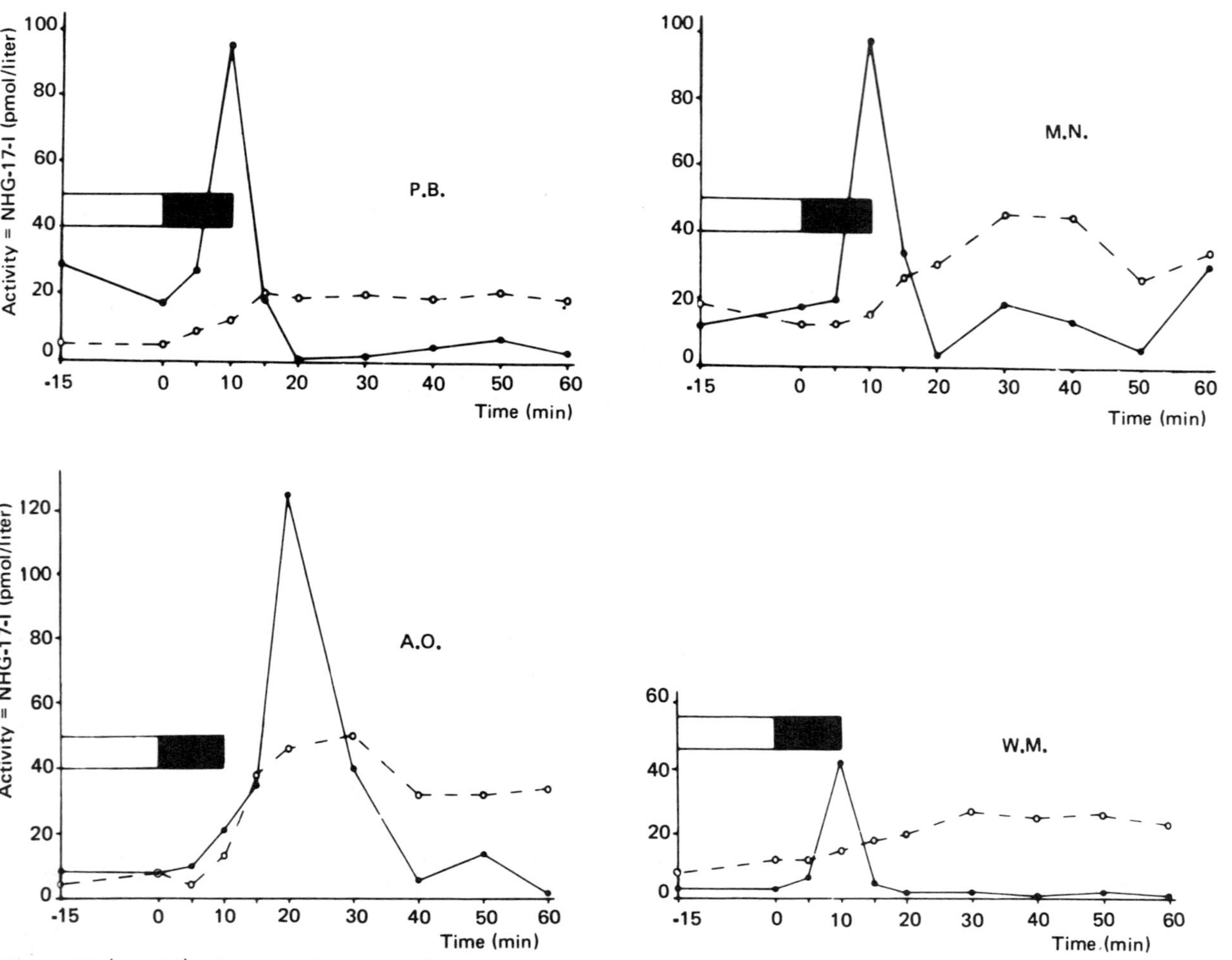

Figure 11 (cont'd)   For legend, see previous page.

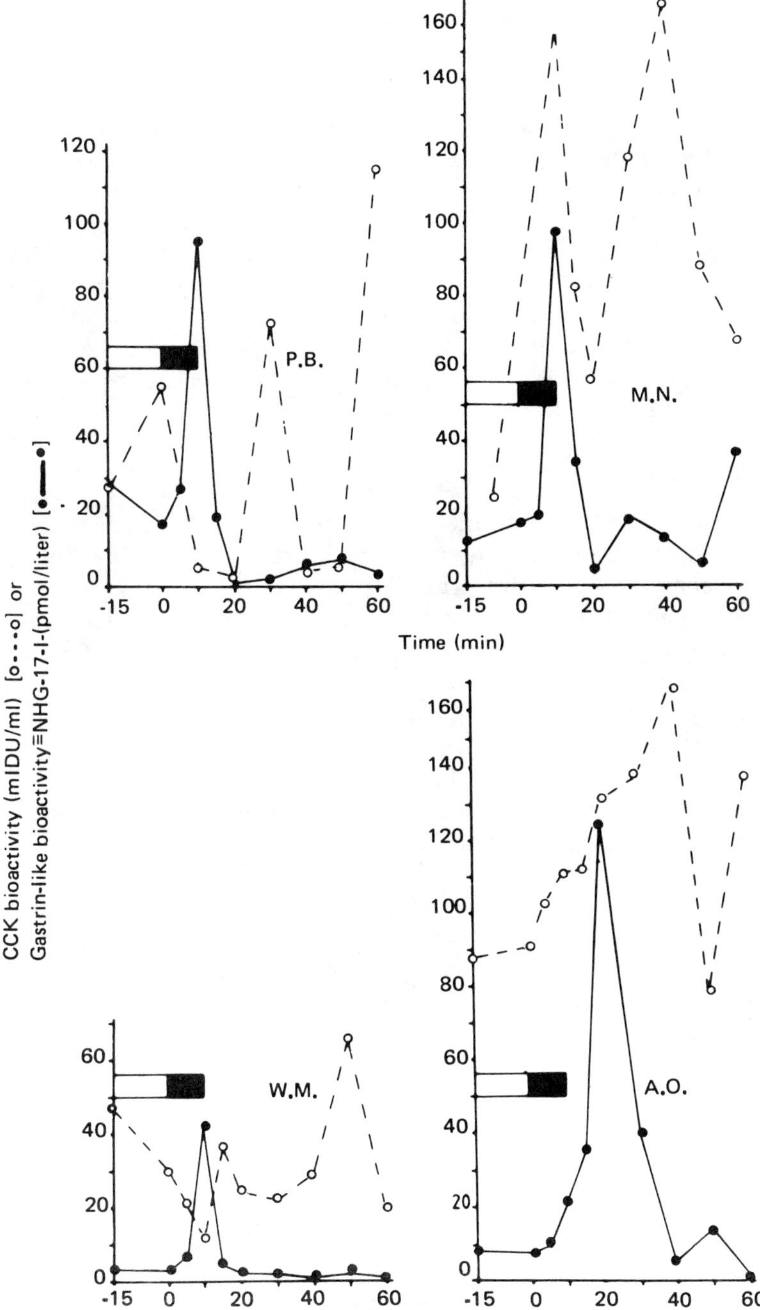

Figure 12 A comparison of the levels of gastrin-like bioactivity and CCK in four duodenal ulcer patients during a test meal. After a fasting period (open box) the patients ate a test meal (closed box). The vertical axis shows both the CCK response (mIDU/ml, open circles) and the gastrin-like bioactivity (pmol/liter, closed circles). The CCK response appears to be biphasic and succeeds the gastrin-like response.

peak at 20 min and persistent elevation as has been previously described. We now know that this is mainly produced by G-34, and Taylor et al. (1979) have now shown that these previously described higher postprandial responses in peptic ulcer patients are due largely to G-34, which has less biological activity.

Measurements of CCK levels were done by the rabbit gallbladder muscle strip bioassay (Marshall et al., 1978) and are shown in Figure 12. In general the response is biphasic and succeeds the gastrin-like response. These results provide further validation of the specificity of the cytochemical bioassay as the peaks only coincide in one patient and even then the second CCK peak is not reflected in the gastrin response.

Zollinger-Ellison Syndrome

This syndrome is classically characterized by a clinical triad of severe ulcer disease (often multiple, jejunal, or recurrent ulcers), massive gastric acid hypersecretion, and a non-β islet-cell gastrin-producing tumor of the pancreas (gastrinoma) (Isenberg et al., 1973). When such tumors are suspected they can be provoked to secrete excessive amounts of gastrin (as measured by radioimmunoassay) by diagnostic infusions of calcium or secretin. The detection of this syndrome is the most important clinical application of radioimmunoassay for gastrin. However, in typical cases there is often only moderate hypergastrinemia in the presence of gross elevation of basal and maximal acid secretion. There are also variations in the disease; some patients do not have ulcers at all, and some lack massive acid hypersecretion and may be difficult to distinguish from patients with ordinary duodenal ulceration. There are also anecdotes of cases in which the severity of the ulcers and the progress and spread of the gastrinoma are poorly related to the actual level of radioimmunoreactive gastrin in the serum. In other words, there are obvious discrepancies between the biological activity of gastrin (or similar secretagogues) and gastrin radioimmunoreactivity in this syndrome (Hansky et al., 1974; Temperley and Stagg, 1971). Case studies illustrating this discrepancy and the use of the cytochemical section bioassay will be described below.

Gastrin and Other Diseases

Abnormal gastrin levels have been found in other interesting clinical situations associated with gastric hypersecretion, such as the isolated retained antrum after gastrectomy, pyloric stenosis, and alterations in gastrin metabolism following disease or surgical resection of the small intestine or kidneys. There is also a rare condition of antral G-cell hyperplasia which mimics, but can be distinguished from, the Zollinger-Ellison syndrome (Straus and Yalow, 1974). However, gastrin assays in these situations have been of academic rather than of diagnostic use.

Hypergastrinemia also occurs without acid hypersecretion, particularly in pernicious anemia, atrophic gastritis, and achlorhydria (with or without gastric cancer). Provided the inflammatory process spares the gastric antrum, or there are circulating parietal cell antibodies, then gastrin release is not inhibited and in fact there is an increase in the G cells of the antrum. If there is antral gastritis the serum gastrin is usually normal (Strickland et al., 1971). In a recent study by Loveridge, Bitensky, et al. (1980) it was shown that gastric parietal cell antibodies in serum from pernicious anemia sufferers inhibit parietal cell function. These workers used both isolated gastric mucosa from the bullfrog and cytochemical techniques to demonstrate that the immunoglobulin fraction from sera of patients with pernicious anemia directly inhibited gastric acid secretion. It appeared that the parietal cell antibodies in the serum interfered with the ability of the parietal cells to respond to gastrin stimulation by binding to the gastrin receptor. In view of the trophic action of gastrin the authors suggested that such a blockade might inhibit parietal cell proliferation and maturation, leading to the mucosal atrophy seen in this condition. Thus, despite the hypergastrinemia, there is no acid hypersecretion and the gastric fundic mucosa becomes atrophic. Consequently, to interpret the relevance of hypergastrinemia, it is also necessary to measure acid secretion.

It can be seen from the preceding discussion that the study of the biochemistry and physiology of the gastrins (and other gastrointestinal hormones) now occupies an important position in experimental gastroenterology. However, the clinician feels rather unenthusiastic in the company of biochemists and physiologists working in this field, as their only useful contribution to daily clinical practice, despite all their fascinating findings, has been in the diagnosis and subsequent management of gastrinomas. Apart from this instance there are no generally accepted indications for measuring circulating gastrin and the clinical importance of this examination has still to be decided. Perhaps this failure of gastrin radioimmunoassay to influence clinical management reflects the inability of the technique to provide functional, biologically meaningful information.

THE NEED FOR A CYTOCHEMICAL BIOASSAY OF GASTRIN-LIKE ACTIVITY

In preceding sections I briefly mentioned that the earlier in vivo, functionally specific, biological assays for gastrin based on the perfused rat stomach (Lai, 1964) were not sufficiently sensitive to measure normal fasting levels of gastrin. They were partially successful in the laboratory diagnosis of the Zollinger-Ellison syndrome (Thomson et al., 1970), but both false positives and false negatives were reflections of the insensitivity of these assays (Jesseph et al., 1967).

More recently, radioimmunoassay for gastrin has eclipsed bioassays in most centers. There are, however, two disadvantages to radioimmunological assays.

The first is that, depending on the antisera used, radioimmunoassays measure a composite antigenic activity which bears little relationship to the biological activity present (WHO, 1975). There are many examples of other situations where radioimmunoassay has demonstrated elevated hormone levels but sensitive bioassays have revealed biologically inactive hormone molecules (Ealey, 1979; Petersen et al., 1978). These discrepancies between biological activity and immunoreactivity may arise in several ways and have been discussed elsewhere (Rees, 1979). The second disadvantage is that radioimmunoassays are still one thousand times less sensitive than the equivalent cytochemical bioassays which have this additional advantage of measuring only biologically active hormone molecules.

In the light of these facts it is obviously desirable to possess a suitably sensitive bioassay for gastrin in order to be able to assign a biological potency to a result obtained by radioimmunoassay, especially where the latter gives results contradicted by other clinical evidence. Such an assay would also help to investigate possible excesses of biologically active gastrin or gastrin-like substances in duodenal ulceration (Work Group II, 1975).

An additional advantage provided by the sensitivity of the cytochemical bioassays (and the ability to use plasma in dilutions of the order of $1:10^2$ and $1:10^3$) is the small volumes of blood required for assays. This makes such a technique ideal for measurements of gastrin in infants, children, small laboratory animlas, and dynamic physiological studies. Most of these advantages have already been exploited in other spheres, such as ACTH measurements (Holdaway et al., 1973; Buckingham and Hodges, 1975).

From the results of the cytochemical section bioassay given in earlier sections and the clinical details given below it will be seen that there is a basis for a reappraisal of the role of gastrin radioimmunoassay and of gastrin itself.

ZOLLINGER-ELLISON SYNDROME

The cytochemical bioassay has been applied to the investigation and diagnosis of suspected cases of the Zollinger-Ellison (Z-E) syndrome; abbreviated, illustrative case histories are given below.

Case 1. A 45-year-old man with a 12 year history of a radiologically proven duodenal ulcer presented with a hematemesis. An emergency partial gastrectomy was performed, and recovery was uneventful. He developed severe upper abdominal pain and vomiting 2 years later. Radiology showed retention in the gastric remnant, jejunitis, and recurrent ulcers in both the afferent and efferent jejunal loops; these were subsequently confirmed by endoscopy.

Fasting plasma gastrin-like bioactivity was 71.5×10^{-12} M (normal range, 1.9-17.6×10^{-12} M). This 10-fold increase in the level of activity strongly

Table 5   Gastrin-like Bioactivity Detected by the Cytochemical
Section Assay in the Plasma of Case 2

Plasma Dilution	Gastrin-like Bioactivity $(M)^a$
$1:10^2$	1.1×10^{-11}
$1:10^3$	1.6×10^{-11}
$1:10^4$	25.0×10^{-11}
$1:10^5$	26.0×10^{-11}

[a]Corrected for dilution.

suggested a gastrinoma. In view of the severity of his symptoms further investigations were not done. At laparotomy a mass was found in the tail of the pancreas and the recurrent ulceration confirmed. A total abdominal gastrectomy and distal pancreatectomy were performed. Histology and immunocytochemistry showed diffuse hyperplasia or microadenomatosis of the G cells. The patient made an excellent recovery.

This was the first case studied using this bioassay and confirmed our ability to detect pathological levels of gastrin-like activity.

Case 2.   This man, aged 37, was first diagnosed as a duodenal ulcer sufferer in 1968 and had undergone a partial gastrectomy in 1975. Since then he had developed stomal ulceration, and investigations in 1976 had shown raised radioimmunoreactive gastrin levels. Exploratory laparotomy at that time confirmed the jejunal ulceration, but no evidence of pancreatic pathology was found. The ulcer was excised and a bilateral selective vagotomy added. Ulceration recurred 2 months after this, and on admission he was complaining of epigastric pain, vomiting, and weight loss. Radiology and endoscopy confirmed jejunal ulceration.

Fasting gastrin levels (measured by radioimmunoassay) at two independent centers were 21 and 17×10^{-12} M (normal less than 50×10^{-12} M), and a secretin provocation test failed to show a typical response suggestive of a gastrinoma. However, his ulceration was aggressive and failed to respond to normal doses of the H_2-receptor blocking agent cimetidine.

Measurement of biological activity revealed plasma levels of 9.25×10^{-12} M; there was no evidence of interference with the assay, and an antigastrin antibody abolished all activity. However, recovery experiments on the plasma showed only 3.6% recovery of exogenously added hormone, suggesting the presence of some factor in the plasma which was inhibiting the gastrin molecule. Utilizing the extreme sensitivity of the cytochemical bioassay we measured serial dilutions of the plasma and the results are shown in Table 5. It appeared that the patient's

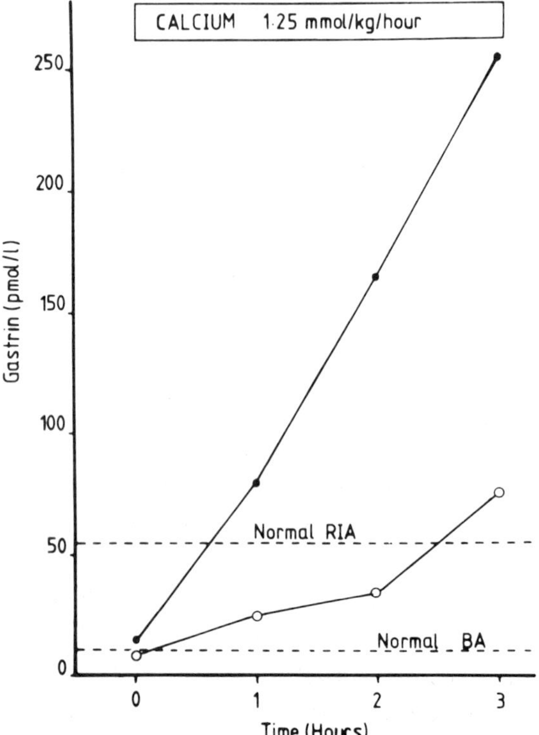

Figure 13   Case 3.   Calcium provocation test. During the 3 hr infusion of calcium the levels of both radioimmunoreactive gastrin (closed circles) and gastrin-like bioactivity (open circles) rose (RIA = radioimmunoactivity; BA = bioactivity).

plasma did in fact contain 25.0×10^{-11} M of gastrin-like activity, but this high level of activity was suppressed, possibly by a factor binding to the gastrin molecules, the binding being labilized at very high dilutions. This might explain the inability of radioimmunoassay to detect these elevated levels, as the antibody might be unable to bind with the hormone in the presence of a blocking agent.

In view of these findings and the strong possibility of the Z-E syndrome, the dose of cimetidine was increased and the ulcers eventually healed.

This case illustrates one way in which discrepancies between immunoactivity and biological activity might arise and a cytochemical technique (i.e., the use of dilution based on the high sensitivity of the system) can be utilized to unravel such a problem.

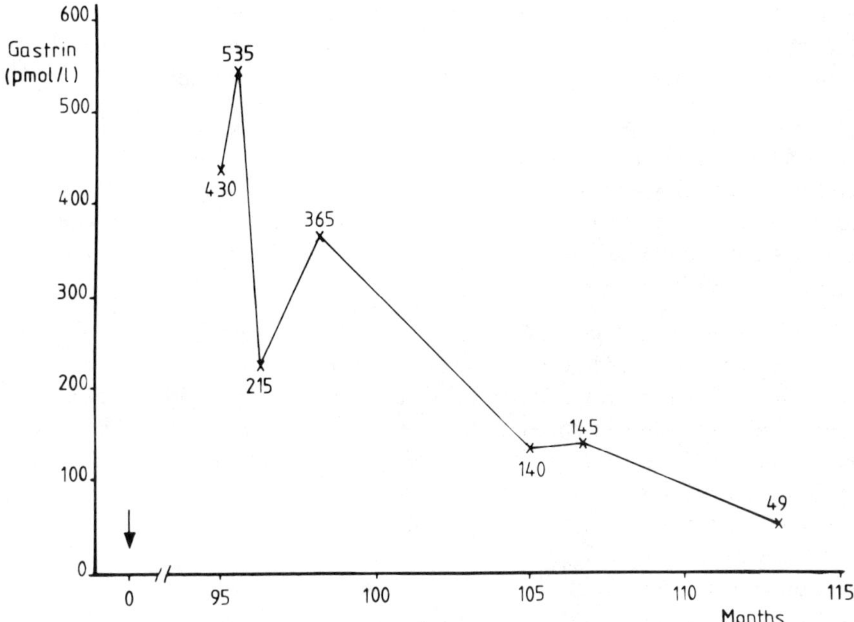

Figure 14   Case 4. Serum immunoreactive gastrin levels measured over the period since her initial surgery (arrow). Plasma for estimation of biological activity was taken when radioimmunoreactivity was 49 pmol/liter and before the final surgery and enucleation of hepatic metastases (see text). (From Hoile, 1982.)

Case 3.   A 52-year-old woman was diagnosed as having primary hyperparathyroidism, and selective venous sampling from her neck veins localized a parathyroid adenoma. Prior to surgery further investigations were done to exclude the multiple endocrine syndrome. These demonstrated elevated levels of fasting serum immunoreactive gastrin (23×10^{-11} M; normal, $3\text{-}55 \times 10^{-12}$ M), but both her basal gastric acid and pentagastrin-stimulated secretion rates were normal. There was no evidence of peptic ulceration. The fasting gastrin-like bioactivity in her plasma at this time was 17.5×10^{-12} M (the upper end of the normal range). Thus, despite a high level of serum radioimmunoreactive gastrin, this patient had normal gastrin biological activity and no evidence of peptic ulceration.

Surgery was undertaken and a benign parathyroid adenoma removed from the predicted site. Postoperatively her calcium levels dropped to the normal range. A calcium provocation test was then performed, and this produced a significant rise in radioimmunoreactive gastrin levels together with a small rise in biological activity (Figure 13); this response is diagnostic of a gastrinoma. Attempts to localize a tumor have so far been unsuccessful, and she has remained well.

It appears that this patient has an occult gastrinoma as part of the multiple endocrine syndrome and that this gastrinoma responds to hypercalcemia by secreting abnormal amounts of gastrin (Walsh and Roth, 1976). It is evident from the history and studies on gastrin-like bioactivity that the gastrin secreted either has a low biological potency or is comprised of very few biologically active molecules or fragments.

Case 4. This 48-year-old female first presented with an epigastric mass which was subsequently found to be a metastatic tumor in the left lobe of the liver. Following left hepatectomy this was shown to be a non-α, non-β islet cell tumor from the pancreas and was proven to be a gastrinoma by immunocytochemistry. She remained well for 8 years before further metastases were identified. Because of pain the largest of these were removed by further surgery, and she was then treated with streptozocin and remained well.

This patient is very interesting for several reasons. First, she had maximal acid secretion with a basal acid output to maximal acid output ratio of 0.9 (a ratio of 0.6 or greater being one criterion for the diagnosis of the Z-E syndrome) and initially raised immunoreactive gastrin levels (Figure 14) but had never shown symptoms or radiological or endoscopic evidence of peptic ulceration. Second, despite the proven presence of multiple gastrinoma deposits in the liver there was no detectable response to a secretin or calcium provocation test, and third, her serum immunoreactive gastrin levels have been steadily falling (Figure 14).

We obtained some freeze-dried plasma from this patient, which had been collected according to our criteria, and measured the gastrin-like biological activity; this was raised at 31.6 pmol/liter (normal range 1.4-18.2 pmol/liter). It can be seen that, although this patient's tumor appears to have changed behavior as far as the production of radioimmunoreactive gastrin molecules is concerned, there continues to be a biological drive for the production of maximal levels of acid. This is presumably a small biologically active gastrin or gastrin-like molecule that is undetectable by radioimmunoassay. It is even more interesting that the mucosal defense mechanisms in this case have been sufficient to prevent ulceration occurring despite the acid production.

CONCLUSIONS

1. There is a growing awareness of the need for functional assays to complement the more analytical or structural assays. In this light, radioimmunoassays and cytochemical bioassays should be regarded as complementary, giving quite different but relevant information. Although radioimmunoassay is suitable for

Table 6 Differing Combinations of Radioimmunological and Biological Gastrin Levels in Relation to the Clinical Presentation

Case	Radio immunoassay	Bioassay	Acid Secretion	Ulcers	Pathology
1	Unknown	Raised	Unknown	Recurrent	Pancreatic G-cell hyperplasia
2	Initially raised, later normal	Raised on dilution of plasma	Unknown	Recurrent	Presumed gastrinoma
3	Raised	Normal	Normal	Absent	Occult gastrinoma
4	Initially raised, later normal	Raised	Raised	Absent	Metastatic gastrinoma

automation and routine laboratory use it cannot discriminate between biologically active and inactive molecules despite the claims of workers using various double antibody and subtraction techniques to measure the contribution to total immunoreactivity of the different molecules; they can only measure the presence of molecules known to have a given bioactivity, and thus, by implication, they assign a level of bioactivity present in the sample measured. This is not the measurement of biological activity.

2. Radioimmunoassay can detect the secretion of hormone fragments and precursors from, for example, nonendocrine tumors such as a gastrinoma, and the detection of these immunoreactive hormones and fragments may only be possible by radioimmunoassay as they may not have biological activity. Thus, the presence of a tumor and even its precise localization can be achieved by radioimmunoassay (Ingemansson et al., 1977). The level of activity and hence the functional significance and clinical relevance of these hormones should then be apportioned by a sensitive bioassay such as the cytochemical bioassay. This is borne out by the clinical studies quoted where the varying combinations of immunological and biological activity are expressed in differing clinical presentations (Table 6).

3. Finally, with regard to duodenal ulceration, my findings of increased fasting gastrin-like bioactivity in patients is completely opposed to the findings of most workers using radioimmunoassay. However, my findings are supported by the work of Stadil and his colleagues (Stadil et al., 1975) who showed that, while total fasting serum gastrin immunoreactivity is alike in normal subjects and duodenal ulcer patients, there are characteristically increased levels of little

gastrin within this total level in the ulcer patients. They concluded that these evaluated levels of little gastrin, with its higher potency, could account for the basal acid hypersecretion seen in many duodenal ulcer patients.

Thus there does appear to be a definite case for the implication of gastrin in the pathogenesis of duodenal ulcer, although a reappraisal of our current knowledge is required in the light of information from the new sensitive cytochemical bioassays. We are still studying the evidence for incriminating gastrin in duodenal ulcer disease, and there can be no verdict as to guilty or not guilty at present. Much of the evidence needs reexamining, and a crucial factor may be whether the gastrin molecules are actually performing the offenses implied by the witnesses' interpretations of laboratory findings.

ACKNOWLEDGMENTS

The hormones used for this work were obtained from the following sources: synthetic human gastrin I (G-17) from the National Institute for Biological Standards and Control. Natural human gastrin I (G-17) and synthetic human big gastrin I (G-34) were donated by Professor R. A. Gregory and Dr. G. Dockray (Department of Physiology, University of Liverpool): Dr. Dockray also performed the radioimmunoassay studies quoted. Pentagastrin was obtained as Peptavlon from Imperial Chemical Industries Limited. Natural porcine cholecystokinin (CCK) and secretin (both 95% pure) were prepared by Professor V. Mutt, GIH Research Unit, Karolinska Institute, Stockholm. The synthetic C-terminal octapeptide of CCK (CCK-OP) was provided by Dr. Miguel Ondetti, Squibb Institute for Medical Research. Rabbit antigastrin antibody (G/R/26-PF) was provided by Dr. B. Morris, Guildford University, Surrey. Dr. C. Marshall (Department of Surgery, Charing Cross Hospital, London) kindly performed the CCK assays.

I offer grateful acknowledgments to the many clinical and scientific colleagues, too numerous to mention individually, who have stimulated my interest in this subject and especially to Professor A. G. Johnson (Sheffield University). I am also grateful to the surgeons and physicians who referred the cases quoted: Mr. K. W. Reynolds (Charing Cross Hospital, London) for Cases 1 and 3; Mr. P. King (St. Stephen's Hospital, London) for Case 2; Mr. T. B. Hugh and Dr. D. Byrnes (St. Vincent's Hospital, Sydney) for Case 4. My thanks are due in particular to Nigel Loveridge who, having patiently taught me the fundamentals of cytochemical technique, has continued to study and improve the assay system and performed some of the assays quoted in this chapter. I also wish to record my gratitude to Rosemary Hughes for her secretarial help and to my wife, who has tolerated domestic chaos during the production of this chapter.

REFERENCES

Baron, J. H. (1973). The rationale of the different operations for peptic ulcer. In *Vagotomy on trial,* A. G. Cox and J. A. Williams (eds.). Heinemann, London, pp. 7-35.

Blair, E. L., Greenwell, J. R., Grund, E. R., Reed, J. D., and Sanders, D. J. (1975). Gastrin response to meals of different composition in normal subjects. *Gut,* 16:766-773.

Bloom, S. R., and Polak, J. M. (1978). Gut hormone overview. In *Gut Hormones,* S. R. Bloom (ed.). Churchill Livingstone, Edinburgh, pp. 3-18.

Bonfils, S., Dubrasquet, M., Accary, J. P., Girodet, J., and Mignon, M. (1978). Bioassay versus gastrin radioimmunoassay in the Zollinger-Ellison syndrome. In *Gut Hormones,* S. R. Bloom (ed.). Churchill Livingstone, Edinburgh, pp. 158-162.

Buckingham, J. C., and Hodges, J. R. (1975). Interrelationships of pituitary and plasma corticotrophin and plasma cortisone during adrenocortical regeneration in the rat. *J. Endocrinol.,* 67:411-417.

Bugat, R., Walsh, J. H., Ippoliti, A., Elashoff, J., and Grossman, M. I. (1976). Detection of a circulating gastric secretagogue in plasma extracts from normogastrinaemic patients with acid hypersecretion. *Gastroenterology,* 71:1114-1116.

Card, W. I., and Marks, I. N. (1960). The relationship between the acid output of the stomach following "maximal" histamine stimulation and the parietal cell mass. *Clin. Sci. Mol. Med.,* 19:147-163.

Chayen, J., Bitensky, L., and Butcher, R. G. (1973). *Practical Histochemistry.* Wiley, New York.

Christofides, N. D., Modlin, I., Sarson, D. L., Albuquerque, R. H., Ghatei, M. A., Adrian, T. E., and Bloom, S. R. (1978). The releases of gastrin, PP, VIP and motilin after an oral water load and atropine in man. *Clin. Sci. Mol. Med.,* 54:12-13P.

Dockray, G. J. (1978). Gastrin overview. In *Gut Hormones,* S. R. Bloom (ed.). Churchill Livingstone, Edinburgh, pp. 129-139.

Dockray, G. J., and Taylor, I. L. (1976). Heptadecapeptide gastrin: Measurement in blood by specific radioimmunoassay. *Gastroenterology,* 71:971-977.

Dockray, G. J., and Walsh, J. H. (1975). Amino-terminal gastrin fragment in serum of Zollinger-Ellison syndrome patients. *Gastroenterology,* 68:222-230.

Ealey, P. A. (1979). The validation of the cytochemical bioassay of thyrotrophin and its application to selected clinical problems. *Ph.D. Thesis,* University of London.

Gregory, R. A. (1976). Heterogeneity of the gastrins in blood and tissue. In *Polypeptide Hormones: Molecular and Cellular Aspects.* CIBA Foundation Symposium, 41:251-265.

Gregory, R. A., and Tracy, H. J. (1964). The constitution and properties of two gastrins extracted from hog antral mucosa. *Gut,* 5:103-117.

Hansky, J., Royle, J. P., and Korman, M. G. (1974). Comparison of bioreactive and immunoreactive gastrin. *Aust. J. Exp. Biol. Med. Sci.,* 52:841-846.

Helander, H. F. (1978). Effects of chronic pentagastrin stimulation on parietal cell size in rat gastric mucosa. *Acta Physiol. Scand.,* Special Suppl.:19-23.

Hoile, R. W. (1979a). The biological activity of gastrin in peptic ulceration.*M.S. Thesis,* University of London.

Hoile, R. W. (1979b). Comparison between levels of gastrin radioimmunoreactivity and gastrin-like bioactivity estimated by a cytochemical section bioassay. *Gut,* 20:A488.

Hoile, R. W. (1982). The technique and clinical application of the cytochemical section bioassay for gastrin-like activity. *Ann. R. Coll. Surg. Engl.,* 64:96-104.

Holdaway, I. M., Rees, L. H., and Landon, J. (1973). Circulating corticotrophin levels in severe hypopituitarism and the neonate. *Lancet,* ii:1170-1172.

Ingemansson, S., Larsson, L. I., Lunderquist, A., and Stadil, F. (1977). Pancreatic vein catheterisation with gastrin assay in normal patients and in patients with the Zollinger-Ellison syndrome. *Am. J. Surg.,* 134:558-563.

Isenberg, J. I., Grossman, M. I., Maxwell, V., and Walsh, J. H. (1975). Increased sensitivity to stimulation of acid secretion by pentagastrin in duodenal ulcer. *J. Clin. Invest.,* 55:330-337.

Isenberg, J. I., Walsh, J. H., and Grossman, M. I. (1973). Zollinger-Ellison syndrome. *Gastroenterology,* 65:140-165.

Jesseph, J. E., Moore, F. T., Endahl, G. L., and Passi, R. B. (1967). Peptic ulceration and enteritis. *Am. J. Surg.,* 113:188-193.

Korman, M. G., Soveny, C., and Hansky, J. (1971). Serum gastrin in duodenal ulcer. *Gut,* 12:899-902.

Lai, K. S. (1964). Studies on gastrin. Part I. A method of biological assay of gastrin. *Gut,* 5:327-333.

Loveridge, N. (1978). A quantitative cytochemical method for carbonic anhydrase activity. *Histochem. J.,* 10:361-372.

Loveridge, N., Bitensky, L., Chayen, J., Hausamen, T. U., Fisher, J. M., Taylor, K. B., Gardner, J. D., Bottazzo, G. F., and Doniach, D. (1980). Inhibition of parietal cell function by human gamma globulin containing gastric parietal cell antibodies. *Clin. Exp. Immunol.,* 41:264-270.

Loveridge, N., Bloom, S. R., Welbourn, R. B., and Chayen, J. (1974). Quantitative cytochemical estimation of the effect of pentagastrin (0.005-5.0 pg/ml) and of plasma on the guinea-pig fundus in-vitro. *Clin. Endocrinol. (Oxf),* 3: 389-396.

Loveridge, N., Hoile, R. W., Johnson, A. G., Gardner, J. D., and Chayen, J. (1980). The cytochemical section bioassay of gastrin-like activity. *J. Immunoassay,* 1:195-209.

Maren, T. H. (1967). Carbonic anhydrase: Chemistry, physiology and inhibition. *Physiol. Rev.*, 47:595-781.

Marshall, C. E., Egberts, E. H., and Johnson, A. G. (1978). An improved method for estimating cholecystokinin in human serum. *J. Endocrinol.*, 79:17-27.

McGuigan, J. E., and Trudeau, W. L. (1973). Differences in rates of gastrin release in normal persons and patients with duodenal ulcer. *N. Engl. J. Med.*, 288:64-66.

Petersen, V. B., McGregor, A. M., Belchetz, P. E., Elkeles, R. S., and Hall, R. (1978). The secretion of thyrotrophin with impaired biological activity in patients with hypothalamic-pituitary disease. *Clin. Endocrinol. (Oxf)*, 8:397-402.

Rees, L. H. (1979). The cytochemical bioassay of polypeptide hormones. In *Quantitative Cytochemistry and Its Applications*, J. R. Pattison, L. Bitensky, and J. Chayen (eds.). Academic, London, pp. 297-309.

Rehfeld, J. F. (1978). Localisation of gastrins to neuro- and adenohypophysis. *Nature*, 271:771-773.

Stadil, F., Rehfeld, J. F., Christiansen, L. A., Malström, J. (1975). Patterns of gastrin components in serum during feeding in normal subjects and duodenal ulcer patients. *Scand. J. Gastroenterol.*, 10:863-868.

Stadil, F., and Stage, J. G. (1979). Gastrinomas as a model for duodenal ulcer disease. In *Gastrins and the Vagus*, J. F. Rehfeld (ed.). Academic, London, pp. 199-210.

Straus, E., and Yalow, R. S. (1974). Differential diagnosis in hyperchlorhydric hypergastrinaemia. *Gastroenterology*, 66:867.

Strickland, R. G., Bhathal, P. S., Korman, M. G., and Hansky, J. (1971). Serum gastrin and the antral mucosa in atrophic gastritis. *Br. Med. J.*, 4:451-453.

Sundler, F., Alumets, J., and Hakanson, R. (1978). Peptides in the gut with a dual distribution in nerves and endocrine cells. In *Gut Hormones*, S. R. Bloom (ed.). Churchill Livingstone, Edinburgh, pp. 406-413.

Taylor, I. L. (1976). Region specific antibodies and the study of gastrin. Ph.D. Thesis, University of Liverpool.

Taylor, I. L., Dockray, G. J., Calam, J., and Walker, R. J. (1979). Big and little gastrin responses to food in normal and ulcer subjects. *Gut*, 20:957-962.

Temperly, J. M., and Stagg, B. H. (1971). Bioassay and radioimmunoassay of plasma gastrin in a case of Zollinger-Ellison syndrome. *Scand. J. Gastroenterol.*, 6:735-738.

Thomson, C. G., Cleaton, I. G. M., and Sircus, S. W. (1970). Experiences with a rat bioassay in the diagnosis of the Z-E syndrome. *Gut*, 11:409-419.

Tracy, H. J., and Gregory, R. A. (1964). Physiological properties of a series of synthetic peptides structurally related to gastrin-I. *Nature*, 204:935-938.

Trowell, O. A. (1959). The culture of mature organs in a synthetic medium. *Exp. Cell Res.*, 16:118-147.

Uvnäs-Wallensten, K., Rehfeld, J. F., Larsson, L. I., and Uvnäs, B. (1977). Heptadecapeptide gastrin in the vagal nerve. *Proc. Natl. Acad. Sci. USA*, 74:5707-5710.

Walsh, J. H., and Grossman, M. I. (1975). Gastrin. *N. Engl. J. Med.*, 292:1324-1332; 1377-1384.

Walsh, J. H., Isenberg, J. I., Arnfield, J., and Maxwell, V. (1976). Clearance and acid-stimulating action of human big and little gastrin in duodenal ulcer subjects. *J. Clin. Invest.*, 57:1125-1131.

Walsh, J. H., and Roth, B. E. (1976). Hormone-secreting tumours of the gastrointestinal tract. In *Recent Advances in Gastroenterology Number 3*, I. A. D. Bouchier (ed.). Churchill Livingstone, Edinburgh, pp. 49-72.

Work Group 11 (1975). Peptic diseases. *Gastroenterology*, 69:1071-1087.

World Health Organisation (1975). 26th Report of the Expert Committee on Biological Standardization. WHO Tech. Rep. Ser. 565, pp. 30-56.

10

Measurement of Antidiuretic Hormone

Peter H. Baylis / Royal Victoria Infirmary and University of Newcastle Upon Tyne, Newcastle Upon Tyne, England

INTRODUCTION

As life emerged from the primordial seas, the need to maintain constancy of tissue composition necessitated the development of homeostatic mechanisms. Conservation and regulation of body water was critical for survival on land, and so evolved antidiuretic hormones which are the principal regulators of water excretion for animals. Arginine vasopressin is the antidiuretic hormone for most mammals, including humans.

Although the concept of constancy of the internal medium had been proposed in the mid-nineteenth century by Claude Bernard (1865), a factor controlling renal water excretion was not discovered until early in the next century. A pressor function was originally ascribed to extracts from the pituitary gland (Oliver and Schäfer, 1895), but by 1901, Magnus and Schäfer had observed oliguria in dogs injected with small extracts from the posterior lobe of the pituitary. Mechanical destruction of the gland resulted in profound polyuria (Schäfer 1909). The precise work of Starling and Verney (1924) demonstrated clearly the antidiuretic effect of posterior pituitary extracts on the isolated kidney, and so a water-regulating factor had been discovered. Over the subsequent decades the mechanisms which control the secretion of this factor, antidiuretic hormone, were elucidated. Changes in blood tonicity appeared to be a major determinant of the release of antidiuretic hormone, mediated by specialized intracranial cells sensitive to osmotic change (Verney, 1947). These observations indicated that a true homeostatic mechanism maintaining constant body water did exist. It is now realized that nonosmotic stimuli may also be important in controlling the secretion of this hormone (Scheiner, 1975).

The physiology of the hormone will be briefly discussed before concentrating in the assay methods available to measure antidiuretic hormone in body fluids, particularly in plasma, and describing the technique, validation, and limitations of a cytochemical bioassay to detect this hormone.

PHYSIOLOGY OF ANTIDIURETIC HORMONE

Chemistry

Arginine vasopressin (AVP), the major mammalian antidiuretic hormone, is a nonapeptide with a molecular weight of 1084. Because three carboxyl groups are amidated it is strongly basic (isoelectric point pH 10.9). Its chemical structure is given in Figure 1. Lysine vasopressin is the other antidiuretic hormone but is confined to Suidae (members of the pig family). The structure of AVP was elucidated by Acher and Chauvet (1954), and the molecule was synthesized in 1958 by du Vigneaud's group. Apart from vasopressin, mammals synthesize a second nonapeptide, oxytocin, the structure of which is similar to the vasopressins but it has no antidiuretic action. Closely related peptides (e.g., vasotocin) are formed in reptiles and birds, and an interesting evolutionary sequence from an ancestral molecule has been postulated by Acher (1974). More recently investigations have concentrated on the relationship between tertiary structures of antidiuretic hormone and its biological function (Forsling, 1979).

Biosynthesis

Neurosecretion

Arginine vasopressin and oxytocin are synthesized in hypothalamic nuclei, each hormone being derived from separate neurons. The major sites are the paraventricular and supraoptic nuclei, although other nuclei (e.g., suprachiasmatic) may also be involved. Each nonapeptide is synthesized with a specific intraneuronal carrier protein called neurophysin (molecular weight 10,000), which in turn is derived from a precursor molecule. AVP and its specific neurophysin (Robinson, 1975) travel from the body of the neuron in the hypothalamus along the axons as discrete neurosecretory granules at a rate of 1-3 mm/hr to three distinct areas of the brain. These three pathways are shown in Figure 2. The major tract terminates in the posterior lobe of the pituitary gland, with the other two supplying the median eminence and the floor of the third ventricle.

Vasopressin is released from the neurohypophysis by calcium-dependent exocytosis. The neurosecretory granule fuses with the axonal membrane causing rupture of the granule with release of AVP, cleaved from neurophysin, into the systemic circulation. Once in the circulation the majority of AVP appears to remain unbound to plasma proteins and neurophysin (Share and Crofton, 1980).

| | 1 | 2 | 3 | 4 | 5 | 6 | 7 | 8 | 9 |

arginine vasopressin: Cys—Tyr—Phe—Glu(NH$_2$) — Asp(NH$_2$) — Cys—Pro—Arg—Gly(NH$_2$)

lysine vasopressin: Cys—Tyr—Phe—Glu(NH$_2$) — Asp(NH$_2$) — Cys—Pro—Lys—Gly(NH$_2$)

oxytocin: Cys—Tyr—Ile—Glu(NH$_2$) — Asp(NH$_2$) — Cys—Pro—Leu—Gly(NH$_2$)

arginine vasotocin: Cys—Tyr—Ile—Glu(NH$_2$) — Asp(NH$_2$) — Cys—Pro—Arg—Gly(NH$_2$)

Figure 1 Amino acid sequence of arginine vasopressin and related posterior pituitary peptides.

Control of Secretion

The major physiological determinant of AVP secretion from minute to minute is plasma tonicity. Jewell and Verney (1957), using indirect urinary methods to estimate antidiuretic hormone activity, localized a group of cells sensitive to osmotic change in the anterior hypothalamus, which were named "osmoreceptors." The first direct confirmation that a rise in blood osmolality was associated with an increase in circulating antidiuretic activity came from Baratz and Ingraham (1959), using a rat bioassay. However, precise methods of measuring AVP were lacking until the development of more sensitive radioimmunoassay techniques.

Recent work has defined the exquisitely sensitive relationship between plasma osmolality and AVP secretion (Robertson et al., 1977). Rises in plasma osmolality of less than 1% cause significant increases in AVP release and consequent antidiuresis. Measurement of plasma AVP and osmolality during infusion of hypertonic saline demonstrates a direct correlation between these two variables in healthy adults (Figure 3). Simple linear regression analysis of plasma AVP (pAVP) and osmolality (pOs) defines the regression line, pAVP = 0.63 (pOs − 284), correlation coefficient r = +0.80, P < 0.001 (Baylis and Robertson, 1980). The slope of this equation, 0.63, is a measure of the sensitivity of the osmoreceptor and AVP-releasing unit. The abscissal intercept, 284 mmol/kg, defines the theoretical threshold for AVP secretion. Whether AVP secretion is indeed completely switched off at low plasma osmolalities remains controversial (Bie, 1980). We must await the development of more sensitive AVP assay techniques before this question can be answered.

Another important aspect of osmoreceptor function is the variation in response to different blood solutes. Although Verney (1947) found very little

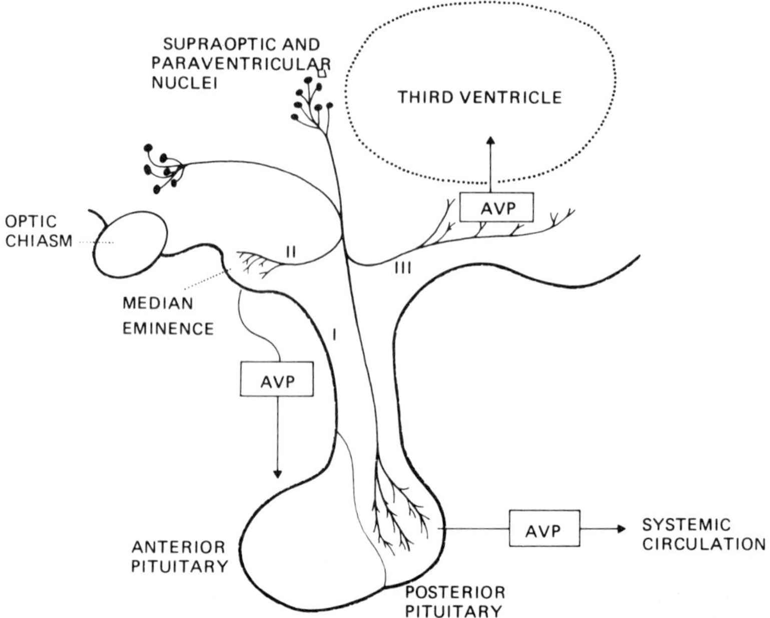

Figure 2 Neuronal pathways of arginine vasopressin secretion in the hypothalamus and posterior pituitary.

difference in antidiuresis produced by a variety of solutes, recent studies have
shown that sodium and mannitol are extremely potent solutes releasing AVP,
that urea is less effective, and that glucose fails to stimulate hormone secretion
(Robertson et al., 1977). Clearly, the cells that are sensitive to solute change are
not uniform in their response as osmoreceptors.

It has been well recognized for many years that profound hypotension and
hypovalemia cause the release of large quantities of AVP. Reduction in blood
pressure or volume of the order of 10-15% is required to initiate AVP release,
but thereafter there is an exponential release of AVP in response to further
hypotension/hypovolemia (Robertson et al., 1977). Thus, under normal physiological circumstances, baroregulatory mechanisms probably play a minor role in
control of AVP secretion.

Many other factors (e.g., nausea, pain, hypoglycemia) have been reported to
stimulate AVP release, but their relevance to the physiological control of AVP
secretion remains to be determined.

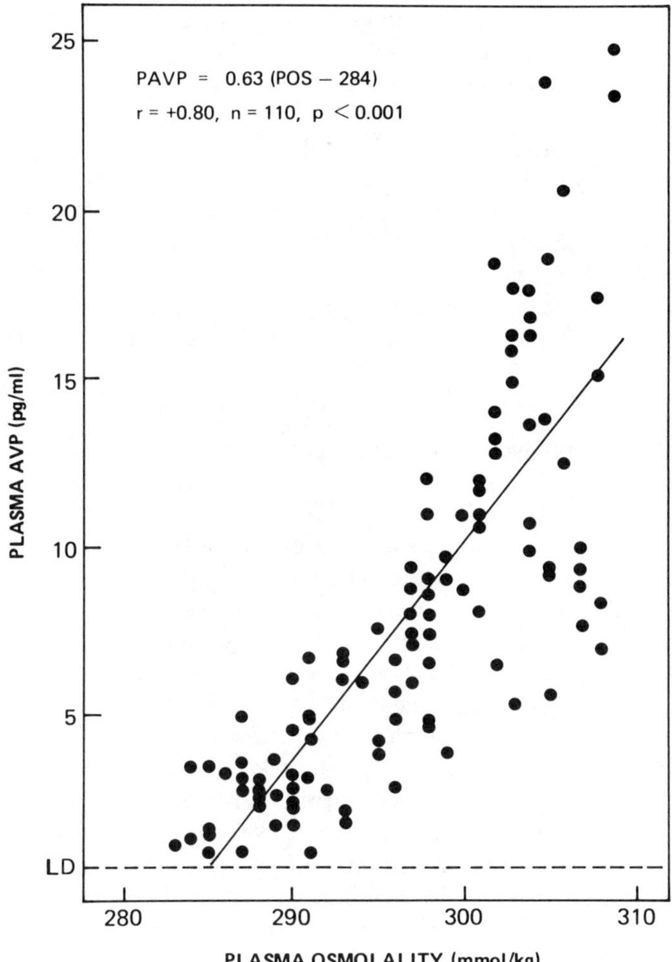

Figure 3   Relationship between plasma arginine vasopressin (AVP) and plasma osmolality during infusion of hypertonic saline in normal subjects. LD represents the limit of detection of the immunoassay. (From Baylis and Robertson, 1980).

Metabolism

The plasma half-life of endogenous AVP is extremely short, of the order of 5-10 min. Total clearance of AVP from plasma is extremely rapid so that changes in plasma AVP concentration are determined principally by changes in hormone secretion rates (Lauson, 1974). The liver and kidneys are the major sites for AVP clearance, renal clearance accounting for 56% of the total AVP clearance, while the liver accounts for about 15%. Plasma itself, except in the pregnant female, possesses little degrading activity, but erythrocytes appear to contain enzymes capable of AVP degradation (Robertson, Roth, et al., 1973). The brain and uterus also metabolize AVP.

Enzymic cleavage of the molecule has been identified at four main sites (Walter and Simmons, 1977). The major site of cleavage is between proline and arginine, at positions 7 and 8, respectively (Figure 1). A second enzyme appears to split the terminal glycinamide from the $Arg-Gly-NH_2$ fragment. Cleavage of the disulfide bond breaks the molecular ring, allowing a fourth enzyme to split the cysteinyl-tyrosyl bond. Many analogs of AVP have how been synthesized, some of which resist N- and C-terminal destruction and consequently have pro-longed plasma half-lives.

Action

In mammals, the major target organ of AVP is the nephron; the collecting tubules are the principal recognized sites of action. The antidiuretic effect depends on the development of a concentration gradient across the tubular cell. The countercurrent system generates a hyperosmolar interstitial renal medulla, and hyposmolar urine enters the distal nephron (Rector, 1977). In the presence of AVP, the permeability of distal tubular cells to water is increased, thus allowing water to move from the tubular lumen into the medullary tissue and consequently concentrating the urine. Permeability to sodium and urea is also influenced by AVP via mechanisms apparently independent of water transport.

The precise intracellular mechanisms involved in AVP-dependent water permeability have yet to be defined. The hormone binds only to the contraluminal tubular surface and activates an adenyl cyclase system to generate cyclic AMP. Phosphorylation of specific protein kinases then occurs, which subsequently may act on the cellular membrane to increase water permeability (Dousa and Barnes, 1977). Other workers, however, have speculated that rearrangement of intracellular microtubules and possibly microfilaments may mediate the action of AVP (Iyengar et al., 1976). Although prostaglandins may modulate the antidiuretic action of AVP at an intracellular level, again little is known about the precise mechanism.

The other actions of AVP, under physiological conditions, are minor in comparison to its renal effect. At high blood concentrations, the hormone contracts vascular smooth muscle, particularly in the splanchnic bed. Recent studies by Hems et al. (1976) have shown that hepatic glycogen phosphorylase activity increases to raise hepatic glucose output after AVP perfusion.

Extremely high concentrations of AVP have been found in the portal veins supplying the anterior pituitary (Zimmerman and Robinson, 1976). The hormone originates from the median eminence (Figure 2) and may act as one of the stimulators of adrenocorticotropin or growth hormone release. Finally, one of

the more controversial roles ascribed to AVP is its consolidating effect on behavior in animals and memory in humans.

MEASUREMENT

Introduction

During the past few decades a considerable body of knowledge about AVP has
accumulated. Its structure, chemical properties, physiological function, and its
role in disturbances of water balance have been defined to some extent. In
studying the function of AVP in health and disease, this hormone has had an
apparent advantage over many other hormones, because a crude estimation of
the hormone's action can be readily assessed by observing changes in urine flow
and concentration. Other hormones do not have such an accessible target organ
as the kidney, the excretions of which are so easily attainable and convenient to
study. In consequence, by observing urinary indices the gross physiology of the
antidiuretic action of AVP was appreciated early in the history of the hormone.
However, although AVP remains a major factor regulating water balance, there
are numerous AVP-independent factors that influence urine diluting and concentrating mechanisms (Berliner and Bennett, 1967). There is now clear evidence
that changes in the distribution of intrarenal blood and glomerular filtration rate
alter urine concentration irrespective of plasma AVP concentration. Even rats
totally deficient in AVP (Brattleboro rat) may generate concentrated urine after
prolonged dehydration. It is therefore unsatisfactory to merely observe changes
in urine flow and concentration to assess systemic AVP concentrations, and
direct methods of hormones measurement are necessary.

Once methods to measure AVP directly were developed there was a sudden
increase in our understanding of the physiological control of secretion of AVP
and of the regulation of water balance. As yet little work has been performed in
applying these techniques to the investigation of clinical problems associated
with disturbances in water regulation. One area of particular diagnostic difficulty may occur in establishing the cause of dilutional hyponatremia in an individual patient. The concept of water intoxication with consequent hyponatremia
caused by persistently elevated plasma AVP and continued drinking was proposed by Schwartz and coworkers in 1957. Other workers have challenged the
concept (Thomas et al., 1978), but precise studies on AVP osmoregulation have
defined various functional abnormalities of AVP secretion in this group of hyponatremic patients (Robertson, 1979). It would appear that some individuals
secrete AVP in a random fashion unrelated to plasma osmolality; others reset
their osmoregulatory line to the left of normal (Figure 4) so that AVP is released
about a lower set point of plasma osmolality. A third group show normal osmoregulation at high plasma osmolalities but fail to switch off AVP release. Finally,

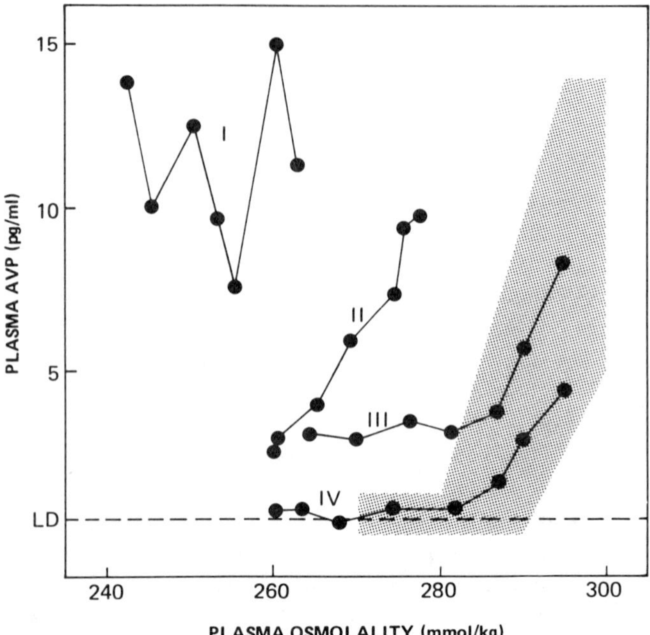

Figure 4   Relationship of plasma arginine vasopressin (AVP) to plasma osmolality during hypertonic saline infusion. Four types of AVP response are noted in patients with the syndrome of inappropriate antidiuresis. (From Robertson et al., 1976. Reprinted from *Kidney International* with permission.)

a few patients appear to have entirely normal osmoregulation despite fulfilling the original criteria for the syndrome of inappropriate antidiuretic hormone secretion (Schwartz et al., 1957). The various patterns of abnormal AVP secretion are related, to some extent, to specific underlying causes. Thus, from both physiological and pathophysiological viewpoints, precise measurement of circulating AVP is fundamental to our understanding of the regulation of water balance in both health and disease.

Indirect urinary methods of assessing the antidiuretic action of AVP are totally inappropriate for the measurement of this hormone in tissues or in body fluids other than circulating blood. Using specific radioimmunoassays, AVP has been detected in urine, cerebrospinal fluid, extrahypothalamic brain tissue, and a variety of neoplasms. The concentration of AVP in some of these fluids or tissues is often extremely low so that highly sensitive methods are required to measure the hormone.

A major obstacle to the direct measurement of AVP, particularly in blood, has been the remarkably low physiological concentration of the hormone. It is now generally agreed that AVP is largely unbound to any plasma proteins and that it circulates in concentrations of the order of 1-10 pg/ml, i.e., approximately 1-10 pmol/liter. This concentration is considerably lower than many other peptide hormones and, in consequence, extremely sensitive assays are required to detect blood AVP levels under physiological conditions. Two assay techniques have been used to solve this problem—one a bioassay, the other a radioimmunoassay—but each has inherent limitations, which will be described below.

Bioassay

Bioassays for this hormone were developed some considerable time before immunoassays. There are basically two bioassay techniques available today, the rat pressor assay and the rat antidiuretic assay. Historically, the pressor assay was developed first. Because it is about 200-500 times less sensitive than the antidiuretic assay it is very limited in its application and cannot be used for the measurement of AVP in plasma.

The pressor assays currently in use are based on the method of Dekanski (1952). The principle is simple. The left common carotid artery of male albino rats, anesthetized with urethane, is catheterized. The catheter is connected to a blood pressure recording device. The sympathetic nervous system is blocked by phenoxybenzamine and the parasympathetic system blocked by bilateral cervical vagotomy. Standard preparations of AVP or samples are injected into a jugular vein, and the pressor effect of the hormone is detected by changes in systemic blood pressure. The sensitivity of a good preparation is of the order of 1-2 nmol AVP per liter, which is approximately 1000 times greater than physiological concentrations of plasma AVP. A further significant limitation to this method concerns the volume of injected solution. Even an injected volume of 0.5 ml may induce a nonspecific pressor response. Despite the report that the use of pithed rats enhances the sensitivity of the assay, the pressor assay is far too insensitive to investigate the function of AVP under normal circumstances.

All the antidiuretic assays for AVP are considerably more sensitive than the pressor assays. The modern bioassays are based upon the original description by Jeffers et al. (1942). Male rats are hydrated with water (50 ml/kg body weight) administered intragastrically by means of a stomach tube and then anesthetized with 12% ethanol (60 ml/kg body weight) orally (Share, 1973). In this way, endogenous secretion of AVP is suppressed by both the water load and the alcohol. A jugular vein is catheterized to administer standard solutions of AVP or samples, and a catheter is positioned in the urinary bladder. The antidiuretic effect of AVP in samples can then be detected by observing changes in urinary flow, urine osmolality, or urine conductivity. Throughout the assay, the animal preparation must be continuously infused via an intragastric tube with ethanol and water to replace urinary losses.

The antidiuretic bioassay is subject to numerous nonspecific factors influencing the antidiuresis. Standards and test solutions should be administered under identical conditions, containing 0.9% NaCl, 0.03% acetic acid, and 0.1% bovine plasma albumin (Share, 1973). Injection of volumes larger than 1 ml may induce nonspecific antidiuresis. Often the response to the first injection of AVP appears to be aberrant. Even in the best hands, this assay can be capricious and frustrating, and there are times when reliable assays cannot be achieved. Many variations to the above procedure have been described in attempts to overcome the difficulties of the assay (Czacskes et al., 1964; Forsling et al., 1968).

The majority of workers find that the sensitivity of the antidiuretic assay is of the order of 5-10 pmol AVP per liter. Although this is very close to apparent normal physiological blood concentrations, greater sensitivity is required to confidently and accurately measure blood levels. Greater sensitivity may be attained by the use of younger rats (Forsling et al., 1968), by using rats with hereditary deficiency of AVP (Sawyer and Valtin, 1967), or by the continuous infusion of low concentrations of AVP during the assay (Forsling et al., 1967).

The other logical solution to increasing the sensitivity of the assay is to concentrate the samples by extraction. Many workers extract their plasma samples and, in consequence, claim that they can detect concentrations of AVP as low as 1 pmol/liter in plasma. Two basic techniques have been employed to extract AVP from plasma and other body fluids. One involves precipitation of proteins by acetone, ammonium sulfate, or trichloracetic acid, and the other involves the absorption of peptides, including AVP, onto glass beads with subsequent elution using aqueous acetone.

Radioimmunoassay

Although the principles of immunoassay were described over 20 years ago, it was not until about 10 years ago that assays were developed that could reliably measure AVP in body fluids. Two major problems were encountered that hindered progress. Antisera of high affinity were required to detect the low concentrations of AVP in body fluids. Even when this had been achieved, considerable difficulties were discovered in applying the assay to the measurement of plasma AVP because of nonspecific interference by large molecular weight proteins.

In spite of its low molecular weight, AVP may be rendered antigenic in mammalian species under certain circumstances. Some workers have successfully produced antisera to AVP by immunizing rabbits with natural or synthetic AVP or LVP emulsified in Freund's complete adjuvant (Roth et al., 1966; Wu and Rockey, 1969), while others have conjugated the nonapeptide to bovine or

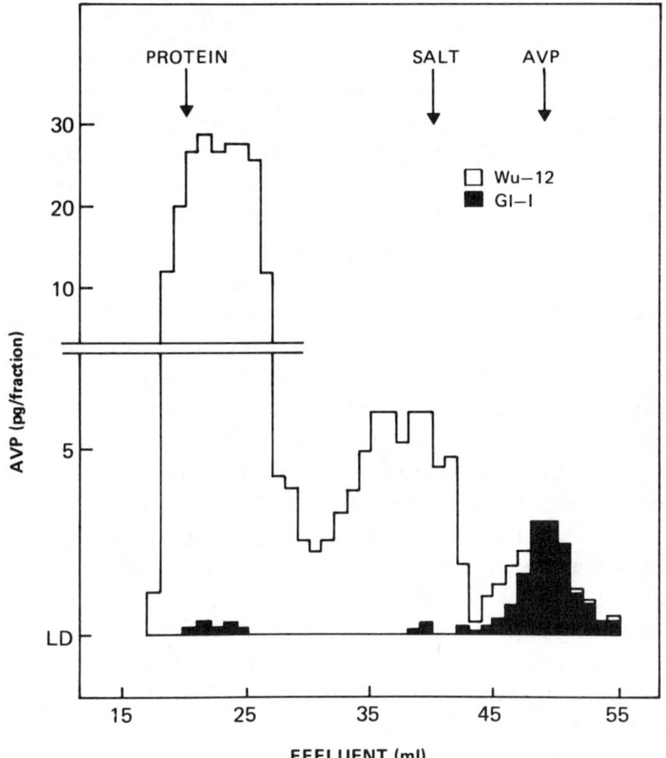

Figure 5   Immunoreactivity as detected by two arginine vasopressin antisera in eluates of plasma filtered on Sephadex G-25. (From Robertson , Roth, et al. 1973.)

human serum albumin using carbodiimide or glutaraldehyde as coupling agents (Permutt et al., 1966; Oyama et al., 1971; Skowsky and Fisher, 1972). Although antisera can be produced quite readily, most are not suitable for use in the measurement of AVP in human plasma, since a suitable antiserum must have very high affinity for AVP to detect the extremely low circulating levels in the human. Nevertheless, numerous groups around the world have now succeeded in raising antisera of sufficiently high affinity to measure AVP in human plasma (Robertson, Mahn, et al., 1973; Hayward et al., 1976; Thomas and Lee, 1976; Baylis and Heath, 1977).

Antisera to AVP vary in their susceptibility to nonspecific interference. This phenomenon was elegantly demonstrated by Robertson and coworkers in 1973 (Figure 5) (Robertson, Roth, et al., 1973). Plasma was chromatographed on a column of Sephadex G-25 and immunoreactive AVP was measured by two different antisera. One antiserum, Wu-12, apparently detected AVP immunoreactivity in fractions associated with the void and salt volumes as well as in the fractions known to contain AVP by the chromatography of synthetic AVP. The G1-I antiserum detected very little AVP immunoreactivity in the first two sets of fractions, suggesting that Wu-12 antiserum was more susceptible to interference by protein and salt. It must be realized that physiological changes in plasma AVP concentration are reflected by changes in the quantity of the AVP immunoreactivity only in fractions after the salt volume and not in any other fractions. This interference can be overcome, to some extent, by extracting all samples to reduce substantial quantities of protein and salt. However, because the readily available extraction methods fail to remove all proteins and salt, antisera that show little interference are clearly preferable. Many workers describing AVP immunoassays neglect to report their studies on interferences, which may explain in part some of the relatively high values of plasma AVP obtained by a few groups (reviewed by Edwards, 1979). Two methods are available to remove protein from plasma. It may be precipitated by the addition of cold acetone (Robertson, Mahr, et al., 1973), or the small peptides may be absorbed onto glass beads, so that the remaining plasma can be discarded (Beardwell, 1971). An obvious further advantage of extracting the plasma is the potential to concentrate the peptides. When antiserum of relatively low affinity is used, the sample has been as large as 50 ml (Beardwell, 1971), but the recently described immunoassays now require only 1-2 ml plasma (Hayward et al., 1976; Robertson, Mahr, et al., 1973).

Other aspects of the AVP immunoassay present no particular problems. Iodination of AVP is commonly performed by the chloramine-T method (Greenwood et al., 1964). To avoid reduction of the disulfide bond within the AVP molecule, some groups prefer to halt the iodination reaction by the addition of serum albumin instead of metabisulfite (Robertson, Roth, et al., 1973). The labeled AVP can be readily purified by fractionation on Sephadex G-25 to give monoiodinated AVP.

Despite these problems, the radioimmunoassay of AVP is considerably more sensitive than the classic bioassay. The best immunoassays can detect plasma AVP concentrations of the order of 0.5-1 pg/ml. These levels are believed to be at the lower end of the normal physiological range. Nonspecific interference remains a potential hazard. And one caveat still needs emphasis: The AVP radioimmunoassay requires continuous attention, is exceedingly fussy, and is not as robust as the immunoassays of many other hormones. Finally, like all other immunoassays, the immunological activity may differ substantially from the biological activity of the hormone.

THE CYTOCHEMICAL BIOASSAY

From the above, the reader can readily appreciate some of the problems of measuring AVP, particularly in plasma. Because AVP concentrations are so low, bioassays and immunoassays have been extended to their limits of sensitivity. Both techniques have a problem with nonspecific interference, and interassay variation remains a particular difficulty with antidiuretic bioassays. Some workers have attempted to circumvent the problems associated with plasma measurements by turning to other fluids. Measurement of AVP in urine is feasible and relatively easy by immunoassay (Miller and Moses, 1971), but urinary AVP excretion is a poor reflection of plasma hormone concentrations and, under some circumstances, provides an erroneous impression of AVP physiology and pathophysiology (reviewed by Robertson, 1977).

Another approach to the measurement of this nonapeptide is required. Cytochemical bioassay techniques have become available over the past decade (Chayen et al., 1976). This technique offers attractive advantages over the more classic methods of AVP measurement. It appears to be an extremely sensitive method, and therefore might overcome one of the major difficulties encountered with the current assays. Furthermore, it might dispense with the need to extract samples. Biological specificity, in contrast to immunological specificity, would be defined.

The Enzyme Reaction

A hormone shows its activity by causing a specific effect on target cells, which is mediated by a sequence of intracellular biological changes. The result is recognized as a specific change in cell function, for example, the alteration in the synthesis of a chemical or protein. The kidney is the most important target tissue for AVP, and therefore it is probable that the hormone has considerable effects on biochemical reactions in some cells of the nephron. It therefore seemed appropriate to investigate the effect of AVP on the enzyme activities of the renal medulla. Preliminary studies in waterloaded and dehydrated rats (i.e., rats with low and high circulating AVP, respectively) indicated that considerable changes in the activities of a number of enzymes in the rat renal medulla had occurred following these maneuvers. One of these, sodium-potassium dependent, magnesium-activated adenosine triphosphatase, (Na^+-K^+)-ATPase, was studied in detail since considerable changes in the activity of this enzyme were detected in the thick ascending limb of the loop of Henle.

Several workers have sought a cytochemical method to demonstrate the activity of this enzyme. The original methods were based on lead ions trapping the phosphate liberated by ATPase activity (Wachstein and Meisel, 1957). Unfortunately, the lead ions caused nonenzymatic hydrolysis of the enzyme substrate, ATP, and also inhibited (Na^+-K^+)-ATPase activity. Consequently, this method fell into disrepute (Beeuwkes and Rosen, 1975). Alternative techniques to demonstrate ATPase activity were devised. Strontium was used as the trapping agent instead of lead and ATP was substituted by a p-nitrophenyl phosphate as the substrate for the ATPase (Ernst, 1972). Other modifications have been made by Guth and Albers (1974). A major advance in the ability to detect changes in (Na^+-K^+)-ATPase activity has come with the preparation of a new stable lead complex in which the lead atom is surrounded by an ammonium citrate-acetate complex. The complex allows free phosphate to approach and react with the central lead atom while at the same time it prevents the lead from hydrolyzing ATP and inhibiting the ATPase (Chayen et al., 1981).

The reaction method that has been developed is based upon the original description by Schwartz et al. (1971), but the concentrations of reagents have been increased fourfold to overcome the low diffusibility of the chemicals in the stabliziler, Polypep 5115. The constituents of the reaction medium, prepared in a 40% solution (w/v) of Polypep 5115 (Sigma) in 200 mmol/liter Tris buffer with 1 mmol/liter sodium acetate, pH 7.4, were as follows: sodium chloride (410 mmol/liter, 24 mg/ml), potassium chloride (37.5 mmol/liter, 2.8 mg/ml), magnesium chloride (20 mmol/liter, 4 mg/ml), ATP (16.5 mmol/liter, 10 mg/ml). Lead ammonium citrate-acetate was dissolved in a small volume of dilute ammonia before being added to the above solution to give a final concentration of 32 mg/ml. The pH was adjusted to 7.5 by the addition of sodium and potassium hydroxides in the ratio of 10:1 (2.5 mmol/liter) or 0.1 mol/liter hydrochloric acid. The (Na^+-K^+)-ATPase activity, defined as ouabain-sensitive ATPase activity, was determined by the difference between the activities of total and ouabain-insensitive ATPase. The reaction medium for the latter required the addition of ouabain octahydrate (400 mmol/liter, 0.3 mg/ml).

The procedure of the preparation of tissue for reaction follows that described by Chayen et al. (1976). Wistar rats were killed by nitrogen asphyxiation. The kidneys were removed immediately and cut into segments which were chilled to $-70°C$ in n-hexane (free from aromatic hydrocarbons, boiling range $67-70°C$). The segments were mounted on chucks ready for sectioning at 10 μm in a Bright's cryostat with a cabinet temperature of $-25°C$. The sections were taken off the cutting blade onto glass slides, ready to measure ATPase activity.

The enzyme reaction was performed at $37°C$. Small Perspex rings were placed on the slides to surround each section completely. The reaction medium was poured into the rings and allowed to remain for 15 min. For alternate sections, reaction medium containing ouabain was required. Thereafter, the sections were washed in 200 mmol/liter Tris buffer, pH 7.4, several times and finally rinsed with distilled water. The sections were removed from the $37°C$ environment, to be immersed, at room temperature, in water that had been saturated with hydrogen sulfide. Final rinsing was with distilled water. The sections were dried with blotting paper and mounted in Farrants' medium.

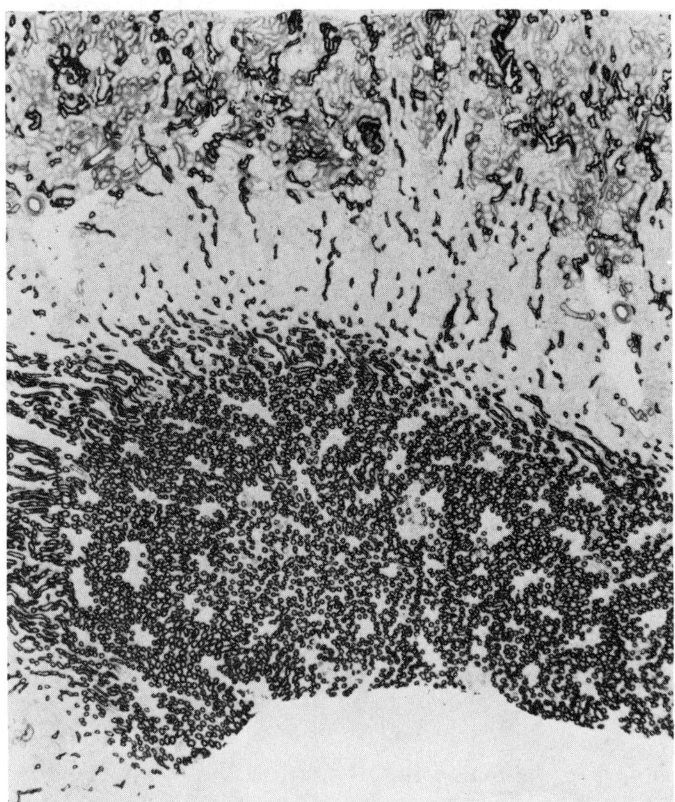

Figure 6   Distribution of total ATPase activity in a rat kidney section (×21).

In later studies (and for Figure 6), the sections were immersed for 5 min in 40% Polypep 5115 in the Tris buffer containing 100 mmol/liter of potassium acetate, pH 7.5, to remove free phosphate from the sections. They were then reacted in the full reaction medium (as above). To inhibit alkaline phosphatase activity, L-p-bromotetramisole oxalate (0.4 mmol/liter; Aldrich) was added to both the reaction media (media with and without ouabain).

Figure 6 shows the distribution of ATPase activity in a section from a rat kidney. Total ATPase activity was distributed throughout the kidney, with highest activities in the proximal convoluted tubules, the medullary thick ascending limbs of the loop of Henle, and the distal convoluted tubules. Using a microchemical assay method, Katz et al. (1979) reported similar distribution of ATPase activity in nephrons from rat, mouse, and rabbit.

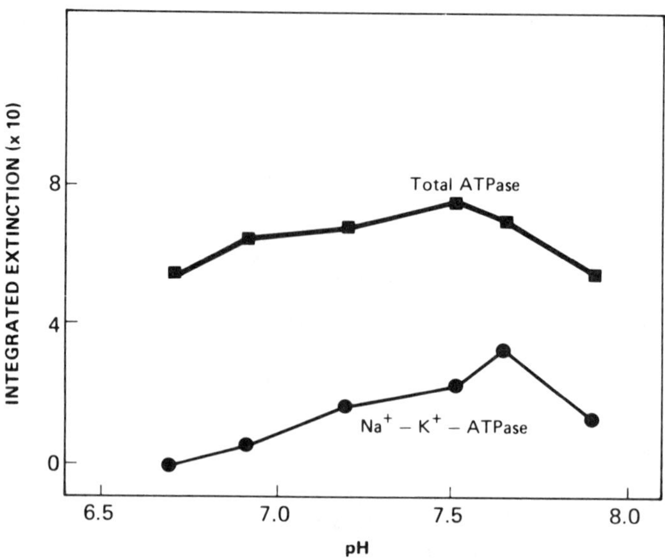

Figure 7 The effect of pH changes on the activity of total and ouabain-sensitive (Na^+-K^+)-ATPases in the rat kidney, expressed as the mean integrated extinction recorded in 20 renal tubular cells of the ascending limb of Henle's loop.

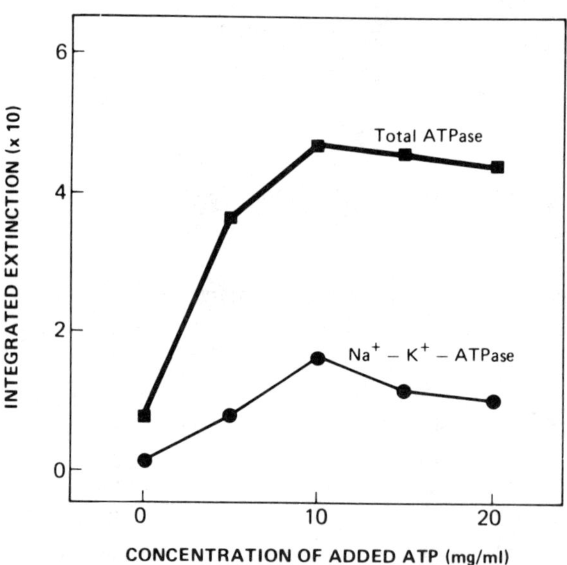

Figure 8 The effect of adenosine triphosphate concentration on the activity of total and ouabain-sensitive (Na^+-K^+)-ATPases in the ascending limb of Henle's loop of the rat.

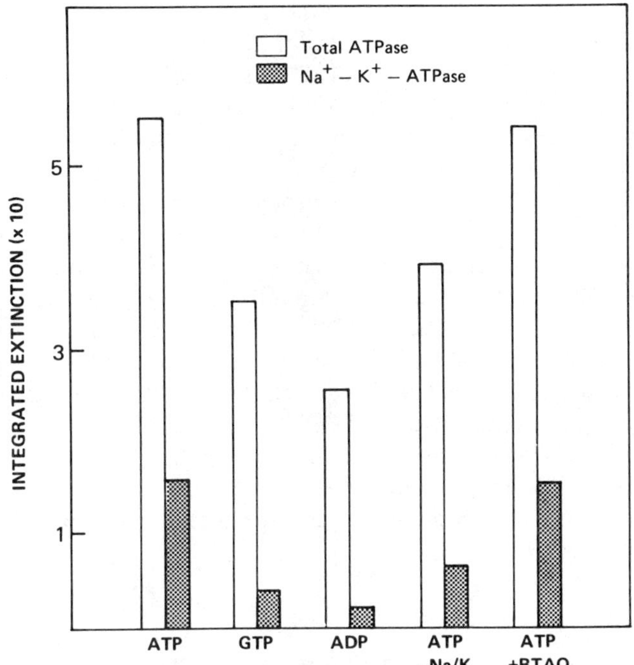

Figure 9   The activities of total and ouabain-sensitive (Na^+-K^+)-ATPases for different substrates in the ascending limb of Henle's loop of the rat. ATP, GTP, ADP, and BTAO represent adenosine triphosphate, guanosine triphosphate, adenosine diphosphate, and bromotetramisole, respectively.

The reaction product, which is presumed to be lead sulfide, was quantified by means of a scanning and integrating microdensitometer (Vickers M85, at 550 nm) expressing the enzyme activity as mean integrated extinction (Chayen et al., 1976). In so doing, a precise measure was made of the enzyme activity.

Optimal pH for the ouabain-sensitive reaction in the thick ascending limb was 7.65 (Figure 7). The addition of ATP at varying concentrations (0-20 mg/ml) defined the peak activity of total and (Na^+-K^+)-ATPase at an ATP concentration of 10 mg/ml in this region of the nephron (Figure 8). Increases in concentration of the trapping agent, lead ammonium citrate-acetate , up to 32 mg/ml, caused an increase in enzyme activity in the thick ascending limb. Thereafter, enzyme activity rose no further and, furthermore, it was not inhibited by higher trapping-agent concentrations. ATPase activity was linear with section thickness over the range 5-20 μm, and with time up to 20 min. ATPase activity remained stable for at least 48 hr.

The new method for measuring (Na^+-K^+)-ATPase activity is also specific.
Figure 9 shows the results of a series of experiments designed to define its specificity in the thick ascending limb. Substitution of ATP by trisodium guanosine
5′-phosphate (17 mmol/liter) or by adenosine diphosphate (16.5 mmol/liter)
stimulated little activity. Removal of sodium and potassium chlorides from the
reaction mixture depressed enzyme activity; the residual ouabain-sensitive
(Na^+-K^+)-ATPase activity was probably due to endogenous ions remaining in
the tissue sections and to sodium and potassium hydroxides required to adjust
the pH of the reaction medium. The alkaline phsophatase inhibitor, L-p-bromotetramisole oxalate (400 mmol/liter) did not affect the APTase activity. It would
therefore appear that the reaction is specific for ATP. The proportion of ATPase
activity that can be attributed to (Na^+-K^+)-ATPase along the nephron, as measured by this method (Chayen et al., 1981), is in broad agreement with the values
obtained by microchemical assay (Katz et al., 1979).

With the development of a reliable, sensitive, and specific method to measure
ouabain-sensitive (Na^+-K^+)-ATPase activity in the nephron, it may be possible
to quantify the effect of a number of hormones on this enzyme. One hormone
that has been investigated is AVP.

Assay of AVP

The method of the cytochemical bioassay of AVP is based upon the changes in
activity of (Na^+-K^+)-ATPase in the thick ascending limb of the loop of Henle in
the rat kidney, and its development rests solely on the enzyme reaction
described in the preceding section. The technical procedure follows the general
method of cytochemical assay described in detail by Chayen et al. (1976). Only
brief details of the AVP assay will be given.

Female Wistar rats (250-300 g) were allowed free access to food and water
before they were killed by asphyxiation with nitrogen. The following procedures
were performed at $37°C$. After removal of the kidneys, the renal capsule was
stripped and each kidney was bisected longitudinally. Three segments were cut
from each half of the kidney in such a way that the segments included cortical
and medullary tissue. Each segment was placed on a defatted lens tissue which
lay upon a metal grid in a vitreosil dish containing Trowell's T8 medium, pH
7.6. The cortical surface of the segment was placed on the lens tissue. The dish
was placed in a culture pot. The air was displaced by a mixture of gases, 95%
oxygen and 5%carbon dioxide, and the pots were sealed with lanolin. Kidney
segments remained in culture at $37°C$ for 5 hr.

After maintenance culture, the vitreosil dishes were removed from the culture
pots and the medium was discarded. Solutions of synthetic AVP (400 μ/mg) or
plasma diluted suitably with Trowell's T8 medium were poured carefully over
the segments. These solutions were allowed to remain in contact with the

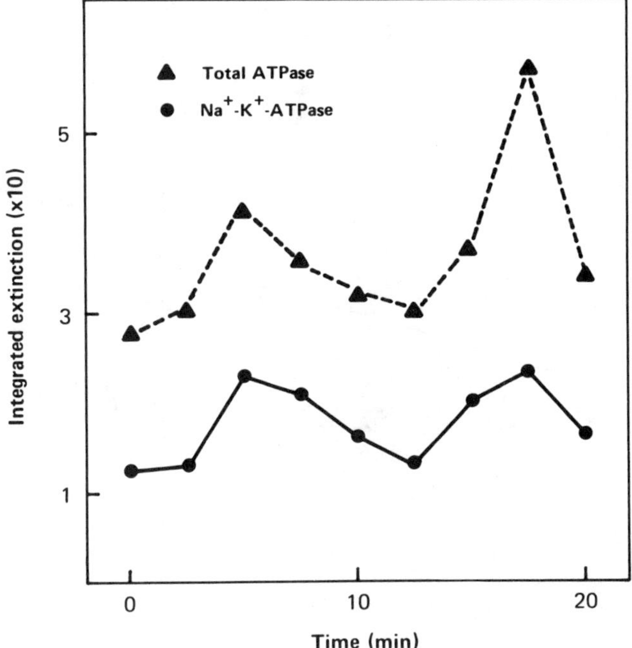

Figure 10   The time course of total and ouabain-sensitive (Na$^+$-K$^+$)-ATPase response exposed to synthetic arginine vasopressin (0.02 pg/ml), in the rat ascending limb of Henle's loop.

segments for specific time periods, before the segments were chilled in n-hexane and prepared for sectioning and reacting as described above.

The activities of total and ouabain-sensitive (Na$^+$-K$^+$)-ATPase in the thick ascending limb, which were stimulated by AVP diluted in Trowell's T8 medium to 0.02 pg/ml, varied with respect to time (Figure 10). Two peak enzyme activities were observed, one following 5 min and the other following 17.5 min exposure to the hormone. The response of the two ATPases ran parallel to each other. Alteration of the AVP concentration within the range 0.002-2.0 pg/ml had no significant effect on the timing of the peak ATPase response. Addition of Trowell's T8 medium alone to cultured segments had no effect on ATPase activity.

Arginine vasopressin gave a linear log dose response over the range 0.002-2.0 pg/ml after exposure to the hormone for 17.5 min (Figure 11). Concentrations of AVP greater than 2.0 pg/ml caused inhibition of both total and (Na$^+$-K$^+$)-ATPase activities, and concentrations less than 0.002 pg/ml failed to further

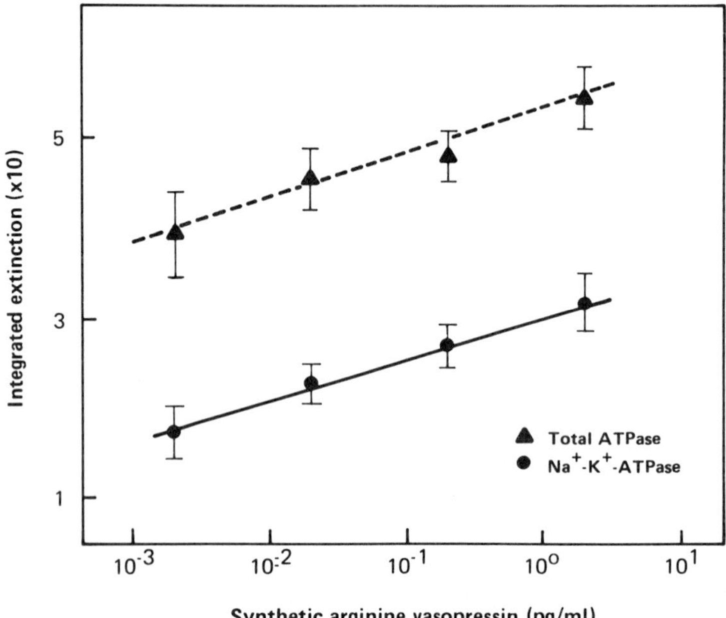

Figure 11 The dose response of total and ouabain-sensitive $(Na^+$-$K^+)$-ATPase activities in the rat renal tubular cells of the ascending limb of Henle's loop induced by increasing concentration of synthetic arginine vasopressin (mean ± SD).

lower either enzyme activity. The correlation coefficient for the points on the $(Na^+$-$K^+)$-ATPase line was +0.99 and for the total ATPase was +0.97. The inter-assay coefficient of variation of the individual concentrations of AVP ranged from 6.3 to 14.4% for $(Na^+$-$K^+)$-ATPase and 4.7 to 16.9% for total ATPase activities.

These observations indicate that there is an AVP-sensitive ATPase located in the rat renal medulla. Although the collecting duct is believed to be the main target site of AVP in the nephron, this region has very little ATPase activity, but the thick ascending limb of Henle's loop does have substantial $(Na^+$-$K^+)$-ATPase activity (Katz et al., 1979). It is possible, however, that AVP may be acting on sites other than the collecting duct. The morphology of the cells in the renal medulla, which show AVP-sensitive ATPase activity, is characteristic of the thick ascending limb of the loop of Henle. Studies by Imbert-Teboul et al. (1978) demonstrated two regions in the rat nephron that had membrane-associated adenyl cyclase activities sensitive to AVP. The collecting duct was one site, but

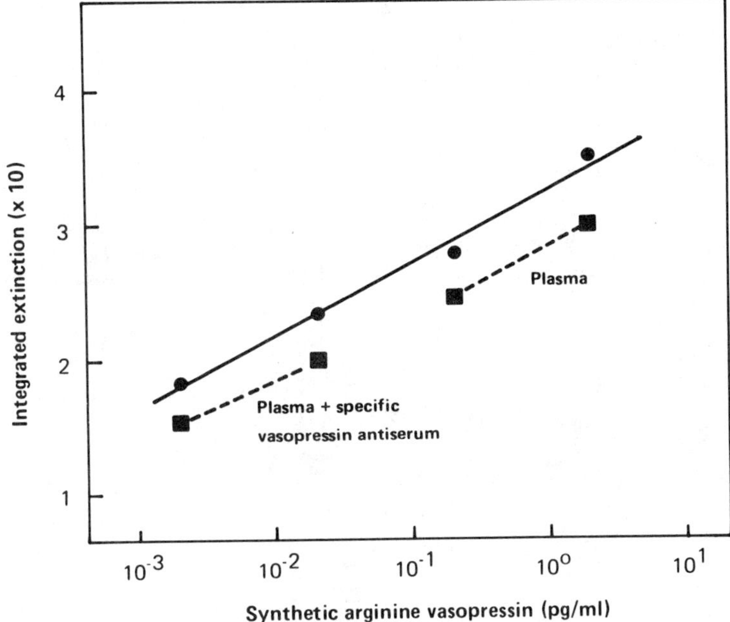

Figure 12   The response of ouabain-sensitive $(Na^+\text{-}K^+)$-ATPase activity to human plasma (filled squares) at dilutions 1:20 and 1:200 gave values parallel to the standard curve (closed circles). Addition of specific arginine vasopressin antiserum to plasma reduced enzyme activity.

significant quantities of adenyl cyclase activity were also observed in the thick ascending limb of Henle's loop. AVP-sensitive adenyl cyclase was not found elsewhere in the nephron. Recent studies have confirmed the presence of an AVP-sensitive adenyl cyclase in the thick ascending limb of the rat nephron (Jackson et al., 1980).

From a physiological point of view, there is considerable logic for a $(Na^+\text{-}K^+)$-ATPase located in the thick ascending limb to be responsive to changes in AVP concentration. This part of the nephron is impermeable to water. Therefore, the pumping of sodium ions from the lumen of the nephron into the renal interstitial tissue by $(Na^+\text{-}K^+)$-ATPase stimulated by AVP will not be accompanied by water, and consequently the tonicity of the interstitial tissue will increase. This, in turn, will raise the osmotic gradient across the collecting duct, which is dependent upon hypotonic urine within the duct and the hypertonic interstitial renal tissue. This sequence of events would augment the action of AVP on the

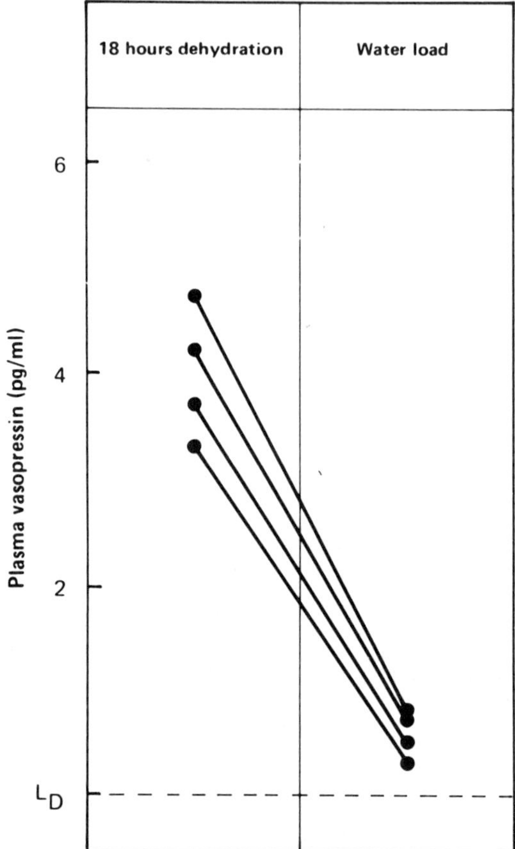

Figure 13   Results of plasma arginine vasopressin, measured by changes in ouabain-sensitive (Na^+-K^+)-ATPase activity in the rat ascending limb of Henle's loop, obtained from four normal subjects after 18 hr dehydration and following a water load of 20 ml per kilogram body weight.

collecting tubule by increasing the osmotic gradient at a time when this tubule becomes more permeable to water under the direct action of AVP.

Physiological Validation of the Bioassay

The addition of plasma from a dehydrated healthy individual, which had been diluted 1:20 and 1:200 with Trowell's T8 medium, to cultured renal segments gave (Na^+-K^+)-ATPase responses parallel to the standard curve (Figure 12).

Antibody specific to AVP (Baylis and Heath, 1977), when added to an aliquot of the same plasma, reduced the effect of AVP on (Na^+-K^+)-ATPase activity by over 99%. The absolute value of plasma AVP concentration fell from 6.7 to 0.05 pg/ml after addition of antibody (Figure 12).

Physiological validation of the cytochemical bioassay of AVP can be obtained by observing the changes in plasma AVP concentration following fluid restriction and the ingestion of a water load. Such maneuvers will cause marked changes in plasma osmolality, which, in turn, alter the secretion of AVP. The results of plasma AVP concentration from four healthy individuals who were fluid restricted for 16 hr and then drank 20 ml of water per kilogram body weight are shown in Figure 13. All subjects had a considerable fall in plasma AVP concentration after the oral water load. The fall in plasma osmolality from 293 ± 3 (mean ± SE) to 284 ± 2 mmol/kg was associated with a mean reduction in plasma AVP from 4.0 ± 0.3 to 0.6 ± 0.1 pg/ml. Immunoreactive plasma AVP was measured in the same plasma samples by a well-established radioimmunoassay (Baylis and Heath, 1977); plasma immunoreactive AVP in the dehydrated subjects was 4.5 ± 0.5 pg/ml, and fell to 1.7 ± 0.2 pg/ml after the water load.

Although there is some discrepancy between the plasma AVP concentrations measured by radioimmunoassay and cytochemical assay, particularly in the plasma samples from hydrated subjects, it is considerably less than has been reported with some other hormones. Fenton et al. (1978) demonstrated a substantial difference in the values of serum parathyroid hormone obtained by immunoassay and cytochemical bioassay. The major discrepancy in the AVP results appears to be in the samples from the hypotonic subjects. It is probable that the immunoassay is detecting either inactive fragments of the hormone or that it is subject to some degree of nonspecific interference as previously described (Figure 6).

Application of the AVP Bioassay

Although only preliminary studies on the cytochemical bioassay of AVP have so far been completed, and, indeed, further validation studies are required, there is considerable evidence that AVP can be measured using cytochemical techniques. The method is exquisitely sensitive and able to detect concentrations of AVP far beyond the capability of any described immunoassay or classic bioassay. It has also overcome the need to extract plasma samples, which is a problem that has bedeviled the majority of immunoassays. In its present form, the AVP cytochmical assay is time consuming and technically difficult and can only measure a few samples in each assay. With regard to the latter point, the immunoassay has an advantage over the current cytochemical technique. However, by changing from a segment to a section assay in which the hormone or plasma solutions are allowed to react with sections instead of segments of renal tissue, a greater number of samples could be handled in one assay.

The cytochemical bioassay of AVP is ideally suited to the investigation of physiological and pathophysiological situations in which plasma AVP may be low. An excellent example would be the study of the role of AVP in hyponatremic states. There is considerable controversy about the importance of the hormone in the pathogenesis of some types of hyponatremia (Thomas et al., 1978). In addition, some physiological questions have remained unanswered because techniques to reliably measure low plasma AVP concentrations have been unavailable. Does the secretion of AVP stop with decreasing plasma osmolality? Is there pulsatile release of the hormone under physiological conditions? Are there AVP fragments which are immunologically active but have no biological actions? The answers to these questions can now be sought.

A further application of the cytochemical assay will be in multiple sample experiments, particularly those using small animals. At present, the smallest sample size in the best AVP immunoassays and classic bioassays is 1 ml. In contrast, with the cytochemical bioassay, since samples have to be diluted by a factor of 10 or 100 so that their effects on $(Na^+\text{-}K^+)$-ATPase activity can be related to the AVP dose-response curve, it will be possible to assay AVP in 10-100 μl aliquots.

Immunoassay of any hormone lacks biological specificity of the hormone. The AVP cytochemical assay confers biological specificity on the measurements. This should be extremely helpful in defining the nature of AVP and related peptides in tissues and in tumors suspected of synthesizing and releasing AVP. Indeed, the scope of a sensitive, biologically specific assay for AVP is immense.

CONCLUSIONS

Since the beginning of the twentieth century, physiologists have realized the existence of a specific antidiuretic factor secreted by the posterior lobe of the pituitary gland, which had its principal action on the kidney. In most mammals, this factor in a nonapeptide called arginine vasopressin. Its gross physiological effects were studied with ease by observing changes in urine concentrations in response to the hormone. As with many other hormones, significant advances in knowledge came only with improved methods of detection. Classic bioassays and radioimmunoassays of AVP exhibit numerous difficulties. Nonspecific interference, poor sensitivity, and bad reproducibility are some of the factors that have troubled many workers. Although extraction of samples overcomes some of the problems, it is only a partial solution.

Preliminary studies on the cytochemical bioassay of AVP strongly suggest that these problems may now be solved. A new method to detect changes in $(NA^+\text{-}K^+)$-ATPase activity has been developed, and forms the basis of the assay.

An AVP-sensitive (Na$^+$-K$^+$)-ATPase has been located in the thick ascending limb on Henle's loop in the rat nephron. It appears to be specific for AVP because antibody to the hormone reduces the effect of AVP on the enzyme activity by 99%. Absolute plasma AVP concentrations measured cytochemically are similar to those measured by radioimmunoassay. Changes in plasma osmolality cause the anticipated changes in plasma AVP measured by cytochemical bioassay. The new assay is more sensitive than any other AVP assay by a factor of 100. Although further work is needed to confirm the method's specificity for AVP, results so far suggest that the new cytochemical bioassay for AVP will be of great value in the investigation of vasopressin function in health and disease.

REFERENCES

Acher, R. (1974). Chemistry of the neurohypophysial hormones: An example of molecular evolution. In *Handbook of Physiology,* Section 7, *Endocrinology,* Vol. 4, Pt. 1, E. Knobil and W. H. Sawyer (eds.). American Physiology Society, Bethesda, Maryland, pp. 119-130.

Acher, R., and Chauvet, J. (1954). La structure de la vasopressine de boeuf. *Biochim. Biophys. Acta,* 14:421-429.

Baratz, R. A., and Ingraham, R. C. (1959). Sensitive bioassay method for measuring antidiuretic hormone in mammalian plasma. *Proc. Soc. Exp. Biol. Med.,* 100:296-299.

Baylis, P. H., and Heath, D. A. (1977). The development of a radioimmunoassay for measurement of human plasma arginine vasopressin. *Clin. Endocrinol. (Oxf),* 7:91-102.

Baylis, P. H., and Robertson, G. L. (1980). Vasopressin response to hypertonic saline infusion to assess posterior pituitary function. *J. Roy. Soc. Med.,* 73: 255-260.

Beardwell, C. G. (1971). Radioimmunoassay of arginine vasopressin in human plasma. *J. Clin. Endocrinol. Metab.,* 33:254-260.

Beeuwkes, R., and Rosen, S. (1975). Renal sodium-potassium adenosine triphosphatase optical localization and x-ray microanalysis. *J. Histochem. Cytochem.,* 23:828-839.

Berliner, R. W., and Bennett, C. M. (1967). Concentration of urine in the mammalian kidney. *Am. J. Med.,* 42:777-789.

Bernard, C. (1865). *Introduction à l'étûde de la medicine expérimentale.* Bailliere, Paris.

Bie, P. (1980). Osmoreceptors, vasopressin, and control of renal water excretion. *Physiol. Rev.,* 60:961-1048.

Chayen, J., Daly, J. R., Loveridge, N., and Bitensky, L. (1976). The cytochemical bioassay of hormones. *Recent Prog. Hormone Res.,* 32:33-79.

Chayen, J., Frost, G. T. B., Dodds, R. A., Bitensky, L., Pitchfork, J., Baylis, P. H., and Barrnett, R. J. (1981). The use of a hidden metal-capture reagent for the measurement of Na^+-K^+-ATPase: A new concept in cytochemistry. *Histochemistry*, 71:533-541.

Czaczkes, J. W., Kleeman, C. R., and Koenig, M. (1964). Physiologic studies of antidiuretic hormone by its direct measurement in human plasma. *J. Clin. Invest.*, 43:1625-1640.

Dekanski, J. (1952). The quantitative assay of vasopressin. *Br. J. Pharmacol.*, 7:567-572.

Dousa, T. P., and Barnes, L. D. (1977). Regulation of protein kinase by vasopressin in renal medulla in situ. *Am. J. Physiol.*, 232:F50-F57.

du Vigneaud, V., Gish, D. T., Katsoyannis, P. G., and Hess, G. P. (1958). Synthesis of the pressor-antidiuretic hormone, arginine vasopressin. *J. Am. Chem. Soc.*, 80:3355-3358.

Edwards, C. R. W. (1979). Vasopressin. In *Hormones in Blood*, C. H. Gray and V. H. T. James (eds.). Academic, London, pp. 423-450.

Ernst, S. A. (1972). Transport adenosine triphosphatase cytochemistry. I. Biochemical characterization of a cytochemical medium for the ultrastructural localization of ouabain-sensitive, potassium-dependent phosphatase activity in the avian salt gland. *J. Histochem. Cytochem.*, 20:13-22.

Fenton, S., Somers, S., and Heath, D. A. (1978). Preliminary studies with a sensitive cytochemical assay for parathyroid hormone. *Clin. Endocrinol. (Oxf)*, 9:381-384.

Forsling, M. L. (1979). Chemistry and structure-activity relationships. In *Antidiuretic Hormone*, Vol. 3, M. L. Forsling (ed.). Churchill Livingstone, Edinburgh, pp. 14-21.

Forsling, M. L., Jones, J. J., and Lee, J. (1967). Intravenous infusion of vasopressin to increase the sensitivity of its assay in the ethanol, water-loaded rat. *Nature (Lond.)*, 215:433-434.

Forsling, M. L., Jones, J. J., and Lee, J. (1968). Factors influencing the sensitivity of the rat to vasopressin. *J. Physiol. (Lond.)*, 196:495-505.

Greenwood, F. C., Hunter, W. M., and Glover, J. S. (1963). The preparation of 131I-labelled human growth hormone of high specific radioactivity. *Biochem. J.*, 89:114-123.

Guth, L., and Albers, R. W. (1974). Histochemical demonstration of $(Na^+$-$K^+)$-activated adenosine triphosphatase. *J. Histochem. Cytochem.*, 22:320-326.

Hayward, J. N., Pavasuthipaisit, K., Perez-Lopez, F. R., and Sofroniew, M. V. (1976). Radioimmunoassay of arginine vasopressin in Rhesus monkey plasma. *Endocrinology*, 98:975-981.

Hems, D. A., Rodrigues, L. M., and Whitton, P. D. (1976). Glycogen phosphorylase, glucose output and vasoconstriction in the perfused rat liver. *Biochem. J.*, 160:367-374.

Imbert-Teboul, M., Charbardès, D., Montégut, M., Clique, A., and Morel, F. (1978). Vasopressin-dependent adenyl cyclase activities in the rat kidney medulla: evidence for two separate sites of action. *Endocrinology*, 102:1254-1261.

Iyengar, R., Lepper, M. G., and Mailman, D. S. (1976). Involvement of microtubules and microfilaments in the action of vasopressin in the canine renal medulla. *J. Supramol. Struct.*, 5:521-530.

Jackson, B. A., Edwards, R. M., and Dousa, T. P. (1980). Lithium-induced polyuria: Effect of lithium on adenylate cyclase and adenosine $3',5'$-monophosphate phosphodiesterase in medullary ascending limb of Henle's loop and medullary collecting tubules. *Endocrinology*, 107:1693-1698.

Jeffers, W. A., Livezey, M. M., and Austin, J. H. (1942). A method for demonstrating an antidiuretic action of minute amounts of pitressin: Statistical analysis of results. *Proc. Soc. Exp. Biol. Med.*, 50:184-188.

Jewell, P. A., and Verney, E. B. (1957). An experimental attempt to determine the site of the neurohypophyseal osmoreceptors in the dog. *Philos. Trans. R. Soc. Lond. [Biol.]* 240:197-324.

Katz, A. I., Doucet, A., and Morel, F. (1979). Na^+-K^+-ATPase activity along the rabbit, rat, and mouse nephron. *Am. J. Physiol.*, 237:F114-F120.

Lauson, H. D. (1974). Metabolism of the neurohypophysial hormones. In *Handbook of Physiology*, Section 7, *Endocrinology*, Vol. 4, Pt. 1, E. Knobil and W. H. Sawyer (eds.). American Physiological Society, Bethesda, Maryland, pp. 287-393.

Magnus, R., and Schäfer, E. A. (1901). The action of pituitary extracts upon the kidney. *J. Physiol. (Lond.)*, 27:ix-x.

Miller, M., and Moses, A. M. (1971). Radioimmunoassay of urinary antidiuretic hormone with application to study of the Brattleboro rat. *Endocrinology*, 88: 1389-1396.

Oliver, G., and Schäfer, E. A. (1895). On the physiological actions of the pituitary body and certain other glandular organs. *J. Physiol. (Lond.)*, 18:277-279.

Oyama, S. N., Kagan, A., and Glick, S. M. (1971). Radioimmunoassay of vasopressin: Application to unextracted urine. *J. Clin. Endocrinol. Metab.*, 33: 739-744.

Permutt, M. A., Parker, C. W., and Utiger, R. D. (1966). Immunochemical studies with lysine vasopressin. *Endocrinology*, 78:809-814.

Rector, F. C., Jr. (1977). Renal concentrating mechanisms. In *Disturbances in Body Fluid Osmolality*, T. E. Andreoli, J. J. Grantham, and F. C. Rector, Jr. (eds.). American Physiological Society, Bethesda, Maryland, pp. 179-196.

Robertson, G. L. (1977). The regulation of vasopressin function in health and disease. *Recent Prog. Hormone Res.*, 33:333-385.

Robertson, G. L. (1979). Physiopathology of ADH secretion. In *Clinical Neuro-endocrinology*, G. Tolis, F. Labrie, J. B. Martin, and F. Naftolin (eds.). Raven, New York, pp. 247-260.

Robertson, G. L., Athar, S., and Shelton, R. L. (1977). Osmotic control of vasopressin function. In *Disturbances in Body Fluid Osmolality*, T. E. Andreoli, J. J. Grantham, and F. C. Rector, Jr. (eds.). American Physiological Society, Bethesda, Maryland, pp. 125-148.

Robertson, G. L., Mahr, E. A., Athar, S., and Sinha, T. (1973). Development and clinical application of a new method for the radioimmunoassay of arginine vasopressin in human plasma. *J. Clin. Invest.*, 52:2340-2352.

Robertson, G. L., Roth, J., Beardwell, C. G., Klein, L. A., Petersen, M. J., and Gordon, P. (1973). Vasopressin: Radioimmunoassay. In *Methods in Investigative and Diagnostic Endocrinology*, Vol. 2a, S. A. Berson and R. S. Yalow (eds.). North-Holland, Amsterdam, pp. 656-668.

Robertson, G. L., Shelton, R. L., and Athar, S. (1976). The osmoregulation of vasopressin. *Kidney Int.*, 10:25-37.

Robinson, A. G. (1975). Isolation, assay, and secretion of individual human neurophysins. *J. Clin. Invest.*, 55:360-367.

Roth, J., Glick, S. M., Klein, L. A., and Petersen, M. J. (1966). Specific antibody to vasopressin in man. *J. Clin. Endocrinol. Metab.*, 26:671-675.

Sawyer, W. H., and Valtin, H. (1967). Antidiuretic responses of rats with hereditary hypothalamic diabetes insipidus to vasopressin, oxytocin and nicotine. *Endocrinology*, 80:207-210.

Schäfer, E. A. (1909). The functions of the pituitary body. *Proc. R. Soc. Lond. [Biol.]*, 81:442-468.

Scheiner, E. (1975). The relationship of antidiuretic hormone to the control of volume and tonicity in the human. *Adv. Clin. Chem.*, 17:1-52.

Schwartz, A., Nagano, K., Kakao, M., Lindenmayer, G. E., and Allan, J. C. (1971). The sodium- and potassium-activated adenosine triphosphatase system. In *Methods in Pharmacology*, Vol. 1, A. Schwartz (ed.). Appleton-Century-Crofts, New York, pp. 361-388.

Schwartz, W. B., Bennett, W., Curelop, S., and Bartter, F. C. (1957). A syndrome of renal sodium loss and hyponatraemia probably resulting from inappropriate secretion of antidiuretic hormone. *Am. J. Med.*, 23:529-542.

Share, L. (1973). Vasopressin: Bioassay and extraction from blood. In *Methods in Investigative and Diagnostic Endocrinology*, Vol. 2A, S. A. Berson and R. S. Yalow (eds.). North-Holland, Amsterdam, pp. 652-656.

Share, L., and Crofton, J. T. (1980). Is vasopressin in the circulation of the dog freely filterable? *J. Endocrinol.*, 86:501-509.

Skowsky, W. R., and Fisher, D. A. (1972). The use of thyroglobulin to induce antigencity to small molecules. *J. Lab. Clin. Med.*, 80:134-144.

Starling, E. H., and Verney, E. B. (1924). The secretion of urine as studied on the isolated kidney. *Proc. R. Soc. Lond. [Biol.]*, 97:321-363.

Thomas, T. H., and Lee, M. R. (1976). The specificity of antisera for the radio-immunoassay of arginine vasopressin in human plasma and urine during water loading and dehydration. *Clin. Sci. Mol. Med.*, 51:525-536.

Thomas, T. H., Morgan, D. B., Swaminathan, R., Ball, S. G., and Lee, M. R. (1978). Severe hyponatraemia. *Lancet*, i:621-624.

Verney, E. B. (1947). The antidiuretic hormone and the factors which determine its release. *Proc. R. Soc. Lond. [Biol.]*, 135:25-106.

Wachstein, M., and Meisel, E. (1957). Histochemistry of hepatic phosphatases at a physiologic pH. *Am. J. Clin. Pathol.*, 27:13-23.

Walter, R., and Simmons, W. H. (1977). Metabolism of neurohypophysial hormones. Considerations from a molecular viewpoint. In *Neurohypophysis*, M. Moses and L. Share (eds.). Karger, Basel, pp. 167-188.

Wu, W. H., and Rockey, J. H. (1969). Antivasopressin antibody. Characterization of high-affinity rabbit antibody with limited association constant heterogeneity. *Biochemistry*, 8:2719-2728.

Zimmerman, E. A., and Robinson, A. G. (1976). Hypothalamic neurons secreting vasopressin and neurophysin. *Kidney Int.*, 10:12-24.

11

Parathyroid Hormone

G. Neil Kent* / M.R.C. Clinical Research Centre, Harrow, Middlesex, England; Kennedy Institute of Rheumatology, London, England; and National Institute for Medical Research, London, England

Joan M. Zanelli / National Institute for Biological Standards and Control, London, England

Calcium homeostasis is essential for life because calcium is involved in cell function and communication, muscle contraction, nerve transmission, blood fluidity, and exocrine and endocrine secretion, as well as the strength of the internal skeleton. The principal tissues and humoral factors responsible for the complex regulatory systems include the parathyroid gland and parathyroid hormone; the liver, kidney, and vitamin D metabolites; bone and intestine; and other peptides, glycoproteins, and steroids, such as calcitonin, prolactin, growth hormone, and estrogens. The many aspects of these systems have been the subject of considerable study and review (Habener and Potts, 1980; Parsons and Zanelli, 1980; Cohn and MacGregor, 1981). The aim of this chapter is to summarize current views on the nature and biological function of parathyroid hormone (PTH) and to describe how cytochemical techniques may be of particular value in elucidating the role of PTH and its interaction with its target issues involved in calcium homeostasis.

NATURE OF THE HORMONE

The predominant form of PTH isolated from bovine (B), porcine (P), and human (H) parathyroid glands is a highly conserved straight-chain 84 amino

*Present affiliation: Sir Charles Gairdner Hospital, Queen Elizabeth II Medical Centre, Nedlands, Western Australia

Table 1 Amino Acid Sequences of Bovine and Human Parathyroid Hormone

```
          1                                    10                          17
Bovine    H2N-Ala-Val-Ser-Glu-Ileu-Gln-Phe-Met-His-Asn-Leu-Gly-Lys-His-Leu-Ser-Ser-
Human         Ser-                         -Leu-                              -Asn-
          18    20                               30              34
Bovinen   Met-Gly-Arg-Val-Glu-Trp-Leu-Arg-Lys-Lys-Leu-Gln-Asp-Val-His-Asn-Phe-
Human

          35              40                              50
Bovine    Val-Ala-Leu-Gly-Ala-Ser-Ileu-Ala-Tyr-Arg-Asp-Gly-Ser-Ser-Gln-Arg-Pro-
Human                       -Pro-Leu    -Pro-        -Ala-Gly-

          52                      60                      68
Bovine    Arg-Lys-Lys-Glu-Asp-Asn-Val-Leu-Val-Glu-Ser-His-Gln-Lys-Ser-Leu-Gly-
Human                                                -Glu-
          70                              80      84
Bovine    Glu-Ala-Asp-Lys-Ala-Asp-Val-Asp-Val-Leu-Ileu-Lys-Ala-Lys-Pro-Gln-COOH
Human                               -Asn-    -Thr-        -Ser-
```

Source: Habener and Potts (1980).

acid polypeptide containing no intrachain disulfide bonds (Table 1) (Habener and Potts, 1980). Like a number of other secretory proteins (Steiner, 1976), PTH is synthesized within the cells in a much larger form (preproPTH) that contains a 31 amino acid adduct on the amino terminus (N-terminus) of the PTH molecule (Kemper et al., 1974). This short-lived form of PTH is converted to proPTH and transferred into the cisternal space of the endoplasmic reticulum (Habener et al., 1976). ProPTH contains only a basic hexapeptide adduct on the N-terminal (Hamilton et al., 1974), and this is removed prior to packaging of PTH into secretory granules (Nakagami et al., 1971; Habener et al., 1979).

The fate of PTH, stored within the secretory granules of the parathyroid cell, is dependent, mainly, on the concentration of extracellular ionized calcium. The exact mechanisms by which the secretory response is mediated, which may be through membrane receptors, such as β-adrenergic receptors or systems involving cyclic nucleotides (Brown and Aurbach, 1981) or phospholipids (Klahr et al., 1982), are the subjects of current investigations. The parathyroid gland stores relatively little PTH (Chu et al., 1973) but has a relatively high rate of biosynthesis (Sherwood et al., 1968) and a high rate of intracellular degradation of the hormone (Chu et al., 1973; Habener et al., 1975). Hypocalcemic stimulation of the gland results in increased secretion of PTH that may be due, during the initial phase, to a rapid decrease in the rate of intracellular degradation followed later by an increase in the biosynthetic rate.

The nature of the secreted PTH is not clear. It is now apparent that intact PTH (1-84) is not the only secreted form of the hormone and that carboxy-terminal (C-terminal) and N-terminal fragments may also be secreted. The formation of these fragments probably occurs within the secretory granules of the cell. All studies have shown a greater proportion of C-terminal to N-terminal fragments, indicating further degradation of the N-terminal fragments within the cell. The physiological roles both of the secreted and of the peripherally generated fragments of PTH will be discussed later.

BIOLOGICAL ACTIVITY

The classic biological activities (both in vivo and in vitro) of intact 1-84 appear to reside in the N-terminal third of PTH (Parsons and Zanelli, 1980; Habener and Potts, 1980). The C-terminal two-thirds of the PTH molecule is believed to be devoid of bioactivity, although a cytochemical response to C-region PTH has been demonstrated (Arber et al., 1980). However, in recent studies on the internal homology of bovine preproPTH, there is a suggestion that bovine preproPTH may be a result of gene duplication, since the best "homology" appeared between sequence $-23 \to +29$ and sequence $+30 \to +80$ (Cohn et al., 1979). Computer analysis for homologies in the cloned complementing DNA (cDNA) copy of bovine PTH messenger RNA has not provided any support for this hypothesis

of gene duplication (Kemper et al., 1981). This latter study, however, relies on
the nucleotide sequence of the cDNA copy of PTH messenger RNA and may not
necessarily reflect the nucleotide sequence of genomic PTH because of the excision of introns from the transcribed copy of the PTH gene. The possibility of
gene replication needs further study since it may provide a basis for an explanation of the apparent biological activity of the carboxyl-terminal region of PTH.

METABOLISM OF PTH

The immunoheterogeneous nature of PTH in the circulation (Berson and Yalow,
1968) is a result both of the secretion and of the peripheral generation of C- and
N-region fragments of PTH that presumably will have different biological activities and metabolic fates (Bennett and McMartin, 1979; Parsons and Zanelli,
1980; Cohn and MacGregor, 1981; Martin, Hruska, et al., 1981; Neuman et al.,
1981). It is now clear from many studies on the metabolism of PTH within the
circulation that intact PTH (1-84) is only found in significant amounts in the
venous effluent of the parathyroid glands, the concentrations being very low, or
undetectable, in peripheral circulation. The liver (Canterbury et al., 1975) and
the kidney (Hruska et al., 1975) are the principle organs involved in the peripheral metabolism of PTH. The metabolism of PTH (1-84) by liver and kidney is
sufficient to result in a very short circulating half-life of approximately 2-4 min
(Silverman and Yalow, 1973; Blum et al., 1974), although the biological half-
life may be even shorter, as discussed later. Biologically inert fragments containing the mid- and C-regions (with relatively long half-lives) account for the greatest concentration of PTH-derived peptides in the peripheral circulation measured by immunoassay. Amino-terminal materials, which may have biological
activity, would appear in much lower concentrations in the peripheral circulation because their half-lives are much shorter—2-10 min as against half-lives of
1 hr to 2 days for carboxyl-terminal fragments (Yalow, 1978). This has resulted
in great difficulty in establishing and defining the normal levels of bioactive PTH
in the peripheral circulation, a subject which has been reviewed in some detail
(Parsons and Potts, 1972; Parsons and Zanelli, 1980). Parsons concluded that the
plasma concentration of PTH was estimated to be approximately 10^{-11} g/ml
[approximately 10^{-12} M or 25 μU/ml expressed in terms of bovine PTH (1-84)] .

PHYSIOLOGICAL ROLE OF PTH

Total parathyroidectomy results in rapid and prolonged hypocalcemia, although
this is not necessarily immediately fatal. It is generally agreed that the parathyroid glands play a central role in the maintenance of the plasma calcium concentration. The calcemic effect of PTH involves several organs and endocrine systems that have different response characteristics. The rapidly responding component consists of the kidney, and to a lesser extent, bone, while the slower

component involves bone and, mediated by vitamin D, the intestine. In their review, Parsons and Zanelli (1980) classified the dose- and time-dependent effects of PTH on these organs as either anabolic or catabolic to the skeleton. The anabolic effects are effected by enhanced renal absorption of calcium, increased intestinal absorption of calcium, and an increased rate of bone formation, whereas the catabolic effects are, mainly, increased bone destruction and to a lesser extent, an effect secondary to renal phosphaturia. The relative physiological importance of these interactions is difficult to determine because of the large variations in the nature of effector and response parameters, i.e., dose dependence, half-lives of the biologically active forms of PTH, lag time, and the magnitude and half-life of the different responses. For example, the catabolic or osteolytic response in bone requires relatively high concentrations of PTH and is characterized by a long lag time and large magnitude and long half-life of response. In contrast to the catabolic action in bone, the renal action, to conserve calcium by increasing reabsorption of calcium, occurs at a much lower concentration of PTH and with a short lag time, large magnitude of response, and small half-life of response.

Kidney

Parathyroid hormone acts rapidly on the kidney to increase reabsorption of calcium, decrease reabsorption of phosphate, and stimulate hydroxylation systems required in the conversion of vitamin D metabolites to active forms. It also increases the excretion of cyclic AMP, not only through an increase in the glomerular filtration of cyclic AMP, but from increased nephrogenous production of cyclic AMP (Kaminsky et al., 1970). Following the initial observations by Chase and Aurbach (1968) that PTH stimulated adenylate cyclase in preparations of renal plasma membranes, studies on isolated nephrons in systems in vitro have achieved considerable progress in defining the regions of the nephron that express this response to intact PTH or to its synthetic N-terminal fragment PTH (1-34), as reviewed by Morel (1981). In the isolated human nephron, synthetic HPTH (1-34) stimulated adenylate cyclase to the greatest extent in the proximal convoluted tubule and to a lesser extent in the straight proximal tubule, medullary and cortical portions of the thick ascending limb, and the early portion of the distal convoluted tubule (Chabardès et al., 1980). Species differences were apparent in that there were varied proportional responses to synthetic BPTH or HPTH (1-34) in regions of the isolated rabbit, rat, or mouse nephron; however, the same regions of the isolated nephrons were involved and the biological significance of these variations is not known. Cytochemical techniques, which relate histological and topobiochemical features, have been used to "map" responses to physiological concentrations of PTH in populations of guinea pig nephrons (Chambers, Schäfer, et al., 1978). This report shows good correlations between the functional responses (i.e., cellular enzyme activities associated with phosphaturia, calcium transport, bicarbonate excretion, hydroxylation mechanisms) and the studies on the activation of membrane adenylate cyclase.

The exact relationships between the in vitro model systems in which the receptor-linked adenylate cyclase is the PTH response parameter and the sites in nephron populations associated with the in vivo renal responses are still under active study. It was widely held that cyclic $3'4'$-adenosine monophosphate (cyclic AMP) mediated the phosphaturic response to raised levels of PTH, but pharmacological studies (Parsons et al., 1975) have shown that it is possible to dissociate these events. Certainly the net effect of PTH upon the kidney is to induce phosphaturia, mild natriuresis, hypocalciuria, and enhanced excretion of bicarbonate and secretion of cyclic AMP into the urine.

Bone

The effects of PTH on bone have been widely studied, and as mentioned earlier can be classified as either anabolic or catabolic with regard to the skeleton (Parsons, 1976).

However, there is great difficulty in discriminating physiological, pathological, and pharmacological effects of PTH on the skeleton because of the widely varying experimental techniques, dosages, and forms of PTH (or its synthetic fragments) used either in vivo or in vitro, a problem faced by previous reviewers (Hekkelman et al., 1975; Parsons et al., 1975). They postulated from both the in vivo and in vitro studies that the anabolic action of PTH occurs at lower concentrations of PTH than the catabolic actions and that it may be associated only with the N-terminal region of PTH.

The anabolic effects of PTH have been discussed elsewhere (Parsons, 1976; Parsons and Zanelli, 1980), and have long been supported by the observation of a decrease in osteoblastic number in hypoparathyroidism (Jowsey et al., 1958). The overall effect of PTH is to enhance osteoblastic function by increasing the number and activity of osteoblasts, and it is possible that this effect may be direct rather than indirect. This anabolic effect of relatively low doses of the synthetic N-region fragment HPTH (1-34) has been studied following long-term administration to either osteoporotic women (Reeve et al., 1980) or to greyhounds (Podbesek et al., 1981). In the greyhound study, daily injections of HPTH (1-34) produced a significant increase in osteoid formation as well as the number of resorption surfaces and in the skeletal calcium accretion rate, plasma alkaline phosphatase activity, and intestinal calcium absorption. The effects were not seen when the HPTH (1-34) was given by continuous subcutaneous infusion. The human osteoporosis trial has demonstrated an anabolic effect of PTH in terms of an increase in trabecular bone volume in response to daily injections. In this case, trabecular bone volume seemed to be increased at the expense of cortical bone since there was no concomitant increase in skeletal calcium accretion rate. It is not known whether the lack of increase in intestinal absorption of calcium during treatment with HPTH (1-34) seen in the human, compared with a definite increase in greyhounds, represents species differences or is a part of the etiology of osteoporosis. The variation in response of different areas of bone to PTH has been previously reported in an experimental model (Yonaga, 1978). In this instance, in which surgically hypoparathyroid rats were given PTH, compact bone recovered at a faster rate than the spongiosa.

The catabolic actions of bovine PTH (1-84) at pharmacological doses have been most widely studied and have formed the basis of some of the in vivo and in vitro bioassay systems for PTH (Zanelli and Parsons, 1980). The interest in the origins of bone cells, and in their definitions according to morphological and functional criteria (Owen, 1980), has been the subject of considerable investigation and reinvestigation. Although detailed review of the current hypotheses is beyond the scope of this chapter, it should be noted that statements concerning direct or indirect actions of intact PTH, or of the synthetic amino region fragment PTH (1-34) on osteoclasts, osteocytes, and osteoblasts may need to be modified in due course. The catabolic actions are thought to be mediated by osteoclasts and to a lesser extent by osteocytes (Parsons and Zanelli, 1980). PTH has a short-term osteolytic effect, probably elicited by activation of resident osteoclasts and osteocytes, and a long-term effect, probably involving recruitment of more osteoclasts. It is difficult to determine which of the pharmacological effects of PTH are important in the physiological regulation of bone metabolism and resorption by PTH. An increase in cyclic AMP production is a response known to be elicited rapidly and is followed by the biochemical activation of osteoclasts resulting in the release of collagenase and acid phosphatase, as well as other lysosomal enzymes, other proteinases, lactate, and citrate (Raisz, 1976; Ibbotson et al., 1980; Parsons and Zanelli, 1980). There are several early morphological changes in osteoclasts, including an increase in the number of nuclei and the development of clear zones and ruffled borders (Holtrop and Raisz, 1979). Ultimately, the cellular changes lead to loss of both matrix and mineral phase of the affected areas of bone.

Intestine

The effect of PTH on the absorption of calcium from the intestine is now assumed to be indirect and is mediated by metabolites of vitamin D (Haussler and McCain, 1977; DeLuca et al., 1979; Norman, 1979; Fraser, 1980). The regulation of vitamin D metabolism in the kidney by PTH is detailed in the reviews cited above; it is generally believed that PTH stimulates the production of 1,25-dihydroxyvitamin D3.

Table 2 Structure-Function Relationships for PTH and Its Analog in Bioassay and Membrane Binding Systems (Relative Molar Potencies; 95% Confidence Limits)

| | In Vivo Bioassays (assay half-maximal dose) | | | In Vitro Bioassays (assay half-maximal dose) | | | | | | | | |
Peptide and Species[b]	Dog/Rat Hypotensive (Dog 5 U/kg) (Rat 50 U/kg)	Rat Hypercalcemia (Munson Assay)	Chick Hypercalcemia (5U/chick) (0.2nmole/chick)	Guinea-Pig Renal Cytochemical Bioassay (3×10^{-12} M)	Rat Renal Adenylate Cyclase (5×10^{-7} M)	GTP-Enhanced Dog Renal Adenylate Cyclase (1.5×10^{-10} M)	Dog Renal Membrane Binding (3×10^{-9} M)	Human Renal Cortical Binding (6×10^{-9} M)	Rat Calvarial Cyclic AMP Production (2×10^{-8} M)	Perfused Dog Tibia Cycle AMP Production (2.5×10^{-9} M)	Mouse/Rat Calvarial Resorption (1×10^{-9} M)	Mouse Calvarial Collagen Synthesis (3×10^{-9} M)
Bovine												
1-84	100[4]	100[8]	100[1,2]	100[12]	100[1,2,5,6]	100[15]	100[7]		100[8]	<10[13]	100[8]	100[10]
1-34[k]	100[4]	3[8]	130(90-190)[1,2]	73(46-115)[12]	80(60-120)[1,2]97[5,6]	100[15]	100[7]	100[11]	20-50[8,9,k]	100[13]	100[g,8]	100[10]
H2O2-Ox 1-84[d]	0[4]		+[j,4]	<10[18]	3[16]							
H2O2-Ox 1-34	0[4]	0[16]	100[j,4]				<0.5[7]			<5[13]		
1-31			74[2]		10[2]							
1-30							6[7]			100[h]		3[10]
1-28			<1[1]		5[e,1,2]		1[7]					1[10]
1-27	+[f,4]		<1[1]		2[e,1,2]							
1-26			<1[1]		<0.3[1,2,6]	<1[16]						
pro PTH 1-40(-6-+34)			20(13-30)[1] 13[2]		5(4-6)[1,2]							
(desamino[1])1-34		100[l,1]	55(40-75)[1]	0.21(0.13-0.33)[12]	<0.1[e,2,6]							
2-34			65(45-90)[1,2]	0.06(0.04-0.09)[12]	2[3,1,2]				10-30[8]		0.4[8]	
3-34			<0.5[1,2]	0[1,12]	<0.3[1,2,6]		10[7]		0[8]	<2[13]	0[8]	0.2[e,10]
(ser[1])1-34			60(50-70)[1]		20(12-40)[1,2]							
(sar[1])1-34[c]			110(70-180)[1]		8(7-11)[1,2]							
(D-Ala[1])1-34			300(200-430)[1]		55(45-60)[1,2]							
(Tyr[1])1-34			70(50-100)[1]		9(7-11)[1,2]							
1-34 amide			330(210-530)[1]		230(200-270)[1,2]							
(D-Ala[1])1-34 amide			280(200-400)[1]		100(90-110)[1]							
(Tyr[34])1-34 amide					300[3]							
(D-Tyr[34])1-34 amide					270[3]							
(D-Tyr[34])2-34 amide					3[3]							
(D-Val[2],D-Tyr[34])1-34 amide					2[3]							
(D-Val[2],D-Tyr[34])2-34 amide					1[3]							
(Nle[8],Nle[18],Tyr[34])1-34 amide[m]			33[16]		190[16]		200[7]			200[17]	200[17]	
(Nle[8],Nle[18],Tyr[34])3-34 amide			<1[16]	0[a,12]	<1[16]	0[a,15]	200[7]			50[13]	0[a,16]	
1-12 + 13-34			<0.5[1,2]		<0.3[1,2]							
24-34	10[4]											
24-28	1[4]											
25-27	0[4]											
53-84 (bovine or human)			<1[16]	0[i,12]	<0.3[16]		0[7]		0[14]			0[14]
Human												
(Ser[1])1-34	+[f,4]		100(70-130)[1]		30(25-35)[n,1]							
(Ala[1])1-34			75(60-100)[1]		70(55-90)[n,1,2]			100[11]				
1-84					12[2]	100[15]		100[11]				

[a]Acts as a competitive inhibitor.

[b]Numbers refer to amino acid position in the sequence of the native hormone from amino to carboxy end of the molecule (see Table 1). Abbreviations used for amino acids are standard three-letter codes.

[c]Sarcosine.

[d]Hydrogen peroxide-oxidized PTH.

[e]Nonparallel log dose-response curve.

[f]Insufficient material for potency estimate although hypotensive activity present.

[g]Only after 48 hr in culture.

[h]Cathepsin B fragments of BPTH (1-84) containin a mixture of (1-27) and (1-30).

[i]Has no activity on 6-8 min peak but active on 14-16 min peak.

[j]In the Japanese quail hypercalcemia assay (1-34) and H_2O_2-oxidized (1-34). are equipotent, whereas the potency of H_2O_2-oxidized (1-84) is much less than native (1-84).

[k]In a study on the histology of cultured rat embryonic radii the maximal effect produced by the native (1-84) was twice that seen with synthetic (1-34). Synthetic (1-34) produced increased maturation of cartilage and increased number of active osteoblasts, effects *opposite* to those of (1-84); i.e., (1-34) was anabolic to bone.

[l]In 100 g parathyroidectomized rats plasma and urinary calcium and phosphate are measured 90 min after an IV injection. Desamino (1-34) is equipotent to (1-34) in hypercalcemic action but unlike (1-34) causes no increase in plasma phosphate. They were equipotent in decreasing urinary excretion of calcium, but desamino (1-34) was much more potent in producing phosphaturia.

[m]Nle is norleucine.

[n]Of the human and bovine (1-34) peptides tested by Goltzman, Callahan, et al. (1975) in renal adenylate cyclase assays with membranes prepared from kidneys of rat, cow, dog, human, and chicken, those with Ala in position 1 are full agonists in all systems. Peptides with Ser in position 1 are full agonists in the mammalian systems but only partial agonists in the chick membrane system (see also Martin et al., 1974).

1. Parsons et al., 1975.
2. Tregear et al., 1973; 1974.
3. Coltrera et al., 1980.
4. Pang et al., 1981.
5. Goltzman, Callahan, et al., 1975.
6. Goltzman, Petytremann, et al., 1975.
7. Segre et al., 1979.
8. Herrmann-Erlee et al., 1976.
9. Gaillard et al., 1976.
10. Bringhurst and Potts, 1981.
11. Nissenson, Teitelbaum, et al., 1981.
12. Arber et al., 1980, 1981
13. Martin, Bellorin-Font, et al., 1981.
14. Hekkelman et al., 1981.
15. Nissenson, Abbott, et al., 1981.
16. Rosenblatt, 1981.
17. Raisz, Lorenzo, et al., 1979.
18. Chambers et al., 1976.
19. Munson, 1961.

STRUCTURE-FUNCTION RELATIONSHIPS

Research into the metabolism of the hormonal peptides secreted by the parathyroid glands has been limited by the low concentrations of these molecules in the circulation and by the lack of sensitivity and specificity of immunoassay systems. The studies on the cleavage sites of exogenously administered radio-iodinated bovine PTH (1-84) (Segre et al., 1974) and the preparation of synthetic PTH peptides and analogs (Tregear et al., 1973, 1974; Rosenblatt and Potts, 1977) have provided the foundation of our knowledge of the structure-activity relationships for PTH. All currently known bioactivities of PTH are associated with the N-terminal third of the molecule, and most of the recent studies have concentrated on this region of the molecule (Table 2). In nearly all the bioassay systems shown in Table 2, the synthetic N-terminal fragment (1-34) of PTH is equivalent, on a molar basis, to the intact (1-84) molecule, and (1-34) is the minimum peptide sequence that retains full biological activity (Tregear et al., 1973). In studies with the two most extensively used assays, i.e., the in vivo chick hypercalcemia assay and the in vitro rat renal membrane adenylate cyclase assay, a number of differences can be seen. This adenylate cyclase assay shows considerable discrimination between the human and bovine (1-34) fragment and seems very dependent on the integrity of the N-terminal amino acid; a minor change, such as substitution of the alanine by serine (the N-terminal amino acid in the human hormone) leaves only 25% residual activation of adenylate cyclase compared to the synthetic BPTH (1-34). Indeed, the simple loss of the N-terminal amino group (i.e., propionyl instead of alanyl residue) reduces activity by more than 99%. However, this change results in a loss of only about 50% of the in vivo hypercalcemic effect. In summary, shortening of the peptide from the N-terminus causes a much earlier loss of activity in the in vitro renal membrane adenylate cyclase assay than in the in vivo hypercalcemia assay. The minimum continuous sequence required for activation of renal adenylate cyclase is (2-26) (Tregear et al., 1973; Goltzman et al., 1978). On the other hand, the in vivo hypercalcemia assay is more sensitive to amino acid depletion from the C-terminus of the (1-34). It is noteworthy that the potency estimates acquired with the cytochemical bioassay and with the adenylate cyclase assay, both in vitro renal target tissue systems, show greater agreement with each other than with the in vivo assay, where the hypercalcemia is believed to be primarily a response of the chick bone (Parsons et al., 1973).

The results obtained with renal membrane binding assay indicate that this type of system is less sensitive to shortening from the N-terminus than from the C-terminus; thus, while (3-34) has virtually no biological activity in other systems, it still shows considerable ability to bind to renal membranes. This indicates that (3-34) has sufficient structural integrity to bind to the renal receptor but is not able to activate adenylate cyclase. Such studies have provided the basis for the design and synthesis of analogs of PTH that have enhanced agonist or antagonist properties (Rosenblatt and Potts, 1981; Mahaffey et al., 1979; Rosenblatt et al., 1977). The potency can be enhanced by substituting the C-terminal (position 34) phenylalanine with tyrosine (i.e., addition of a single hydroxyl functional group) and by amidating the functional carboxyl group of this residue. Furthermore, some of these new synthetic analogs of HPTH have been produced with norleucine replacing the methionine residues in the native sequence, since oxidation of methionine residues results in reduced solubility of the hormone and loss of some hormonal bioactivities (Table 2) (Rasmussen and Craig, 1962; Tashjian et al., 1964; Pang et al., 1981). The analog [Nle^8, Nle^{18},Tyr^{34}] BPTH (1-34) amide is reported to have approximately twice the activity of synthetic BPTH (1-34) (Rosenblatt and Potts, 1977), while deletion of the two amino acids at the N-terminus produces an inhibitor [Nle^8,Nle^{18}, Tyr^{34}] BPTH (3-34) amide, which has a PTH-receptor binding avidity equal to or greater than either BPTH (1-84) or BPTH (1-34) (Segre et al., 1979). Such lines of study have provided information on the structural requirements of the (1-34) fragment of PTH needed for receptor binding (antagonist studies) and receptor activation (agonist studies).

In most studies, bovine PTH (1-84) and (1-34) were equipotent (Table 2). One exception to this generalization was reported by Martin et al. (1978; Martin, Hruska, et al., 1981), who studied the peripheral metabolism and activity of BPTH (1-34) and (1-84) in the isolated perfused tibia of the dog. In this system, which monitored concentrations of immunoreactive PTH in afferent and efferent perfusate, there was an arteriovenous difference of 40% after infusion of BPTH (1-34) but no demonstrable arteriovenous difference in either intact BPTH (1-84) or oxidized BPTH (1-34). Infusion of BPTH (1-34) produced a large increase in the concentration of cyclic AMP in the venous effluent from the tibia, but only a slight increase was recorded even with large doses of the intact BPTH (1-84). Surprisingly, although the BPTH (3-34) acted as a potent competitive inhibitor when given in combination with the BPTH (1-34), the substituted analog, [Nle^8,Nle^{18},Tyr^{34}] BPTH (3-34) amide did not act as an inhibitor but as an agonist, with approximately half the activity of the PTH (1-34). These authors concluded that in the particular case of the canine skeleton, the peripheral cleavage of BPTH (1-84) may be a prerequisite for the expression of bioactivity. This may help to explain the earlier observations that intact BPTH (1-84) had to be "processed" by passage through the peripheral circulation, before producing an arteriovenous difference in the concentration of calcium of blood perfusing the cat tibia (Parsons and Robinson, 1968).

WHY MEASURE CIRCULATING LEVELS OF PTH?

So far, we have not focused attention on the concentrations of PTH required to elicit a certain response, but there are obviously many situations in which both the nature and the amount of bioactive PTH need to be known. The simplest cases are in the determination of primary hyperfunction and hypofunction of the parathyroid gland. To date, diagnostic discrimination of immunoreactive forms of PTH in peripheral blood has been satisfactory for some clinical purposes. It is well known that the actual numerical values for blood levels of immunoreactive PTH in terms of different standards by different laboratories differ widely (Zanelli and Gaines Das, 1980; 1983). Furthermore, the relative values may be inappropriate to the degree of biological disturbances as assessed by other biochemical or clinical indices (Goltzman, 1978). A greater disparity between immunoassay values and biological effects of PTH occurs in such situations as hyperparathyroidism secondary to renal failure, associated with the painful condition of renal osteodystrophy, in which clearance of PTH is drastically reduced (Slatopolsky et al., 1980). The biological situation becomes paradoxical in pseudohypoparathyroidism type I in which there are episodes of hypocalcemia and a loss of renal responsiveness to pharmacological doses of BPTH administered exogenously despite elevated plasma levels of immunoassayable PTH. There is also a great need to define the effects on the parathyroid gland of a number of physiological effectors, e.g., neurological or neuroendocrine effects on circadian or pulsatile secretion; gut hormones as well as other hormones; and glandular activity during pregnancy, lactation, and fetal life. The application of the cytochemical bioassay system to such problems, and comparisons with other PTH assay systems, will be discussed in the remaining sections of this chapter.

ASSAYS OF CIRCULATING PTH: TECHNIQUES, VALIDATION, PROBLEMS

There are currently only three assay systems that have sufficient sensitivity to assay PTH in clinical samples, i.e., serum or plasma. These are immunoassay; a receptor bioassay (the guanylate nucleotide-amplified renal adenylate cyclase systems); and the cytochemical bioassay. Only the cytochemical bioassay is capable of measuring PTH in all normal subjects.

Immunoassay

There is no doubt that during the last two decades, immunoassay systems have provided the basic information for current views on the physiology and pathology of the parathyroid gland and on the nature of the circulating forms of PTH.

There are many reviews on many different immunoassay systems, with particular emphasis on the clinical interpretation of values derived from these techniques (Arnaud et al., 1974; Bouillon, 1978; Raisz, Yajnik et al., 1979; Habener and Segre, 1979; Hawker and Di Bella, 1980; Martin et al., 1980). Detailed discussion of immunoassay systems and their biological relevance is outside the subject of this chapter, but it is generally accepted that immunoassay systems detect and measure variable composites of antigenic determinants which may not have the structural requirements necessary for expression of biological activities.

The first generation of clinical immunoassays, most of which were based on radioimmunoassay rather than immunoradiometric methods, used reagents derived from bovine PTH, the only species of extracted PTH available in sufficient quantity, and they depended upon cross-reactivity with human PTH. Once the chemical structure and amino acid sequences of the bovine and human PTH molecules had been established (Keutmann, 1974), limited amounts of peptide fragments of PTH molecules, prepared by enzymatic cleavage or synthesized, were available for research into the specificity of immunoassay systems with respect to N- and C-regions of the PTH molecule.

The most widely used immunoassay systems for routine clinical purposes, based on antisera raised to bovine, porcine, or human PTH, in guinea pigs, rabbits, chickens, goats, and sheep, have been reported as having specificity for the mid and carboxyl regions of PTH. The apparent tendency for experimental animals, such as guinea pigs, chickens, goats, and sheep, to generate antibodies that recognize intact, N-, mid-, or C-regions, may indicate that species-specific antigenic determinants are associated with these regions of the PTH molecule. However, there are also practical reasons for selecting antisera which recognize the C-region. The cleavage products of PTH which lack the N-terminal third have a longer half-life than the intact secreted hormone, and thus the accumulated cleavage products are more readily detected and measured (see reviews cited above).

A second generation of immunoassays has been developed using antisera specifically directed to antigenic determinants within the N-terminal one-third of the PTH molecule, i.e., that region of the molecule associated with the known biological activities of the intact PTH (1-84) molecule (Desplan et al., 1977; Manning et al., 1980; Segre et al., 1981). Although the low circulating concentrations of N-region immunoreactivities, either in the form of the intact PTH (1-84) or as an N-region cleavage product, impose great demands upon the sensitivity of such immunoassay systems, they are proving to be of value in clinical research studies into the acute secretion and metabolism of PTH.

A new generation of region-specific immunoassays is now emerging (Marx et al., 1981; Mallette et al., 1981; Manning et al., 1981) because a variety of different synthetic peptide fragments of defined sequences within the N-, mid-, and C-regions have become commercially available for such purposes from international manufacturers (Bachem Fine Chemicals, Torrance, California; Beckman Products, Palo Alto, California; Peninsula Laboratories, San Carols, California, and the Protein Research Foundation, Osaka, Japan).

Different region-specific systems are now also commercially available as complete immunoassay kits (Immuno Nuclear Corporation, Stillwater, Minnesota). Specificity of immunoassay systems can be expected to be more closely defined with the introduction of monoclonal antibodies.

In summary, the increasing availability of region-specific immunoassay systems will enable rapid progress to be made in comparing and contrasting the relative proportions of the circulating metabolized forms of immunoreactive PTH in clinically characterized groups of patients.

Receptor Bioassay: Guanyl Nucleotide-Amplified Renal Adenylate Cyclase Systems

A new in vitro bioassay of PTH has recently been developed using a PTH activation of adenylate cyclase in canine renal cortical membranes (Abbott et al., 1980; Nissenson, Abbott, et al., 1981). The sensitivity to the PTH has been dramatically enhanced by the addition of the hydrolysis-resistant analog of guanosine triphosphate, $5'$-guanylimidodiphosphate (GTP analog). This assay is reported to be sensitive to as little as 30 pg/ml (10^{-11} M), using a commercial preparation of BPTH (1-34) as standard, and is highly specific for PTH. Synthetic BPTH (1-34) is equipotent (on a molar basis) with both human and bovine PTH (1-84). The assay has a detection limit of 300 pg/ml HPTH (1-84) (3×10^{-11} M) in the presence of 10% serum. Thus, although it does not have sufficient sensitivity to measure bioactive PTH in the peripheral circulation of normal individuals, it can detect and measure PTH in parathyroid venous effluent serum and in peripheral serum from some patients with hyperparathyroidism secondary to renal failure. A brief report from a separate group (Avioli et al., 1981) claims that this assay can be made more sensitive and can measure PTH in peripheral serum of normal subjects (normal range 25-96 pg equivalents/ml BPTH (1-34); 6-24 pM) although the lower point of the normal range is at the detection limit of the assay. An explanation of the methods by which this increase in sensitivity has been obtained has not been published.

A preliminary report of a GTP analog-amplified system using human instead of canine membranes confirms that bioactivity in peripheral serum from human subjects can be detected and measured (Niepel et al., 1981; personal communication). Clinical studies using this homologous system will be awaited with interest.

More detailed considerations of the biological implications of these membrane receptor bioassay systems will be discussed after the following section.

The Cytochemical Bioassay

Measurement of circulating PTH by cytochemical bioassay (Chambers, Dunham, et al., 1978) was developed using techniques first used in the cytochemical bioassay of ACTH (Chayen et al., 1972) and further exploited for other polypeptide hormones (Chayen, 1980). Results using this bioassay for PTH have been reported by at least four different groups (Chambers, Dunham, et al., 1978; Fenton et al., 1978; Goltzman et al., 1980) and by Posillico et al. (1981), who used a modified technique. The specific details and methods required for this assay have been discussed elsewhere (Chapter 3 of this book; Chambers, Dunham, et al., 1978; Chayen, 1980). In brief, the assay currently in use in most laboratories is carried out with female Hartley strain guinea pigs (350-450 g) which have had their diet supplemented for at least 7 days with a multiple vitamin preparation (e.g., Adexolin; Glaxo Laboratories, U.K.) in their drinking water. After killing the animals by asphyxiation in nitrogen, the kidneys are removed, decapsulated, and halved sagitally. Two segments are cut from each pole of each half kidney so that a total of 16 segments from two kidneys is

---

Figure 1 (Overleaf) Photomicrographs of serial cryostat sections (10 μm) of guinea pig kidney used for a PTH cytochemical bioassay. Sections were reacted for (a) alkaline phosphatase activity, 5 min reaction at 22°C using the Gomori-Takamatsu procedure (as outlined in Chayen et al., 1973) or, (b) G6PD, 10 min reaction using method detailed in the text. Two glomeruli (G) can be seen with associated macula densa (MD) and arteriole (A). Proximal convoluted (P) tubules can be seen mainly in the lower right part of the photograph, and distal convoluted (D) tubules can be seen in the upper left part of the photograph. Alkaline phosphatase reactivity is found in the brush-border area of the proximal convoluted tubules. There is little G6PD activity in the glomeruli or arteriole and more in the macula densa and proximal convoluted tubules. The formazan reaction product on the P tubules is asymmetrically distributed within both the cell and the tubule, with the greatest amount of formazan being associated with the glomerular side of the tubule cell. Distal convoluted tubules show the highest G6PD activity in this section, and it is evenly distributed throughout the cytoplasm of the cells. This is a very characteristic pattern of G6PD activity in D tubules, with the nuclear area of each cell appearing as areas of low-formazan production. Cells within these tubules would be measured for a PTH cytochemical bioassay. The sections were photographed at ×300 magnification.

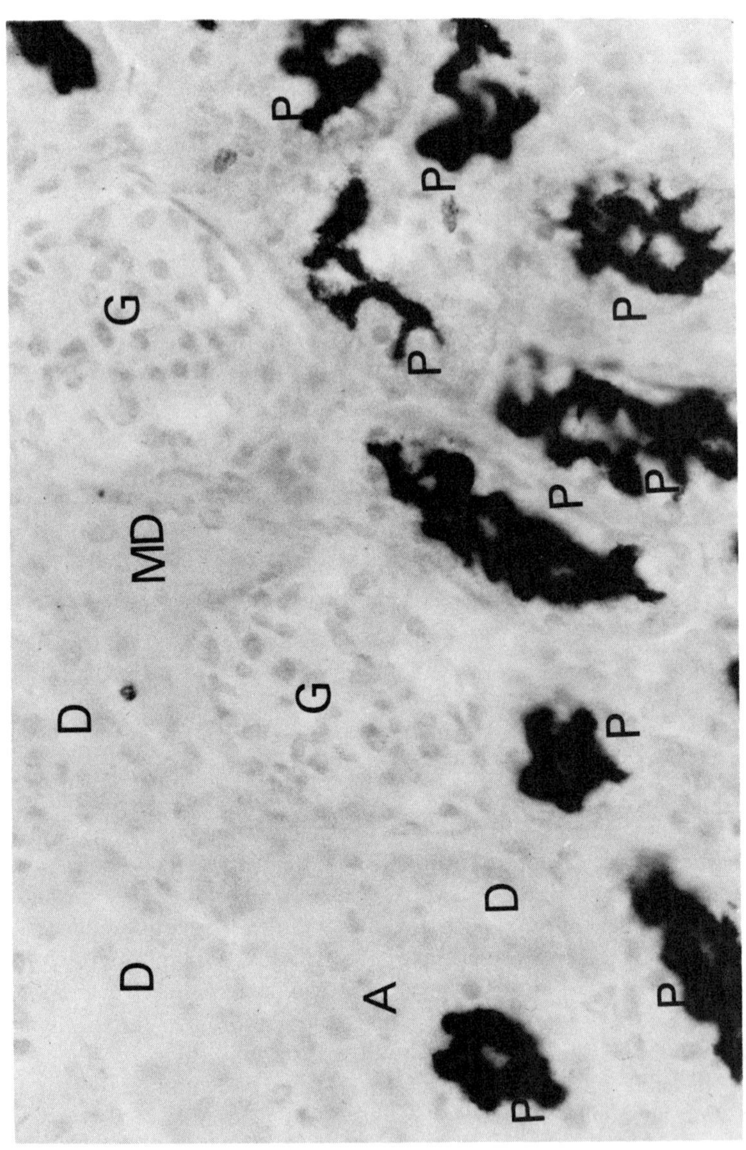

Figure 1a

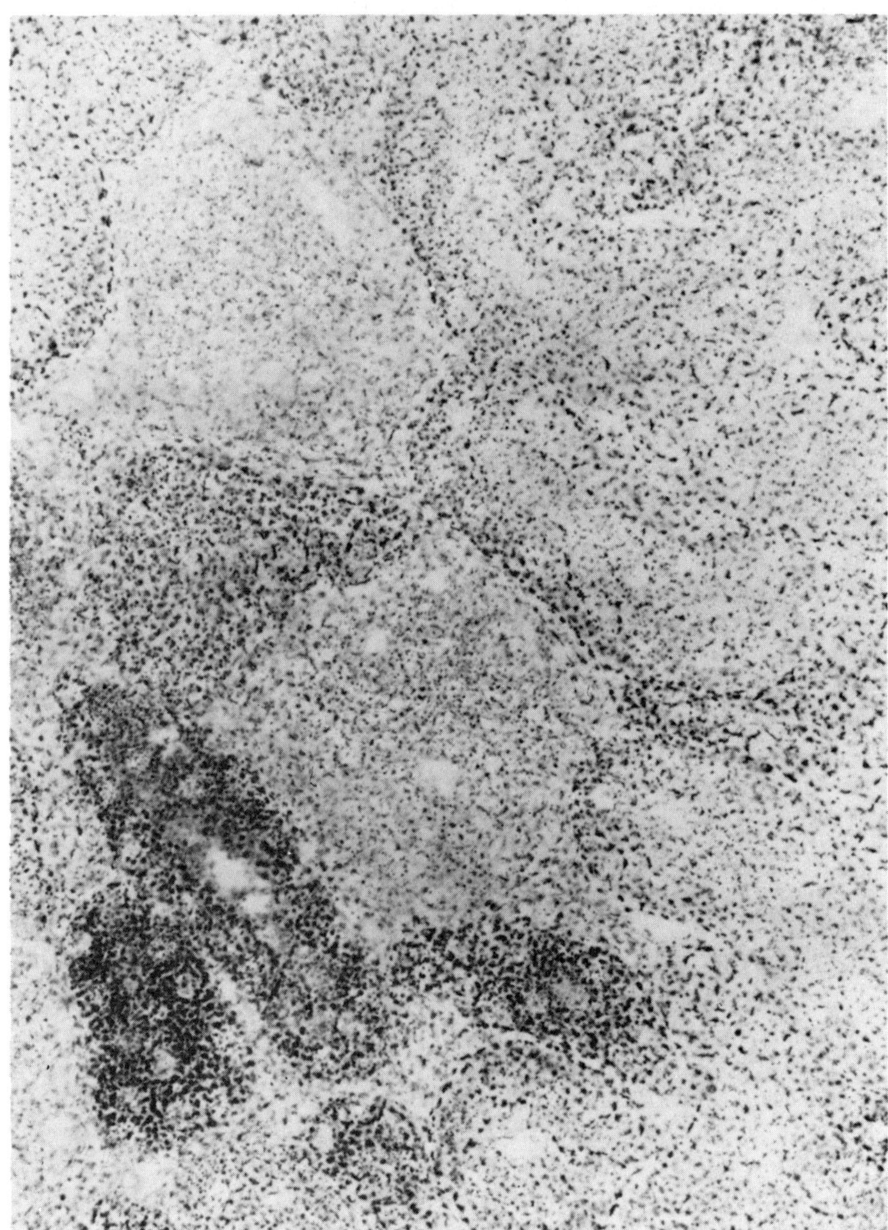

Figure 1b

available for culture. The segments are cultured individually (cortex down) on
defatted lens tissue on stainless steel grids in vitreosil dishes containing Trowell's T8 medium (pH 7.6) (Trowell, 1959) at a level up to the top of the grid.
The dishes are placed in culture chambers at 37°C in an atmosphere of 95%
O_2:5% CO_2 for up to 5 hr. After this time, the medium is removed and replaced
with fresh medium for a wash period of 8 min. This wash solution is then replaced with fresh medium (pH 7.6) containing either a dilution of standard
(nominally 0.001-1 pg/ml BPTH (1-84)) PTH or a dilution of a sample of
plasma, which is poured over the segment and left for a fixed period of time,
usually 6-8 min. Each segment is then removed from the medium, rapidly
chilled to -70°C in n-hexane (aromatic hydrocarbon free), and maintained at
-70°C in a slurry of alcohol-powdered solid carbon dioxide. The chilled segments are stored in dry, stoppered tubes at -70°C and are sectioned for quantitative cytochemistry within 72 hr of chilling.

Sections, 16 μm thick, are cut from each segment in a cryostat maintained
at less than -25°C, with the haft of the knife packed in solid carbon dioxide,
according to techniques previously described (see Chapter 3) (Chayen 1980).
Glucose-6-phosphate dehydrogenase (G6PD) activity in each section is quantified as the amount of colored formazan precipitated within the target cells
after reacting the tissue sections for a predetermined time, usually 10 min,
in a reaction medium. This medium consists of 30% GO_4-grade polyvinyl alcohol (colloid stabilizer) in 50 mM glycylglycine buffer, pH 8.0, containing 5 mM
neotetrazolium, 5 mM glucose-6-phosphate, 3 mM nicotinamide adenine dinucleotide phosphate (NADP), 10 mM potassium cyanide (KCN), and 0.67 mM
phenazine methosulfate. The reaction medium is purged of oxygen by bubbling
with nitrogen gas and the reaction carried out in a chamber containing an atmosphere of nitrogen. In this reaction, G6PD activity produces reduced NADP,
and these reducing equivalents are stoichiometrically transferred to the neotetrazolium salt (2 mole hydrogen per mole of neotetrazolium) via the intermediate hydrogen acceptor, phenazine methosulfate. Reduction of neotetrazolium
produces a colored formazan precipitate. Sections are mounted in Farrants'
medium, and the amount of formazan deposited in cells, identified as those of
the distal convoluted tubules, associated with glomeruli in the outer cortex (see
Figure 1), is measured with a Vickers M85 (or M85a) scanning and integrating
microdensitometer. Usual settings for these machines are X40 objective, spot
size 1 (0.5 μm), wavelength 585 nm, bandwidth 20, and an A3 mask (15 μm
diameter under these conditions). One cell in each of ten different distal tubules
(each tubule being associated with a separate glomerulus) is measured in both of
the duplicate sections of all the kidney segments. The mean integrated extinction for each sample of segment is calculated from these 20 readings by subtract-

ing a background value and dividing by an experimentally determined integration value for a neutral density filter at these particular machine settings (i.e., spot size, mask, objective, and wavelength).

With this assay system, there is a dose-related increase in G6PD activity in distal convoluted tubules with concentrations of BPTH (1-84) between 5×10^{-16} M and 5×10^{-13} M. By this assay, the range of blood plasma levels of bioactive PTH in normal subjects, expressed in terms of an ampuled house standard of highly purified bovine PTH (NIBSC code 77/533) is 2.5-75 μIU/ml, or approximately 1-30 pg equivalent/ml, or 0.1-3 pM (Chambers, Dunham, et al., 1978; Fenton et al., 1978; Goltzman et al., 1980).

Assay Validation

In addition to the classic criteria of statistical validity, biological specificity must be critically examined from two aspects—first, in terms of "the hormone" as defined by its classic properties and, second, in terms of the bioactivity in a sample of biological fluid.

As described by Chambers, Dunham, et al. (1978), Fenton et al. (1978), and Goltzman et al. (1980) the cytochemical bioassay for parathyroid hormone has been established using highly purified PTH extracted from bovine parathyroid tissue. The bovine PTH in the ampuled preparations used by the groups was identified, characterized, and calibrated by classic procedures, i.e., amino acid composition analysis, end-terminal analysis (Keutmann, 1974), in vivo biological activity in the chick hypercalcemia of Parsons et al. (1973), and in vitro biological activity in the renal membrane adenylate cyclase system (Marcus and Aurbach, 1969). Despite the scarcity of human PTH, ampules containing samples of hormone, purified after extraction of human parathyroid adenomata and characterized by chemical assay, immunoassay, and in vitro bioassay systems, have been available for time-dependent and dose-dependent study by cytochemical bioassay (Zanelli and Gaines-Das, 1980; 1983). Studies comparing cytochemical responses to multiple dilutions of both bovine and human PTH show that the extracted hormone from both species produce highly significant linear log dose responses that are parallel to each other. Studies on specificity included testing the response to the preparation of bovine PTH after it had been subjected to controlled oxidation (Zanelli, O'Hare, et al., 1980). The product had the characteristic retention time in high-performance liquid chromatography (Zanelli, O'Hare, et al., 1980). In the in vivo chick hypercalcemia bioassay, this material had lost 90% of its bioactivity; a similar loss was recorded in the cytochemical bioassay (Chambers et al., 1976). Oxidation of human PTH also caused loss of bioactivity. The specificity was also tested as follows: (1) incubation of bovine and human PTH with an antiserum used in clinical immunoassay and known to recognize antigenic determinants within the N-region of the PTH molecule abolished the cytochemical response (Kent, unpublished results); (2) incubation of human PTH with a synthetic peptide [Nle[8],Nle[18],Tyr[34]] BPTH (3-34) amide, known to act as a competitive antagonist for PTH in the classic in vitro renal membrane adenylate cyclase bioassay (Rosenblatt et al., 1977), also inhibited the cytochemical response (Arber et al., 1981;) (3) application of pharmacological concentrations of other substances known to act on kidney tissue, or known to stimulate membrane adenylate cyclase in other tissue, e.g., human and salmon calcitonins, arginine vasopressin, adrenocorticotropic hormone, insulin, glucagon, 1,25-dihydroxycholecalciferol, prostaglandin E_2, and epinephrine, showed that even supraphysiological concentrations of these substances did not mimic the effect of PTH in stimulating G6PD activity. On the other hand, cyclic AMP or its dibutyryl derivative (at concentrations of 10^{-11}-10^{-9} M) mimicked the stimulation of G6PD activity produced by PTH (Goltzman et al., 1981).

The validity of the assay of PTH-like bioactivity in plasma has been assessed as follows. First, plasma, including that from patients with primary hyperparathyroidism, assayed at multiple dilutions, has elicited a dose-dependent response curve which is not significantly nonparallel to that produced by the standard preparation of PTH. Second, plasma from patients who have undergone total thyroparathyroidectomy and who are clinically hypoparathyroid has not elicited a response in the assay. It should be noted that patients with either idiopathic hypoparathyroidism, or hypocalcemia following the sometimes less than total parathyroidectomy for removal of a parathyroid adenoma, may have low, but not necessarily absent, concentrations of circulating bioPTH; the cytochemical bioassay system is readily able to measure such concentrations which are undetectable by immunoassay. Third, it has been possible to "neutralize" the PTH-like plasma bioactivity by preincubation with antiserum under conditions known to abolish the bioactivity of the standard preparation. Fourth, there is quantitative recovery of standard PTH added to the plasma sample before assay; this test is to exclude interference by endogenous substances which may prevent or enhance expression of the cytochemical response to PTH. Fifth, the addition of a known competitive antagonist to human PTH such as the [Nle[8],Nle[18], Tyr[34]] BPTH (3-34) amide has abolished the cytochemical response to PTH bioactivity in the plasma.

As with all clinical assay systems, the "activity" detectable in clinical samples may be composite, and include contributions from the secreted hormone and associated active products produced by the normal or abnormal parathyroid tissue or generated by postsecretion events which may also be normal or abnormal. Study of a new group of patients should seek to establish apparent identity between the standard preparation used in the cytochemical bioassay, usually highly purified extracted hormone, and the plasma bioactivity: in addition to the five aspects already listed, such studies should also include investigations of the time course of the response. Should these studies yield results that suggest that the standard and the sample do not appear to be identical, valid assay of the activity in plasma would not be possible (see also Chapters 2 and 3). On the other hand, the discrepant responses could be used to suggest the nature of the active material or that of the interference (Chapter 3).

Comparison of Results Obtained with the Cytochemical Bioassay and the Renal Receptor Assay

The guanyl-nucleotide-amplified adenylate cyclase bioassay, described in detail by Nissenson and his colleagues (Nissenson, Abbott, et al., 1981) has been validated by the criteria defined in the previous section for validation of the cytochemical bioassay. Yet there are major differences in the nature of the bioactivity detected and quantified by the two systems.

First, there appears to be a difference in sensitivity. The limit of detection for the amplified receptor assay for standards prepared from intact bovine PTH, intact human PTH, or synthetic bovine or human (1-34) peptide (equipotent on a molar basis) is 30 pM, with a dose range extending by a factor of at least 10 to concentrations greater than 300 pM. Thus, it seems that the receptor system is more than three orders of magnitude less sensitive than the cytochemical system. On the other hand, the limit of detection and the dose range for immunoassay systems, which recognize intact PTH, and for the amplified receptor assay, in terms of concentrations of standards of extracted hormone, appear to be very similar (i.e., 0.1-10 ng/ml; 10 pM to 1 nM). However when comparing the numerical values, considerable allowance must be made for the problems of standardizing biological peptides (see also Chapter 2).

The amplified receptor system of Nissenson, Abbott et al. (1981) is not sufficiently sensitive to measure bioactive PTH in the peripheral circulation of normal individuals. As can be seen from Table 3, the values for bioactive PTH in the serum from five of eight patients with hyperparathyroidism secondary to renal failure are in reasonable agreement with those reported by three different groups, using the cytochemical bioassay.

Niepel and her colleagues (1981; personal communication) have recently modified the renal adenylate cyclase assay to use human tissue; they report levels of PTH in serum that were detectable in 7 of 11 normal patients and elevated levels in the primary and secondary hyperparathyroid patients (Table 3). The results obtained with this assay are in good agreement with those obtained with the cytochemical bioassay. In the case of Avioli et al. (1981), the discrepancies between the levels reported for normal individuals [<6-24 pM BPTH (1-34) or

Table 3 Comparison of Blood Levels of PTH Measured by Different Bioassay Techniques in Normal Individuals and Primary and Secondary Hyperparathyroid Patients[a]

Bioassay Technique	Normal		Hyperparathyroid		Reference
Cytochemical bioassay[d]	2.9-29	(16)	34-11,000	(13)	Goltzman et al., 1980
	4-33	(8)	41-300	(8)	Fenton et al., 1978
	1.0-10	(4)	29-460	(6)	Nagant, Fischer, Dambascher, Devogelauer, et. al., 1981
GTP-enhanced renal adenylate cyclase	Not measurable[b]		<300-4000[e]	(8)	Nissenson, Abbott, et. al., 1981
	<60-240[f]		Not reported		Avioli et al. 1981
	<23-46	(11)[c,e]	35-220	(7)[e]	Niepel et al. (personal communication)
			<23	(1)[e]	

[a]Numbers in parentheses denote number of patients in the group.

[b]Canine renal cortex.

[c]Human renal cortex.

[d]Results expressed in pg equivalents of bovine PTH (1-84)/ml.

[e]Results expressed in pg equivalents of human PTH (1-84)/ml.

[f]Results converted to pg equivalents of bovine PTH (1-84)/ml assuming that bovine PTH (1-34), used as a standard, is equipotent on a molar basis with bovine PTH (1-84).

It should be noted that the nominal weight of a powder, obtained or purchased as peptide hormone, may contain up to 20% by weight of residues of buffer salts and water; of the remaining 80% of an unknown proportion of peptide may be a nonhormonal contaminant such as the α-chain of hemoglobin (in the case of extracted hormone), or hormone-related error peptides (in the case of synthetic peptides), or intermediate oxidation products of PTH, of varying and unknown degrees of bioactivity which inevitably arise during purification and lyophilization preparative procedures. See also Chapter 2.

<60-240 pg/ml BPTH (1-84)] compared with the normal values of 0.1-3 pM [BPTH (1-34) or 1-30 pg/ml BPTH (1-84)] found with the cytochemical bioassay systems are more marked.

There is insufficient published information on the two respective guanyl nucleotide amplified systems of Avioli et al. (1981) and Niepel et al. (1981) to speculate further on the bases for the increased sensitivity.

However, apart from the superficial differences in apparent sensitivity, the most significant difference between the cytochemical bioassay and the guanyl-nucleotide-amplified adenylate cyclase bioassay is in the form in which the sample must be presented either to the kidney tissue or to the membranes used as the assay system. Prompt processing of blood samples is essential for the cytochemical bioassay, as bioactivity in plasma is rapidly destroyed if it is left in contact with cells. The standard procedure is to collect blood into heparinized tubes in ice; the chilled blood is centrifuged and the plasma separated from cells within 10 min of collection. Plasma is then dispensed in aliquots, "snap frozen" rapidly in dry-ice, and stored at -70°C until assayed. Bioactive PTH is stable in plasma stored at 4°C for 90 min, whereas over 95% of the bioactivity is lost from whole blood stored at 4°C for 30 min and there is an 85% loss of bioactivity in plasma at room temperature for 30 min (Fenton, personal communication; Kent and Reeve, unpublished results). Paradoxically, the adenylate cyclase assay specifies the use of serum prepared from blood stored in ice for 1 hr prior to centrifugation (Nissenson, Abbott, et al., 1981). Anticoagulants, such as heparin or citrate, cause significant interference with this assay and preclude the assay of PTH in plasma. Thus it appears that although both assay systems use kidney cortex as a target tissue, there is a remarkable discrepancy between the assays concerning the apparent stability of bioPTH molecules in blood. This raises the possibility that the two assays respond through different mechanisms or that the cytochemical bioassay has a more stringent structural requirement for PTH than does the adenylate cyclase assay.

Two groups using the cytochemical bioassay (Chambers, Dunham, et al., 1978; Goltzman et al., 1980) and one of those using a guanyl-nucleotide-amplified renal membrane adenylate cyclase assay (Avioli et al., 1981) have attempted to validate the bioassay system physiologically by observing changes in concentrations of bioactive PTH in peripheral circulation relative to changes in peripheral concentrations of calcium produced by in vivo infusion of either calcium salts or chelating agents. A brief report from one group (Chambers et al., 1979) indicated that the level of bioactive PTH in plasma, as measured by cytochemical bioassay, was initially sensitive to small changes of total or ionized plasma calcium within the physiological range. However, if the infusion of calcium or chelating agent was continued to maintain the plasma calcium concentration at the upper or lower range of normal for up to 1 hr, the parathyroid secretory responsiveness, as measured by plasma bioactivity, was not maintained: a rebound response that escaped either the suppression or the stimulation became apparent after 20-30 min. Plasma immunoreactivity, as measured in both N- and C-region immunoassays, did not mirror the minute-to-minute increase in bioPTH concentrations evident at the start of the chelating agent, nor the suppression seen at the start of the calcium infusion. It is possible that the other groups (Goltzman et al., 1980; Avioli et al., 1981) have used sampling at the longer time intervals conventionally used for immunoassay experiments and thus have not observed these early and different response phases. Further information on the nature of the bioactive form or forms of PTH may be obtained, if the immediate short-term response of the parathyroid gland in vivo to small changes in circulating plasma calcium concentrations are studied by the cytochemical bioassay and by the guanyl-nucleotide-amplified renal membrane adenylate cyclase bioassay. Comparison of the calcium-dependent secretory response of parathyroid tissue in vitro has been studied by bioassay (a modified cytochemical bioassay system) and by immunoassay (Posillico et al., 1981; personal communication) and will be discussed in the next section.

Advantages and Disadvantages of the Different Assay Systems

The advantage of the radioimmunoassay and of the renal adenylate cyclase assays is their great throughput, making them ideal for routine clinical use. The disadvantage of both is that they are dependent on the characteristics of the reagents used in these assays (as discussed in Chapter 2). Moreover, the former is liable to measure biologically inactive molecules that have the necessary antigenic determinants; it is also relatively insensitive. To what extent the renal adenylate cyclase assay can be regarded as a true bioassay awaits further evidence, particularly in view of the discordant results obtained for the normal circulating levels of PTH (as discussed above, and shown in Table 3). The cytochemical bioassay is a true bioassay, so that it does not record biologically inactive fragments and metabolites of PTH. Its sensitivity is about one thousandth that of the best of the other two assays. However, it is unlikely to be suitable for use as a routine clinical assay because of the small number of samples that can be measured in each assay, and because it requires skilled operators. Thus, a maximum of only four to six samples can be measured in each assay, and it is difficult to foresee sufficient increase in output, even by the development of a section bioassay, that would justify its use as a routine diagnostic method. However, its use is justified when it is essential to define the concentration of biologically active PTH, as will be discussed in relation to pseudohypoparathyroidism; when low levels or concentrations of hormone have to be measured; or when anomalous results require detailed analysis of the bioactivity, as when the presence of inhibitory substances may be suspected.

APPLICATIONS OF THE CYTOCHEMICAL BIOASSAY OF PTH

In broad terms, the cytochemical bioassay of PTH (and the cytochemical techniques required for the assay) can be used in two different experimental situations: (1) in the study of the physiology and pathophysiology of PTH, both as secreted by the parathyroid gland and as present in the peripheral circulation, and (2) as a technique in studies on the mechanism of action and structure-function relationships of PTH on target tissues.

PTH from the Parathyroids and in the Peripheral Circulation

Studies on the control of secretion of PTH from the parathyroid gland under different experimental conditions have depended almost entirely upon quantitation of the secreted products by PTH immunoassay or measurement of radioactively labeled PTH molecules (Blum et al., 1974; Sherwood et al., 1978; Altenähr et al., 1979; Brown, 1981; Cohn and MacGregor, 1981). It is well established that a considerable amount of the material secreted consists of fragments of PTH and that the proportion of C-terminal fragments to intact hormone varies with changes in calcium concentration; in fact, it is markedly raised when extracellular concentration is increased (Cohn and MacGregor, 1981). This may result from increased intracellular degradation of PTH at high extracellular calcium concentrations which would act as a control mechanism to alter PTH secretion with variations in calcium concentrations without a significant alteration in the rate of biosynthesis of PTH (Cohn, Morrissey, et al., 1981). Thus in most of these studies the ratio of measured immunoreactive PTH secreted at low and high extracellular calcium is in the range of 2-10. More recent work by Posillico et al. (1981; personal communication), with either isolated guinea pig thyroparathyroid complex or superfused human parathyroid adenoma, has shown that the ratio of bioactive PTH (measured by a modified cytochemical bioassay system) at low (0.5 mM) and high (2.5 mM) calcium was 92 and 55, respectively, whereas the ratio for immunoreactive PTH was only 2 (Table 4). Thus, the maximum biosynthetic rate is the same in both assay systems, but with increased calcium concentration in the medium there is a much greater decrease in the secretion of bioactive PTH than in immunoreactive PTH. Moreover, the results obtained with the guinea pig thyroparathyroid complex shows a biphasic secretory response curve of bioactive PTH to varying calcium concentrations. At calcium concentrations of between 1 and 1.87 mM, the secretion of bioactive PTH fell from 75 to 16 pg/ml per min and then declined from 16 to 1 pg/ml per min when the calcium concentration rose from 2 to 2.87 mM. Over both ranges of calcium concentration, the PTH secretory response was sigmoidal in nature. The ratio of secretory rate at the highest and lowest calcium concentrations in each range was 4.6 and 16 for low and high calcium ranges, respectively. Thus, the

Table 4   Secretion of PTH from Isolated Parathyroid Tissues[a]

Medium Ca Concentration (mM)	Human Parathyroid Adenoma		Isolated Guinea Pig Thyroparathyroid Complex
	BioPTH[b]	IPTH[c]	BioPTH[b]
2.5 (high)	23 ± 2	634 ± 75	0.82 ± 0.05
0.5 (Low)	1282 ± 184	1147 ± 43	75.4 ± 1.05
Ratio at low/high calcium	55.7	1.8	92

[a]Biosynthetic rate, pg/ml per min; mean ± SEM.

[b]Bioactive PTH.

[c]Immunoreactive PTH.

Source: Posillico, Anderson, et al. (1981); Posillico, Tyrey, et al. (1981).

parathyroid gland appears to have two different acute secretory response patterns to progressive hypocalcemia, with a changeover region between 1.75 and 2 mM calcium.

The cytochemical bioassay would also be of great value in studies on PTH secretagogues (Morrissey and Cohn, 1979a, b; Brown and Aurbach, 1980). In isolated porcine parathyroid cells incubated with radioactive amino acids, the specific radioactivity of secreted hormone was much greater than that of the intracellular pool of PTH and hypocalcemia caused an increase in secretion from both stored and newly synthesized PTH. The PTH secretagogues, dibutyryl-cAMP and (l)-isoproterenol, stimulated secretion only from the stored pool of hormone. It appears that the gland contains PTH pools that can be mobilized to different degrees by different secretagogues. The bioactivity of this released PTH under these conditions has not been measured, so it is difficult to determine the physiological role of secretagogues, other than calcium. Recent ultrastructural evidence from human parathyroid adenomata maintained in tissue culture (Dietel and Dorn-Quint, 1980) has shown that, in the initial phase of the increased secretion of PTH in response to hypocalcemia, there is marked proliferation of the organelles responsible for the production of PTH (i.e., free ribosomes and endoplasmic reticulum) without any increase in those responsible for hormone secretion (i.e., Golgi complex and secretory granules). This supports the hypothesis of a "bypass" secretion of PTH during acute hypocalcemia (Cohn and MacGregor, 1981). The net effect of the high biosynthetic rate, high rate of intracellular degradation, secretion of fragments, and the possibility of

different secretory pools of PTH within the parathyroid cell demonstrate the
need for determining precisely the amount of bioactive PTH secreted by the
gland in response to the wide range of agents effective in modifying the rate of
secretion of PTH.

Effect of Surgery

Many surgical procedures can produce significant disturbances in blood electrolyte levels, and because of the requirement for a constant level of extracellular
ionized calcium for a number of cellular functions, it is of considerable clinical
importance to understand the interaction of blood ionized calcium levels and the
secretory state of the parathyroid glands during such periods of imposed stress.
This problem was addressed in a recent study by Chambers and his colleagues
(1983) in which arterial blood bioPTH levels were measured in patients before,
during, and after bypass circulation for open heart surgery after cardioplegic
heart arrest. Blood electrolytes, proteins, total and ionized calcium, and pH and
body temperature were also measured. In these patients, blood levels of total
calcium and albumin decreased during heart bypass (probably due to hemodila-
tion) but the level of ionized calcium remained within the normal range, as
did that of bioactive PTH. Blood pH fell after cooling of the body to 30°C or
below. The sharp decline in pH (to below 7.3) during the cooling to 30°C or
above resulted in a marked increase in blood level of bioactive PTH without any
change in blood ionized calcium concentration. If patients were cooled below
30°C the same degree of acidemia produced only a small transient increase in
blood bioactive PTH levels. Thus under these surgical conditions the normal
feedback control of PTH secretion by ionized calcium is uncoupled to such an
extent that acidosis during hypothermia (above 30°C) either causes an increase
in PTH secretion or reduces the rate of peripheral inactivation of the hormone.
The intriguing observation in this study is that despite all the surgical procedures
and changes in body temperature, blood pH, and bioPTH levels the concentration of ionized calcium is still well maintained.

Diurnal Rhythms

The blood level of ionized calcium (Ca^{2+}) is thought to be very constant, although improvements in the accuracy of measurement have shown a significant
diurnal rhythm of this parameter in normal adult males (Markowitz et al., 1981).
Although the maximum change in blood Ca^{2+} concentration was only ±3% of
the mean daily value, it was reproducible in four separate studies, with a peak at
0900-1000 hr and the nadir at approximately 2000 hr. Another report also
showed a similar diurnal change in blood Ca^{2+} concentration with an 8% decline
from 0800 hr to the middle of the afternoon (Sachs et al., 1981). Plasma

phosphate (P) levels showed a diurnal rhythm, but this was greater in mangitude
($\pm 14\%$ of mean daily value) and of the opposite phase to that of the blood Ca^{2+}
rhythm. The nadir occurred at about 1100 hr followed by two peaks, one at
about 1600 hr and a larger peak at 0200-0400 hr (Markowitz et al., 1981).

These studies were preceded by several studies that had reported considerable diurnal variation in the serum levels of immunoreactive PTH (Berson and
Yalow, 1967; Arnaud et al., 1971; Jubiz et al., 1972; Sinha et al., 1975, Parthemore et al., 1978). The immunoassays used in these studies all estimated pre-
dominately the C-terminal region of PTH and presumably C-terminal fragments
and since these have a relatively long plasma half-life, the results obtained with
these assays tend to integrate changes over a period of minutes or hours. In general, there was a peak level of PTH at night (usually during sleep) which declined during the morning. This prompted more detailed studies on blood PTH
levels during sleep (Parthemore et al., 1978; Kripke et al., 1978). More frequent
times of blood sampling during sleep showed an even more complex pattern of
PTH secretion, suggesting episodic secretion of PTH. Not surprisingly, the rapid
changes in PTH secretion were not correlated with changes in blood total calcium concentration, although some of the periods of high secretion rate were
quite sustained (Parthemore et al., 1978). Unfortunately, the rate of decline of
immunoreactive PTH, between samples only minutes apart, was considerably
greater than the plasma PTH half-life after parathyroidectomy in patients with
primary hyperparathyroidism measured with the same immunoassay. Kripke
et al., (1978) did find some correlation between changes in plasma total calcium
concentration and plasma PTH levels during the high-frequency episodes of PTH
secretion but very little interrelationship in the lower frequency range. During
these low-frequency periods both PTH and calcium were better related to the
stage of sleep rather than to each other, suggesting a possible neural involvement
in the control of PTH secretion.

Ultradian rhythms of the blood levels of immunoreactive PTH have been
demonstrated in the anesthetized dog (Fox et al., 1981). The immunoassay used
in this study is reported to detect only intact PTH (1-84) since it has virtually no
affinity for fragments (44-68) and (53-84) of PTH and only minimal affinity for
fragments (1-34) and (28-48) of PTH. Statistically significant cycles of immunoreactive PTH, with an amplitude $\pm 14\%$ of the mean immunoreactive PTH level
and mean period of 12 min (range 10-15 min), were seen in pre- and postcaval
blood samples from normocalcemic dogs. Similar cycles, with a slightly shorter
period of 8.4 min, were also seen in the thyroparathyroid venous effluent, indicating that the cycles seen in the peripheral blood probably result from changes
in the rate of secretion of the hormone. This rhythmicity in blood levels of immunoreactive PTH was abolished during the early part of a 60 min EGTA*-

---

*1,2-Bis-2-aminoethoxyethane-N,N'-tetra-acetic acid.

induced hypocalcemia (when PTH secretion was markedly elevated) but tended
to return toward the end of the EGTA infusion despite persistent hypocalcemia.
The cycles of blood levels of immunoreactive PTH were not correlated to any
measurable changes in blood total calcium concentration or, in a preliminary
study, to the ionized calcium concentration. The precise mechanism that syn-
chronizes all four glands is still unclear.

PTH in Pregnancy and in the Fetus

A preliminary study by Arber and colleagues (personal communication) compared ionized calcium and bioactive PTH (measured by the cytochemical bioassay) in blood samples collected at different times during the day in pregnant
women. The samples, collected every 3 hr for one 24 hr period, in each of four
women in the third trimester of pregnancy, showed a dirunal rhythm in blood
levels of bioactive PTH with values increasing about three- to fivefold from a late
afternoon nadir (about 1600 hr) to a maximum at about 0200-0400 hr. The levels
at the peak were toward the upper limit of the normal range. The radioimmunoassayable levels of PTH, using both N- and C-terminal region-specific assays, on
the same samples were all toward the upper limit of the normal range, but did
not show any significant diurnal variation. Although the results from this small
study suggest a diurnal rhythm in blood levels of bioactive PTH, it will be necessary to repeat these types of studies on normal individuals, to investigate the
possible circadian and ultradian rhythm of bioPTH levels in blood. If these are
established, it will be necessary to take into account the frequency and amplitude of such rhythms in interpreting the results from single samples of blood collected at different times of the day.

The homeostatic mechanisms involved in the physiological calcium stress during pregnancy have been widely studied (Pitkin, 1975; Pitkin et al., 1979). Maternal blood calcium levels tend to fall during late pregnancy and are generally
lower than in the fetal circulation. Fetal hypercalcemia is evident in both the
total and ionized fractions. There is general agreement that there is an increase
in immunoreactive PTH in the maternal circulation during the third trimester
of pregnancy, but several studies have shown that umbilical cord blood levels are
low or undetectable. Such observations led to the hypothesis that the fetal
parathyroid glands are suppressed in utero because of the persistent hypercalcemia. However, there is evidence to suggest that this is not the case: there is
immunocytological evidence of the presence of PTH in human parathyroid
glands as early as 10 weeks of gestation (Leroyer-Alizon et al., 1981); the parathyroid glands produce a substance that can resorb bone in vitro (Scothorne,
1964), and primate fetal responses to induced hypocalcemia are quantitatively
similar to the maternal response (Pitkin et al., 1980). This prompted a detailed
study on the comparative blood levels of bioactive and immunoactive PTH as

Table 5 Biologically Active PTH in the Fetal and Maternal Plasma at Delivery

	Mother (peripheral) venous blood)	Fetus (umbilical cord venous blood)
Calcium, corrected for albumin (mM)	2.33 ± 0.09	2.66 ± 0.05
Phosphate (mM)	1.00 ± 0.06	1.61 ± 0.14
N-PTH immunoassay, normal range < 40-120 pg/ml; 68% undectable	255 ± 84	Undetectable
C-PTH immunoassay, normal range 103-790 pg/ml	470 ± 70	197 ± 47
Bio-PTH cytochemical bioassay, normal range 1-6 pg/ml	8.07 ± 4.13	43.1 ± 18.3

Source: Data from Allgrove, Manning, et al., 1981).

well as calcium and phosphate in the maternal and fetal circulations (Allgrove, Manning, et al., 1981; personal communications). Samples of peripheral venous blood from mothers at delivery and from umbilical cord venous blood of full-term normal infants were collected; the cytochemical bioassay was used to measure bioactive PTH. The results are summarized in Table 5. Plasma calcium and phosphate were significantly elevated in the fetus compared to the mother. Immunoradiometric levels of PTH, measured with the N-terminal assay, were elevated in about half the maternal group, but were undetectable in the fetal circulation. The plasma levels measured with the C-terminal assay were within the normal range for the maternal group, but were significantly higher than the fetal group. In marked contrast, the levels of bioactive PTH were five- to six-fold higher in the fetal circulation than in the maternal circulation, despite the slight though significant increase above normal in the maternal group. Studies on the nature of the bioactivity in fetal plasma indicated that it was PTH, since the activation of G6PD in guinea pig kidney by fetal plasma had a time course identical to that produced by maternal plasma and that pretreatment of fetal plasma samples with PTH antiserum (that used in the N-terminal immunoassay) abolished all the bioactivity. There was a striking positive correlation ($r = 0.83$, $P < 0.01$) between the difference between maternal and fetal plasma calcium concentrations (i.e., the placental calcium gradient) and fetal plasma levels of bioactive PTH.

   This study confirms that mild physiological hyperparathyroidism exists during late pregnancy, which might not be detected by immunoassay because it might perhaps be associated with increased clearance of C-terminal fragments of PTH. In the fetus, however, despite undetectable N-terminal and low C-terminal immunoreactivity, PTH bioactivity is markedly increased. Allgrove and his coworkers calculated that the ratio of mean values of N-terminal PTH (by immunoassay) to bioactive PTH was <2.4:1 (i.e., ~ 40% biologically active) in fetal plasma and 61:1 (i.e., less than 2% biologically active) in the maternal circulation. This difference could be explained either by a relative alteration in the degradation profile of PTH or to changes in the metabolic clearance rate of PTH. Thus, rather than being suppressed, independent studies show that both human and primate fetal parathyroids appear to be very active even in the face of the hypercalcemia. This, together with the significant correlation between fetal plasma PTH bioactivity and the gradient of calcium across the placenta, suggest that the maintenance of the fetal hypercalcemia may be a consequence of fetal parathyroid gland activity, perhaps on the placenta.

   In studies on anesthetized pregnant pigs near term, Care and colleagues (personal communication) found that the fetal blood concentration of calcium (both total and ionized) was considerably higher than in the maternal circulation. Fetal blood levels of immunoreactive calcitonin, and of PTH (measured both by a C-terminal immunoassay and by the cytochemical bioassay) also were higher than in the maternal circulation. The fetal/maternal ratio for immunoreactive PTH was 1.7, but it was 3.4 for bioassayable PTH, indicating a relatively high proportion of bioactive PTH in the fetal circulation rather than a change in the rate of clearance of C-terminal fragments of PTH. Parathyroid hormone secretion, measured by the immunoassay, was incompletely suppressed by induced hypercalcemia in the fetus, suggesting some autonomy of the parathyroid gland. This apparent autonomy may be a result of some other PTH secretogogue in the fetal circulation (e.g., catecholamines, which are released during anesthesia, or the β-blocker dl-propanolol, which caused a transient reduction in fetal plasma immunoreactive PTH concentration). On the other hand, fetal hypercalcemia in the pig, as in the human, may be a consequence of increased fetal parathyroid gland activity. The measurement of the fetal blood levels of bioactive PTH during the period of induced hypercalcemia would help to clarify this situation since there is already evidence for considerable dissociation between the blood levels of PTH measured by the immunoassay and by the cytochemical bioassay.

Half-Life of PTH in the Circulation

The dissociation of circulating immunoreactive and bioactive PTH has recently been studied by three different groups. Fenton and his colleagues (personal communication) and Allgrove and his colleagues (personal communication) followed the decrease in blood levels of PTH, as measured by the cytochemical bioassay and a variety of immunoassays, in patients undergoing parathyroidectomy for primary or secondary hyperparathyroidism. Fenton and his colleagues studied three patients with hypercalcemia, one with primary hyperparathyroidism and two with hyperparathyroidism secondary to renal failure. When plasma PTH was measured by cytochemical bioassay and by two radioimmunoassays, one a C-terminal assay and one with mixed specificity measuring both C- and N-regions of PTH, the PTH levels were found to be elevated in all three patients by the three assay systems used. From the time of removal of the first parathyroid gland, bioassayable levels of PTH fell rapidly to be within or below the normal range by 30 min. The initial rate of decline in the patient with primary hyperparathyroidism indicated a biological half-life of approximately 2 min. The rate of loss of immunoreactivity was considerably slower than the loss of bioactivity. Thus, the C- and N-immunoassay assay levels were within the normal range 60 min after operation in the patient with primary hyperparathyroidism but took 6-24 hr to reach normal values in the patients with renal failure. The blood levels measured with the C-terminal assay were elevated at 2 hr after operation in the primary hyperparathyroid patient but reached normal levels by 24 hr, whereas the two renal failure patients still had markedly elevated levels 24 hr after operation.

In a similar study, Allgrove and colleagues (personal communication) measured the rate of loss of bioactive PTH in two patients with primary hyperparathyroidism at surgery. This group reported a single exponential loss of bioactivity that rapidly fell below the normal range, with half-lives of 1.2 and 0.96 min. In a study on one patient with primary hypomagnesemia, a rapid infusion of magnesium, sufficient to increase plasma Mg into the normal range, provoked a large transient increase in the secretion of PTH and this too declined as a single exponential with a half-life of 1.2 min.

The third study involved the intravenous injection of synthetic human PTH (1-34) into two normal males (Kent, Loveridge, Reeve, and Zanelli, unpublished results). Plasma levels of bioactive and immunoreactive PTH reached peak levels at 2-4 min after injection and declined in a single exponential (with a half-life of 1.5 min) when measured with the cytochemical bioassay. When the same samples were measured with a homologous radioimmunoassay (Zanelli, Rafferty, et al. 1980), there was a double exponential loss of immunoreactivity with the first phase (after allowing for a 3 min mixing period) having a half-life of about 10 min. Thus, the results from all three independent studies are in good agreement that the half-life of circulating bioactive PTH, as measured by the cytochemical bioassay, is 1-2 min.

Table 6 Biochemical Data on Pseudohypoparathyroid and Nutritional Vitamin D-Deficient Patients Before and After Treatment (mean ± SEM)

	Pseudohypoparathyroidism Type 1		Nutritional Vitamin D Deficiency	
	Pretreatment (n = 4)	Posttreatment (n = 7)	Pretreatment (n = 9)	Posttreatment (n = 7)
Plasma Calcium (mM)	1.76 ± 0.1	2.38 ± 0.03	1.94 ± 0.08	2.26 ± 0.02
Bioactive PTH, normal range 1.3-6.9 (pg/ml)	30 ± 8	4 ± 0.9	38 ± 5	4.2 ± 1.1
N-terminal PTH Immunoassay, normal range <40-120* (pg/ml); 68% not detectable	1200 ± 300	1 normal 3 120-200 3 480-1400	600 ± 70	Normal
Ratio $\dfrac{\text{mean bioactive PTH}}{\text{mean immunoreactive PTH}}$	1:40	—	1:15.8	—

*Based on (1-34) PTH.

Source: Data from Allgrove, Manning, et al. (1981 and personal communication).

Pseudohypoparathyroidism Type 1

Pseudohypoparathyroidism (PSP), first described by Albright et al. (1942), is
an inherited disease of primary unresponsiveness to multiple hormones and was
originally characterized by a resistance to the metabolic effects of PTH. In the
untreated situation the patients generally have hypocalcemia, hyperphosphat-
emia, elevated serum immunoreactive PTH, and resistance to the renal phosphaturic and adenylate cyclase effects of exogenously administered bovine
PTH. These and observations with other hormones led to the hypothesis that
the syndrome was a result of a defective receptor response to hormones that involve activation of adenylate cyclase, particularly PTH. This suggestion is supported by in vitro studies that indicate a deficiency in the guanine-nucleotide-
binding regulatory subunit of renal membranes, erythrocytes, and skin fibroblasts from PSP patients (Bourne et al., 1981; Farfel et al., 1981). However, this
view may be too simplistic, since the hypercalcemic and phosphaturic responses
to exogenous PTH can be restored in patients with PSP after treatment with vitamin D (Suh et al., 1970; Stögmann and Fischer, 1975; however, Breslau et al.,
1980 had discordant findings) and in one PSP pateint following the removal of
four hyperplastic parathyroid glands (Nagant and Krane, 1978). This led to two
recent studies on the bioactivity and immunoreactivity of PTH in the blood of
patients with PSP both before and after treatment (Nagant, Fischer, Dambacher, Werder, et al., 1981; Allgrove, Adami, et al., 1981; personal communication).

In the study by Allgrove et al. (Table 6), in both the PSP type I and nutritional vitamin D-deficient groups, the patients had similar degrees of hypocalcemia and elevated levels of PTH by both bioassay and N-terminal immunoassay.
However, the ratio of bioactive to immunoreactive PTH (not corrected for the
different standards used in the two assay systems) ranged from 1:23 to 1:79
for the PSP patients and from 1:7 to 1:29 for the vitamin D-deficient group
(Allgrove, personal communication). The results indicate that in the untreated
PSP patient there is approximately three times as much immunoreactive PTH
per amount of bioactive PTH as there is in the untreated vitamin D-deficient
patients. After treatment, the plasma concentrations of total calcium were
within the normal range for both groups, as were the blood levels of bioactive
PTH. Immunoreactive PTH fell to normal levels after treatment in most of
the vitamin D-deficient patients [although some remained elevated even after
2 years on treatment (Allgrove, personal communication)] but remained elevated in most of the PSP patients although it had decreased from pretreatment
levels.

In the other study (Nagant, Fischer, Dambacher, Devogelauer, 1981), a
greater number of patient groups were investigated (Table 7). Patients with PSP
receiving treatment had normal levels of bioactive PTH but still had paradoxically

Table 7 Comparison of Bioactive and Immunoactive Levels of PTH in Controls and Patients with Various Parathyroid Diseases[a]

	ol	Patient Group			
	Control	Pseudohypo-parathyroid[b]	Primary Hypoparathyroid	Primary Hyper-parathyroid	Secondary Hyper-parathyroid[c]
Bioactive PTH (pg/ml)	8.4 ± 2.6 (9)	12.8 ± 4.3 (11)	0.9 ± 0.15 (4)	24.4 ± 66 (10)	71.2 ± 43.3 (12)
N-terminal PTH Immunoassay (pg/ml)	91.3 ± 2.12 (4)	570 ± 85 (11)	<20	453 ± 85 (10)	418 ± 88 (5)
Ratio $\frac{\text{mean bioactivity}}{\text{mean immunoreactivity}}$	1:1.09	1:44.5	—	1:1.85	1:5.9
Recovery of bioactivity added to plasma (%)	68 ± 4.3 (8)	9.9 ± 3.4 (9)	— —	78 ± 11.7 (3)	77.7 ± 7 (3)
Plasma Ca (mM)	2.38 ± 0.07 (4)	2.29 ± 0.07 (13)	2.23 ± 0.16 (6)	2.87 ± 0.06 (10)	2.07 ± 0.12 (6)

[a]Data are shown for each group as mean ± SEM (number in group).

[b]Patients on treatment with vitamin D.

[c]Nutrition vitamin D deficiency and chronic renal failure.

Source: Data from Nagant et al. (1981) and Loveridge (personal communication).

high levels of N-terminal immunoreactive PTH. The ratio of mean bioactive to immunoreactive PTH in PSP patients was 4 times lower than that in the control group and 24 and 7.5 times lower than the ratio in primary and secondary hyperparathyroid patients, respectively. One interesting observation in this study is that patients with primary hyperparathyroidism had the greatest percentage of PTH immunoreactivity that was bioactive. Thus, the results reported from the two groups for both treated and untreated PSP are in good agreement.

One of the criteria for the assay of PTH bioactivity in plasma samples is the quantitative recovery of biologically active standard PTH added to the plasma sample before assay (see Assay Validation above). Applying this criterion to samples from the patients studied by Nagant and his colleagues, good recovery was found in the control and in both primary and secondary hyperparathyroid groups, but very low (<10%) recovery from plasma from PSP patients (Table 7). This certainly suggests that there is something present in the plasma of PSP patients that "interferes" with the expression of bioactive PTH added to the plasma and therefore probably interferes with the expression of endogenous PTH bioactivity. Recent evidence with plasma from one untreated PSP patient indicated that at least 100 ng/ml of bioactive PTH needed to be added to the plasma sample to achieve a reasonable recovery of bioactive PTH (Loveridge, personal communication). The measured levels of immunoreactive PTH do not immediately suggest that the low recovery values can be accounted for by the presence of antibodies to PTH; however, this possibility cannot yet be excluded.

The results from both groups of workers show a dissociation of bioactive and immunoreactive PTH in PSP type I patients, as would be predicted from the clinical condition of the patients. This supports a hypothesis that the impaired response to PTH in PSP type I may be due in part, or solely, to blocking of the response to PTH, presumably at the receptor level, by a circulating form of PTH that has immunological cross-reactivity with PTH but is devoid of biological activity. Whether this is associated with genetic phenotypes which secrete an aberrant form of PTH or which show alterations in the peripheral metabolism of PTH is not known. Further studies in this complex syndrome, especially involving bioassay techniques for other hormones as well as PTH, are required.

Hypercalcemia of Malignancy

Hypercalcemia is a common complication of malignant disease and appears to result from either direct resorption of bone by tumor metastases or by osteolytic humoral factors secreted by the tumor (Stewart et al., 1980; and as reviewed by Bockman, 1980). The cytochemical bioassay of PTH has been used in studies on the nature of the hypercalcemia factor in plasma (Goltzman et al., 1981) or in the tumor itself (Loveridge, Kent, and Heath, unpublished results) in patients with hypercalcemia of malignancy not of parathyroid origin.

PTH-like bioactivity, measured with the cytochemical bioassay, was increased in the plasma from 10 of 16 patients with hypercalcemia of malignancy who also showed increased nephrogenous cyclic AMP excretion (Goltzman et al., 1981). The blood levels of PTH-like bioactivity were 10-fold higher in patients with elevated nephrogenous cyclic AMP excretion than those with suppressed nephrogenous cyclic AMP excretion. It is interesting to note that these workers found that cyclic AMP, and its dibutyryl derivative, could mimic the G6PD stimulation elicited by PTH in the cytochemical bioassay. In one of the patients studied after the tumor had been resected, there was a fall in plasma PTH-like bioactivity, serum calcium, and nephrogenous cyclic AMP excretion. Parathyroid hormone-like bioactivity could be demonstrated in the tissue culture fluid of this tumor maintained in tissue culture. Plasma from patients with increased PTH bioactivity was fractionated by gel chromatography and had a major component of "bioactivity" which eluted before PTH (i.e., larger in size). This contrasted with the elution profile of bioactive PTH in plasma from a patient with primary hyperparathyroidism where most of the bioactivity coeluted with intact PTH. When plasma from some patients with hypercalcemia of malignancy showing high PTH bioactivity was incubated with a PTH antiserum before bioassay, there was only a very small loss of PTH-like bioactivity, whereas the same antiserum caused >99% loss of bioactivity in plasma from a patient with primary hyperparathyroidism.

Loveridge, Kent, and Heath (personal communication) investigated the possible PTH-like bioactivity in extracts of a carcinoid tumor of the left lung. The patient was severely hypercalcemic (>4 mmol/liter), had radiological evidence of "hyperparathyroid" bone disease (subperiosteal erosion, brown cysts, and chondrocalcinosis), severe osteitis fibrosa in the bone biopsy, undetectable immunoassayable PTH, and, at autopsy, four macroscopically and microscopically normal parathyroid glands. Neutralized acetic acid extracts of lyophilized tumor (approximately 100 mg tumor in 5 ml of extract) were studied for PTH-like bioactivity using the cytochemical bioassay. At a final dilution of $1:10^9$-$1:10^{10}$ the tumor extract gave maximal activation of G6PD in the cytochemical bioassay for PTH and further dilution (to $1:10^{14}$) produced a decrease in G6PD activation parallel to the PTH standard curve. A time course similar to that given by PTH was produced by the tumor extract at these concentrations. Time-course studies with a $1:10^{10}$ dilution of the extract also revealed other PTH-like cytochemical responses, i.e., activation of alkaline phosphatase, NADPH-diaphorase, and glyceraldehyde-3-phosphate dehydrogenase in proximal, pars recta, and distal tubules, respectively. Unlike PTH, the tumor extract did not appear to activate carbonic anhydrase in proximal tubules of the guinea pig kidney. Preincubation of the tumor extract with a PTH antiserum did not significantly inhibit the activation of G6PD by the tumor extract, whereas it did abolish the response to approximately an equivalent amount of bioactive standard PTH. Not surprisingly, the tumor extract did not show any appreciable PTH-like immunoreactivity in several PTH immunoassay systems.

In these two studies, PTH-like bioactivity was found to be present, both in plasma and in tumor extracts. Since its bioactivity could not be neutralized by PTH antisera it does not meet the strict specificity criteria for PTH. The material does appear to be very potent in relation to PTH and/or is of a much smaller molecular weight than PTH, although the evidence from Goltzman et al. (1981) suggests that it is larger than PTH (however, this does not preclude the possibility that it is a small molecule bound to some larger molecule in plasma). The factor is probably not a prostaglandin, since prostaglandins of the E series are not detected by the cytochemical bioassay (Goltzman et al., 1981). Its identity is unknown but it has sufficient PTH-like activity to interact in the PTH cytochemical bioassay.

This assay is thus likely to prove to be of great value in attempts to isolate and identify the active molecular species of the hypercalcemic factor secreted by some tumors.

STRUCTURE-FUNCTION RELATIONSHIPS OF PTH AND FUTURE DEVELOPMENTS OF THE CYTOCHEMICAL BIOASSAY

All known bioactivities of PTH are associated only with the N-terminal region of the molecule (Table 2). In the cytochemical bioassay of PTH, based on the stimulation of G6PD activity, fresh normal plasma and intact highly purified extracted PTH (human and bovine) produced two peaks of enzyme activation at 6-8 and 14-16 min after addition of plasma or hormone (Chambers et al., 1976; Arber et al., 1980). Synthetic human PTH (1-34) produces only the first peak of G6PD activation, whereas synthetic human PTH (53-84) produces only the second peak (Arber et al., 1980). Oxidation of PTH (1-84) with hydrogen peroxide [associated with the loss of classic biological activity; Table 2 (Tashjian et al., 1964)], incubation with the competitive antagonist (Nle^8, Nle^{18},Tyr^{34})BPTH (3-34) amide or incubation of plasma at 37°C for 2 hr results in loss of the 6-8 min peak of G6PD activation but does not abolish the 14-16 min response. Thus, the 6-8 min activation of G6PD is associated only with the N-terminal region of PTH and hence correlates with the biological activity of the molecule measured in other bioassay systems. In contrast, the 14-16 min enzyme activation is associated with the C-terminal region of PTH and is stimulated by fragments of PTH that are apparently devoid of bioactivity in other conventional bioassay systems. The cytochemical response elicited by the PTH fragment (53-84) is dose dependent, can be abolished by incubation with an antiserum used in the immunoassay for carboxyl-region PTH, and thus provides the only current bioassay system for

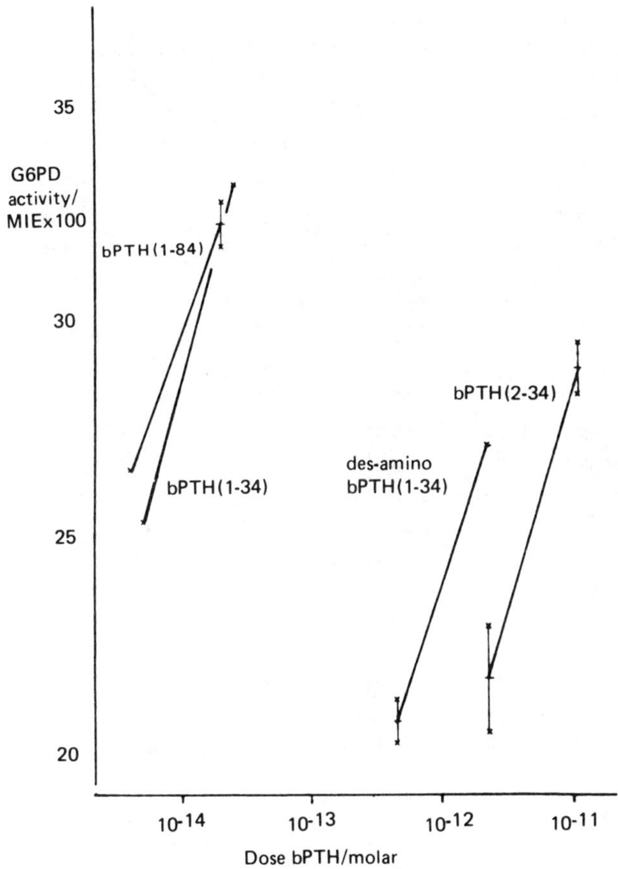

Figure 2   Structure-activity relationships of BPTH in the cytochemical bioassay of PTH measured as glucose-6-phosphate dehydrogenase activity in response to various concentrations of BPTH or its synthetic fragments. Standard PTH cytochemical bioassay conditions were used after preliminary experiments had shown that the time course of activation of G6PD was the same for all the peptides studied. (Data from Arber et al., 1981.)

C-terminal region PTH. This assay may prove to be of great use in the investigations on the biological activity and pathological significance of the C-terminal region of PTH.

As noted earlier, Goltzman et al. (1981) found that either cyclic AMP or its dibutyryl derivative produced a dose-dependent increase in G6PD activation when used in the dose range 10^{-11}-10^{-9} M. This response to cyclic AMP was parallel to the activation produced by PTH, and maximal doses of PTH and cyclic AMP were not additive in activating G6PD, suggesting that a common receptor is involved. The peak time of activation of G6PD was the same for both PTH and cyclic AMP (6 min), but whereas the activity rapidly declined by 8 min in the presence of PTH, G6PD stimulation was maintained (at least to 10 min) in the presence of dibutyryl cAMP. This evidence supports the hypothesis that cyclic AMP is the second messenger for PTH activation of G6PD in the kidney.

A study on the structure-function relationships of synthetic analogs of bovine PTH in the cytochemical bioassay system based on the 6-8 min activation of G6PD has recently been described (Figure 2; Table 2) (Arber, Zanelli, and Parsons, 1981). Native BPTH (1-84) and synthetic BPTH (1-34) are equipotent on a molar basis. Progressive modification of, and deletion from, the N-terminus led to a decrease in biological potency so that the fragment BPTH (3-34) produced no significant response even at molar concentrations 250 times the effective dose of BPTH (1-34). In fact the synthetic analog $(Nle^8,Nle^{18},Tyr^{34})$BPTH (3-34) amide, as already stated, was a competitive inhibitor for both (1-84) and (1-34) PTH, a result seen with other assay systems (Table 2). All the active analogs of bPTH show the 6-8 min activation of G6PD seen with native BPTH (1-84).

The series of potency estimates given in Table 2 shows that the guinea pig cytochemical bioassay has much closer similarities in structure-function requirements to the rat renal adenylate cyclase assay than to the in vivo chick hypercalcemia assay, indicating the need for an alternative cytochemical bioassay system for PTH based on bone. Techniques have recently been developed for cutting nondecalcified cryostat sections of bone suitable for quantitative cytochemistry (Johnstone, 1979). McPartlin et al. (1977; 1978) have shown that it is possible to use the PTH-induced changes in bone alkaline phosphatase in neonatal rat calvaria as a cytochemical bioassay for PTH. The direction and magnitude of the PTH-induced changes in alkaline phosphatase in this system were dependent upon the age of the animal and that there was a dose dependency for PTH 4-7 min after the addition of the hormone.

Several other lines of study have pointed to the apparent dissociation between kidney and bone responses to PTH. There is evidence from Martin, Bellorin-Font, et al. (1981) that (1-34) is active but that (1-84) is inactive in stimulating cyclic AMP production in isolated perfused bone; moreover, [Nle^8, Nle^{18},Tyr^{34}] BPTH (3-34) amide is an agonist in this system but it acts as an antagonist in all the kidney bioassay systems. The variability of the nature of hypoparathyroid bone changes in bone biopsies from PSP patients (Kidd et al., 1980) suggests that it may be possible that the circulating form of PTH in PSP type I patients is modified in such a way that it is active on bone but not active on the kidney. It may be particularly relevant in investigating the etiology of renal osteodystrophy to reconsider the role of the "biologically inert" carboxyl-region PTH.

To conclude this review of application of the cytochemical bioassay of PTH, it is evident that in the years since it was first described, this approach has already contributed significantly to the knowledge of some of the biological characteristics of PTH. We would predict that in the future quantitative cytochemical techniques will be used to give direct insight and discrimination of the complex physiological and pathological interactions of secreted and metabolized forms of PTH with their heterologous and homologous target tissues.

ACKNOWLEDGMENTS

We should like to thank Dr. J. Alaghband-Zadeh, Dr. J. Allgrove, Dr. C. Arber, Professor A. D. Care, Dr. D. Chambers, Mr. S. Fenton, Dr. D. Goltzman, Dr. D. Heath, Dr. N. Loveridge, Miss B. Niepel, and Dr. J. Posillico for providing us with their recent unpublished findings and for their constructive comments on the application of the cytochemical bioassay of parathyroid hormone.

Dr. M. Rosenblatt kindly provided a preprint of his chapter on PTH structure-function relationships and some of the synthetic PTH analogs used in several of the studies in this chapter.

We would also like to thank Dr. J. Chayen and Dr. L. Bitensky for many hours of discussion on cytochemical bioassays.

Dr. Kent would like to thank the Percy Bilton Charity for providing a fellowship to support some of the work reported in this chapter.

Dr. C. Arber was supported by MRC Grant G978/49/SB to Dr. J. Zanelli.

Our thanks to Miss D. Stewart for the preparation of this manuscript.

DEDICATION

This chapter is dedicated to our friend and colleague Dr. John A. Parsons who died in July 1981 and was to have been a coauthor of this chapter. Dr. Parsons, through his enthusiasm and foresight, initiated the collaborative effort that was ultimately responsible for the development and application of the cytochemical bioassay of PTH. His encouragement and enthusiasm resulted in several of the studies reported in this chapter. Dr. Parsons' informed comment and scientific insight in this field will be sadly missed by us all.

REFERENCES

Abbott, S. R., Nissenson, R. A., Teitelbaum, A. P., Clark, O. H., and Arnaud, C. D. (1980). Biologically active parathyroid hormone in human hyperparathyroid serum: Assay and characterization. *Trans. Assoc. Am. Physicians*, 93: 192-202.

Albright, F., Burnett, C. H., Smith, P. M., and Parsons, W. (1942). Pseudohypo-parathyroidism—an example of "Seabright Bentam syndrome." *Endocrinology,* 30:922-932.

Allgrove, J., Adami, S., Jayaweera, P., Chayen, J., and O'Riordan, J. L. H. (1981). Biologically active parathyroid hormone in pseudohypoparathyroidism. *Calcif. Tissue Int.,* 33 (suppl):189.

Allgrove, J., Manning, R. M., Adami, S., Chayen, J., and O'Riordan, J. L. H. (1981). Biologically active parathyroid hormone in foetal and maternal plasma. *Clin. Sci.,* 60:11P.

Altenähr, E., Arps, H., Montz, R., and Dorn, G. (1979). Quantitative ultrastructural and radioimmunologic assessment of parathyroid gland activity in primary hyperparathyroidism. *Lab. Invest.,* 41:303-312.

Arber, C. E., Zanelli, J. M., and Parsons, J. A. (1981). Comparison of the bioactivity of native bovine parathyroid hormone and its aminoterminal analogs in a cytochemical bioassay system. *Endocrinology,* 109(suppl):228. Proceedings of the 63rd meeting of the Endocrine Society.

Arber, C. E., Zanelli, J. M., Parsons, J. A., Bitensky, L., and Chayen, J. (1980). Comparison of the bioactivity of highly purified human parathyroid hormone and of synthetic amino- and carboxyl-region fragments. *J. Endocrinol.,* 85: 55-56P.

Arnaud, D. C., Goldsmith, R. S., Bordier, P. J., and Sizemore, G. W. (1974). Influence of immunoheterorgeneity of circulating parathyroid hormone on results of radioimmunoassays of sera in man. *Am. J. Med.,* 56:785-793.

Arnaud, C. D., Tsau, H. S., and Littledike, T. (1971). Radioimmunoassay of human parathyroid hormone in serum. *J. Clin. Invest.,* 50:21-34.

Avioli, R. C., Greene, A., Bell, N. H., and Epstein, S. (1981). A sensitive and rapid bioassay of human serum parathyroid hormone. *Calcif. Tissue Int.,* 33(suppl.):318.

Bennett, H. P. J., and McMartin, C. (1979). Peptide hormones and their analogues: Distribution, clearance from the circulation, and inactivation in vivo. *Pharmacol. Rev.,* 30:247-292.

Berson, S. A., and Yalow, R. S. (1967). Radioimmunoassay of peptide hormones in plasma. *N. Engl. J. Med.,* 277:640-647.

Berson, S. A., and Yalow, R. S. (1968). Immunochemical heterogeneity of parathyroid hormone in plasma. *J. Clin. Endocrinol. Metab.,* 28:1037-1047.

Blum, J. W., Mayer, G.P., and Potts, J. T., Jr. (1974). Parathyroid hormone responses during spontaneous hypocalcemia and induced hypercalcemia in cows. *Endocrinology,* 95:84-92.

Bockman, R. S. (1980). Hypercalcemia in malignancy. *Clin Endocrinol. Metab.,* 9:317-333.

Bouillon, R. (1978). Clinical applications and technical aspects of parathyroid hormone measurements. *Ric. Clin. Lab.,* 8:75-86.

Bourne, H. R., Kaslow, H. R., Brickman, A. S., and Farfel, Z. (1981). Fibroblast defect in pseudohypoparathyroidism Type I: Reduced activity of receptor-cyclase coupling protein. *J. Clin. Endocrinol. Metab.*, 53:636-640.

Breslau, N. A., Notman, D. S., Canterbury, J. M., and Moses, A. M. (1980). Studies on the attainment of normocalcemia in patients with pseudohypoparathyroidism. *Am. J. Med.*, 68:856-860.

Bringhurst, F. R., and Potts, Jr. (1981). Bone collagen synthesis in vitro: Structure/activity relations among parathyroid hormone fragments and analogues. *Endocrinology*, 108:103-108.

Brown, E. M. (1981). Set point for calcium: Its role in normal and abnormal parathyroid secretion. In *Hormonal Control of Calcium Metabolism*, D. V. Cohn, R. V. Talmage, and J. S. Matthews (eds.). Excerpta Medica, Amsterdam, pp. 35-41.

Brown, E. M., and Aurbach, G. D. (1981). Role of cyclic nucleotides in secretory mechanisms and actions of parathyroid hormone and calcitonin. *Vitam. Horm.* 38:205-256.

Canterbury, J. M., Bricker, L. A., Levey, G. S., Kozlovskis, P. L., Ruiz, E., Zull, J. E., and Reiss, E. (1975). Metabolism of bovine PTH. Immunological and biological characteristics of fragments generated by liver perfusion. *J. Clin. Invest.*, 55:1245-1253.

Chabardès, D., Gagnan-Brunette, M., Imbert-Teboul, M., Gontcharevskia, O., Montégut, M., Clique, A., and Morel, F. (1980). Adenylate cyclase responsiveness to hormones in various portions of the human nephron. *J. Clin. Invest.*, 65:439-448.

Chambers, D. J., Dunham, J., Zanelli, J. M., Parsons, J. A., Bitensky, L., and Chayen, J. (1978). A sensitive bioassay of parathyroid hormone in plasma. *Clin. Endocrinol.* (Oxf), 9:375-379.

Chambers, D. J., Dunham, J., Slavin, B., Quinney, J., Chayen, J., and Braimbridge, M. V. (1982). The effect of ionized calcium, pH and temperature on bioactive parathyroid hormone during and after open-heart surgery. *Ann. Thorac. Surg.* (in press).

Chambers, D. J., Schafer, H., Laugharn, J. A., Jr., Johnstone, J., Zanelli, J. M., Parsons, J. A., Bitensky, L., and Chayen, J. (1978). Dose-related activation by PTH of specific enzymes in various regions of the kidney. In *Endocrinology of Calcium Metabolism*. D. H. Copp and R. V. Talmage (eds.). Excerpta Medica, Amsterdam-Oxford, pp. 216-220.

Chambers, D. J., Tully, G., Rafferty, B., Zanelli, J. M., Chayen, J., and Parsons, J. A. (1979). Acute increase and decrease of biologically active circulating human parathyroid hormone (bioPTH) induced by positive and negative calcium challenges within the physiological range. *Calcif. Tissue Int.*, 27 (suppl): A6.

Chambers, D. J., Zanelli, J. M., Parsons, J. A., and Chayen, J. (1976). Cytochemical responses to the guinea-pig kidney cortex to low concentrations of bovine parathyroid hormone (0.001-0.01 pg/ml). *J. Endocrinol.*, 71:87P.

Chase, L. R., and Aurbach, G. D. (1968). Renal adenylate cyclase: anatomically different sites for parathyroid hormone and vasopressin. *Science*, 159:545-547.

Chayen, J. (1980). *The Cytochemical Bioassay of Polypeptide Hormones.* Springer-Verlag, Berlin.

Chayen, J., Bitensky, L., and Butcher, R. G. (1973). *Practical Histochemistry.* Wiley, London.

Chayen, J., Loveridge, N., and Daly, J. R. (1972). A sensitive bioassay for adrenocorticotrophic hormone in human plasma. *Clin. Endocrinol.* (Oxf), 1:219-233.

Chu, L. L.H., MacGregor, R. R., Anast, C. S., Hamilton, J. W., and Cohn, D. V. (1973). Studies on the biosynthesis of rat parathyroid hormone and proparathyroid hormone: Adaptation of the parathyroid gland to dietary restriction of calcium. *Endocrinology*, 93:915-924.

Cohn, D. V., and MacGregor, R. R. (1981). The biosynthesis, intracellular processing and secretion of parathormone. *Endocrine Rev.*, 2:1-26.

Cohn, D. V., Morrissey, J. J., Hamilton, J. W., Chu, L. L. H., and MacGregor, R. R. (1981). Secretion of parathormone from two independently controlled intracellular pools. In *Hormonal Control of Calcium Metabolism*, D. V. Cohn, R. V. Talmage, and J. L. Matthews (eds.). Excerpta-Medica, Amsterdam-Oxford-Princeton, pp. 28-34.

Cohn, D. V., Smardo, F. L., Jr., and Morrissey, J. J. (1979). Evidence for internal homology in bovine preproparathyroid hormone. *Proc. Natl. Acad. Sci. USA*, 76:1469-1471.

Coltrera, M., Rosenblatt, M., and Potts, J. T., Jr. (1980). Analogues of parathyroid hormone containing D-amino acid: Evaluation of biological activity and stability. *Biochemistry*, 19:4380-4385.

DeLuca, H. F., Paaren, H. E., and Schnoes, H. K. (1979). Vitamin D and calcium metabolism. In *Topics in Current Chemistry*, Vol. 83, M. J. S. Dewar, K. Hafner, E. Heilbronner, S. Ito, J.-M. Lehn, K. Niedenzu, C. W. Rees, K. Schäfer, G. Wittig, and F. L. Boschke (eds.). Springer-Verlag, New York, pp. 1-65.

Desplan, C., Julliene, A., Moukhtar, M. S., and Milhaud, G. (1977). Sensitive assay for biologically active fragment of parathyroid hormone. *Lancet*, ii:198-199.

Dietel, M., and Dorn-Quint, G. (1980). By-pass secretion of human parathyroid adenomas. A particular intracellular form of rapid adaptation to external stimuli. *Lab. Invest.*, 43:116-125.

Farfel, Z., Brothers, V. M., Brickman, A. S., Conte, F., Neer, R., and Bourne, H. R. (1981). Pseudohypoparathyroidism: Inheritance of deficient receptor-cyclase coupling activity. *Proc. Natl. Acad. Sci. USA*, 78:3098-3102.

Fenton, S., Somers, S., and Heath, D. A. (1978). Preliminary studies with a sensitive cytochemical assay for parathyroid hormone. *Clin. Endocrinol.* (Oxf), 9:381-834.

Fox, J., Offord, K. P., Heath, H., III (1981). Episodic secretion of parathyroid hormone in the dog. *Am. J. Physiol.,* 241 (Endocrinol. Metab. 4):E171-E177.

Fraser, D. R. (1980). Regulation of the metabolism of Vitamin D. *Physiol. Rev.* 60:551-613.

Gaillard, P. J., Herrmann-Erlee, M. P. M., and Hekkelman, J. W. (1976). Effects of PTH and some synthetic fragments of embryonic bone in vitro. *Calcif. Tissue Res.,* 21 (suppl):70-74.

Goltzman, D. (1978). Parathyroid hormone and hyperparathyroidism: Current concepts. *Can. J. Surg.,* 21:285-289.

Goltzman, D., Callahan, E. N., Tregear, G. W., and Potts, J. T., Jr. (1975). Studies of bioactive analogues of parathyroid hormone. In *Peptides: Chemistry, Structures and Biology,* R. Walter and J. Meienhofer (eds.). Proceedings of the Fourth American Peptide Symposium, Ann Arbor Science Publishers, Ann Arbor, Michigan, pp. 571-577.

Goltzman, D., Callahan, E. N., Tregear, G. W., and Potts, J. T., Jr. (1978). Influence of guanyl nucleotides on parathyroid hormone-stimulated adenyl cyclase activity in renal cortical membranes. *Endocrinology,* 103:1352-1360.

Goltzman, D., Henderson, B., and Loveridge, N. (1980). Cytochemical bioassay of parathyroid hormone. *J. Clin. Invest.,* 65:1309-1317.

Goltzman, D., Peytremann, A., Callahan, E. N., Tregear, G. W., Potts, J. T., Jr. (1975). Analysis of the requirements for parathyroid hormone action in renal membranes with the use of inhibiting analogues. *J. Biol. Chem.,* 250: 3199-3203.

Goltzman, D., Stewart, A. F., and Broadus, A. E. (1981). Malignancy-associated hypercalcemia: Evaluation with a cytochemical bioassay for parathyroid hormone. *J. Clin. Endocrinol. Metab.* 53:899-904.

Habener, J. F., Amherdt, M., Ravazzol, M., and Orci, L. (1979). Parathyroid hormone biosynthesis. Correlation of conversion of biosynthetic precursors with intracellular protein migration as determined by electron microscopy autoradiography. *J. Cell Biol.,* 80:715-731.

Habener, J. F., Kemper, B., and Potts, J. T., Jr. (1975). Calcium-dependent intracellular degradation of parathyroid hormone: A possible mechanism for the regulation of hormone stores. *Endocrinology,* 97:431-441.

Habener, J. F., and Potts, J. R., Jr. (1980). I. Parathyroid hormone: Recent advances in studies of the chemistry, biosynthesis, control of secretion, metabolism and immunoassay. *Handbuch der inneren Medizin, VI/1A. Knochen. Gelenken. Muskelm.* Springer-Verlag, Berlin, pp. 577-597.

Habener, J. F., Potts, J. T., Jr., and Rich, A. (1976). Pre-proparathyroid hormone. Evidence for an early biosynthetic precursor of proparathyroid hormone. *J. Biol. Chem.*, 251:3893-3899.

Habener, J. F., and Segre, G. V. (1979). Parathyroid hormone radioimmunoassay. *Ann. Intern. Med.*, 91:782-785.

Hamilton, J. W., Niall, H. T., Jacobs, J. W., Keutmann, H. T., Potts, J. T., Jr., and Cohn, D.V. (1974). The N-terminal amino acid sequence of bovine proparathyroid hormone. *Proc. Natl. Acad. Sci. USA*, 71:653-656.

Haussler, M. R., and McCain, T. A. (1977). Basic and clinical concepts related to vitamin D metabolism and action. *N. Engl. J. Med.*, 297:1041-1050.

Hawker, C. D., and DiBella, F. P. (1980). RIA for intact and carboxyl-terminal PTH: Clinical interpretation and diagnostic significance. *Ann. Clin. Lab. Sci.*, 10:76-88.

Hekkelman, J. W., Herrmann-Erlee, M. P. M., Biet, M. J., DiBon, A., Boonekamp, P. M., and Nijweide, P. J. (1981). Studies on the regulatory mechanism of the cyclic AMP response to PTH in bone tissue. In *Hormonal Control of Calcium Metabolism*, D. V. Cohn, R. V. Talmage, and J. L. Matthews (eds.). Excerpta Medica, Amsterdam-Oxford-Princeton, pp. 169-177.

Hekkelman, J. W., Herrmann-Erlee, M. P. M., Heersche, J. N. M., and Gaillard, P. J. (1975). Studies on the mechanism of parathyroid hormone action on embryonic bone in vitro. In *Calcium Regulating Hormones*, R. V. Talmage, M. Owen, and J. A. Parsons (eds.). Excerpta Medica, Amsterdam, pp. 185-195.

Herrmann-Erlee, M. P. M., Heersche, J. N. M., Hekkelman, J. W., Gaillard, P. J., Tregear, G. W., Parsons, J. A., and Potts, J. T., Jr. (1976). Effects on bone in vitro of bovine parathyroid hormone and synthetic fragments representing residues 1-34, 2-34 and 3-34. *Endocrinol. Res. Commun.*, 3:21-35.

Holtrop, M. E., and Raisz, L. H. (1979). Comparison of the effects of 1.25-dihydroxycholecalciferol, prostaglandin E_2 and osteoclast-activating factor with parathyroid hormone on the ultra-structure of osteoclases in cultured long bones of foetal rats. *Calcif. Tissue Int.*, 29:201-205.

Hruska, K. A., Kopelman, R., and Rutherford, W. E. (1975). Metabolism of immunoreactive parathyroid hormone in the dog. The role of kidney and the effects of chronic renal disease. *J. Clin. Invest.*, 56:39-48.

Ibbotson, K. J., D'Souza, S. M., Kanis, J. A., Douglas, D. S., and Russell, R. G.G. (1980). Physiological and pharmacological regulation of bone resorption. *Metab. Bone Dis. Rel. Res.*, 2:177-189.

Johnstone, J. J. A. (1979). The routine sectioning of undecalcified bone for cytochemical studies. *Histochem. J.*, 11:359-365.

Jowsey, J., Rowland, R. E., Marshall, J. H., and McLean, F. C. (1958). The effect of parathyroidectomy on Haversian remodelling of bone. *Endocrinology*, 63:903-908.

Jubiz, W., Canterbury, J. M., Reiss, E., and Tyler, F. H. (1972). Circadian rhythm in serum parathyroid hormone concentration in human subjects: Correlation with serum calcium, phosphatase, albumin and growth hormone levels. *J. Clin. Invest.*, 51:2040-2046.

Kaminsky, N. I., Broadus, A. E., Hardman, J. G., Jones, D. J., Jr., Ball, J. H., Sutherland, E. W., and Liddle, G. W. (1970). Effects of parathyroid hormone on plasma and urinary adenosine 3′,5′-monophosphate in man. *J. Clin. Invest.*, 49:2387-2395.

Kemper, B., Habener, J. F., Mulligan, R. C., and Potts, J. T., Jr. (1974). Pre-proparaparathyroid hormone: a direct translation produce of parathyroid messenger RNA. *Proc. Natl. Acad. Sci. USA*, 71:3731-3735.

Kemper, B., Weaver, C. A., and Gordon, D. F. (1981). Structure and function of bovine parathyroid hormone messenger RNA. In *Hormonal Control of Calcium Metabolism*, D. V. Cohn, R. V. Talmage, and J. L. Matthews (eds.). Excerpta-Medica, Amsterdam-Oxford-Princeton, pp. 19-27.

Keutmann, H. T. (1974). The chemistry of parathyroid hormone. *Clin. Endocrinol. Metab.*, 3:173-197.

Kidd, G. S., Schaaf, M., Adler, R. A., Lassman, M. N., and Wray, H. L. (1980). Skeletal responsiveness in pseudohypoparathyroidism. A spectrum of clinical disease. *Am. J. Med.*, 68:772-781.

Klahr, S., Morrissey, J., and Hruska, K. (1982). Effect of parathyroid hormone on phospholipid metabolism. *Adv. Exp. Med. Biol.* 151:561-575.

Kripke, D. F., Lavie, P., Parker, D., Huey, L., and Deftos, L. J. (1978). Plasma parathyroid hormone and calcium are related to sleep stage cycles. *J. Clin. Endocrinol. Metab.*, 47:1021-1027.

Leroyer-Alizon, E., David, L., Anast, C. S., and Bubois, P. M. (1981). Immunocytological evidence for parathyroid hormone in human fetal parathyroid glands. *J. Clin. Endocrinol. Metab.*, 52:513-516.

Mahaffey, J. E., Rosenblatt, M., Shepard, G. L., and Potts, J. T., Jr. (1979). Parathyroid hormone inhibitors. Determination of minimum sequence requirements. *J. Biol. Chem.*, 254:6496-6498.

Mallette, L. E., Renfro, M., Lemoncelli, J., and Rosenblatt, M. (1981). Radioimmunoassays for the 28-48 region of parathyroid hormone detect intact hormone but not hormone fragments. *Calcif. Tissue Int.*, 33:375-380.

Manning, R. M., Adami, S., Papapoulos, S. E., Gleed, J. H., Hendy, G. N., Rosenblatt, M. and O'Riordan, J. L. H. (1981). A carboxy-terminal specific assay for human parathyroid hormone. *Clin. Endocrinol.* (Oxf), 15:439-449.

Manning, R. M., Hendy, G. N., Papapoulos, S. E., and O'Riordan, J. L. H. (1980). The development of homologous immunological assays for human parathyroid hormone. *J. Endocrinol.*, 85:161-170.

Marcus, R., and Aurbach, G. D. (1969). Bioassay of parathyroid hormone in vitro with a stable preparation of adenyl cyclase from rat kidney. *Endocrinology*, 85:801-810.

Markowitz, H., Rotkin, L., and Rosen, J. F. (1981). Circadian rhythms of blood minerals in humans. *Science*, 213:672-674.

Martin, K. J., Bellorin-Font, E., Freitag, J., Rosenblatt, M., and Slatopolsky, E. (1981). The arterio-venous difference for immunoreactive parathyroid hormone and the production of adenosine 3',5'-monophosphate by isolated perfused bone: Studies with analogues of parathyroid hormone. *Endocrinology*, 109:956-959.

Martin, K. J., Freitag, J. J., Conrades, M. B., Hruska, K. A., Klahr, S., and Slatopolsky, E. (1978). Selective uptake of the synthetic amino terminal fragment of bovine parathyroid hormone by isolated peripheral bone. *J. Clin. Invest.*, 62:256-261.

Martin, K. J., Hruska, K., Bellorin-Font, E., and Slatoplolsky, E. (1981). The peripheral metabolism of parathyroid hormone: Activation or inactivation? In *Hormonal Control of Calcium Metabolism*, D. V. Cohn, R. V. Talmage, and J. L. Matthews (eds.). Excerpta Medica, Amsterdam-Oxford-Princeton, pp. 64-69.

Martin, K. J., Hruska, K., Freitag, J., Bellorin-Font, E., Klahr, S., and Slatopolsky, E. (1980). Clinical ability of radioimmunoassay for parathyroid hormone. *Mineral Electrolyte Metab.*, 3:283-290.

Martin, T. J., Vakakis, N., Eisman, J. A., Livesey, S. J., and Tregear, G. (1974). Chick kidney adenylate cyclase: Sensitivity to parathyroid hormone and synthetic human and bovine peptides. *J. Endocrinol.*, 63:369-375.

Marx, S. J., Sharp, M. F., Krudy, A., Rosenblatt, M., and Mallette, L. E. (1981). RIA for the middle region of human parathyroid hormone: Studies with a radioactive synthetic peptide. *J. Clin. Endocrinol. Metab.*, 53:76-84.

McPartlin, J., Skrabanek, P., and Powell, D. (1978). Early effects of parathyroid hormone on rat calvarian bone alkaline phosphatase. *Endocrinology*, 103: 1573-1578.

McPartlin, J., Skrabanek, P., Powell, D., and Harrington, M. G. (1977). Bone alkaline phosphatase: Quantitative cytochemical characterization and the response to parathyrin in vitro. *Biochem. Soc. Trans.*, 5:1734-1736.

Morel, F. (1981). Sites of hormone action in the mammalian nephron. *Am. J. Physiol.*, 240 (Renal Fluid Electrolyte Physiol. 9):F159-F164.

Morrissey, J. J., and Cohn, D. V. (1979a). Regulation of secretion of parathormone and secretory protein-I from separate intracellular pools by calcium, dibutyryl cAMP and (1)-isoproterenol. *J. Cell. Biol.*, 82:93-102.

Morrissey, J. J., and Cohn, D. V. (1979b). Secretion and degradation of parathormone is a function of intracellular inactivation of hormone pools: Modulation by calcium and dibutyryl cAMP. *J. Cell Biol.*, 83:521-528.

Munson, P. L. (1961). Biological assay of parathyroid hormone. In *The Para-*

thyroids, R. O. Greep and R. V. Talmage (eds.). Thomas, Springfield, Illinois, pp. 94-113.

Nagant de Deuxchaisnes, C., Fischer, J. A., Dambacher, M. A., Devogelaer, J.-P;. Arber, C. E., Zanelli, J. M., Parsons, J. A., Loveridge, N., Bitensky, L., and Chayen, J. (1981). Dissociation of bioactive and immunoreactive parathyroid hormone in pseudohypoparathyroidism Type I. *J. Clin. Endocrinol. Metab.,* 53:1105-1109.

Nagant de Deuxchaisnes, C., Fischer, J. A., Dambacher, M. A., Werder, E., Devogelaer, J. P., De Meyer, R., Loveridge, N., Bitensky, L., and Chayen, J. (1981). Inhibition of cytochemical bioactivity of parathyroid hormone by plasma in pseudohypoparathyroidism Type I (PSPI). *Calcif. Tissue Int.,* 33 (suppl):204.

Nagant de Deuxchaisnes, C., and Krane, S. M. (1978). Hypoparathyroidism. In *Metabolic Bone Disease,* Vol. 2, L. V. Avioli and S. M. Krane (eds.). Academic, New York, pp. 217-445.

Nakagami, K., Warshawsk, H., and Leblond, C. P. (1971). The elaboration of protein and carbohydrate by rat parathyroid cells as revealed by electron microscope radioautography. *J. Cell. Biol.,* 51:596-610.

Neuman, W. F., Schneider, N., and Doolittle, R. (1981). The peripheral metabolism of parathyroid hormone. In *Hormone Control of Calcium Metabolism,* D. V. Cohn, R. V. Talmage, and J. L. Matthews (eds.). Excerpta Medica, Amsterdam-Oxford-Princeton, pp. 55-63.

Niepel, B., Atkinson, M. J., Radeke, H., and Hesch, R.-D. (1981). A sensitive homologous bioassay for human parathyroid hormone (hPTH). *Calcif. Tissue Int.,* 33(suppl):206.

Nissenson, R. A., Abbott, S. R., Teitelbaum, A. P., Clark, O. H., and Arnaud, C. D. (1981). Endogenous biologically active human parathyroid hormone: Measurement by a guanyl nucleotide-amplified renal adenylate cyclase assay. *J. Clin. Endocrinol. Metab.,* 52:840-846.

Nissenson, R. A., Teitelbaum, A. P., Abbott, S. R., Pliam, N., Silve, C., Zitzner, L., Nyiredy, K., and Arnaud, C. D. (1981). Parathyroid hormone receptors in kidney and bone: Relation to adenylate cyclase activation. In *Hormonal Control of Calcium Metabolism,* D. V. Cohn, R. V. Talmage, and J. L. Matthews (eds.). Excerpta Medica, Amsterdam-Oxford-Princeton, pp. 44-54.

Norman, A. W. (1979). *Vitamin D: The Calcium Homeostatic Steroid Hormone.* Academic, New York.

Owen, M. (1980). The origin of bone cells in the post-natal organism. *Arthritis Rheum.,* 23:1073-1080.

Pang, P. K. T., Yang, M. C. M., Dhosla, M. C., and Bumpus, F. M. (1981). Vascular action of parathyroid hormone fragments and analogs. In *Hormonal Control of Calcium Metabolism,* D. V. Cohn, R. V. Talmage and J. L. Matthews (eds.). Excerpta Medica, Amsterdam-Oxford-Princeton, pp. 293-297.

Parsons, J. A. (1976). Parathyroid physiology and the skeleton. In *Biochemistry and Physiology of Bone*, Vol. IV, G. H. Bourne (ed.). Academic, New York, pp. 159-225.

Parsons, J. A., and Potts, J. T., Jr. (1972). Physiology and chemistry of parathyroid hormone in calcium metabolism and bone disease. *Clin. Endocrinol. Metab.*, 1:33-78.

Parsons, J. A., Rafferty, B., Gray, D., Reit, B., Zanelli, J. M., Keutmann, H. T., Tregear, G. W., Callahan, E. N., and Potts, J.T., Jr. (1975). Pharmacology of parathyroid hormone and some of its fragments and analogues. In *Calcium Regulating Hormones*, R. V. Talmage, M. Owen, and J. A. Parsons (eds.). Excerpta Medica, Amsterdam, pp. 33-39.

Parsons, J. A., Reit, B., and Robinson, C. J. (1973). A bioassay for parathyroid hormone using chicks. *Endocrinology*, 92:454-462.

Parsons, J. A., and Robinson, C. J. (1968). A rapid indirect hypercalcaemic action of parathyroid hormone demonstrated in isolated blood-perfused bone. In *Parathyroid Hormone and Thyrocalcitonin (Calcitonin)*, R. V. Talmage and L. F. Bélanger (eds). Excerpta Medica, Amsterdam, pp. 329-331.

Parsons, J. A., and Zanelli, J. M. (1980). Physiological role of the parathyroid glands. *Handbuch der inneren Medizin VI/1A. Knochen. Gelenke. Muskeln.* Springer-Verlag, Berlin, pp. 135-172.

Parthemore, J. G., Roos, B. A., Parker, D. C., Kripke, D. F., Avioli, L. V., and Deftos, L. J. (1978). Assessment of acute and chronic changes in parathyroid hormone secretion by a radioimmunoassay with predominant specificity for the carboxy-terminal region of the molecule. *J. Clin. Endocrinol. Metab.*, 47: 284-289.

Pitkin, R. (1975). Calcium metabolism in pregnancy: A review. *Am. J. Obstet. Gynecol.*, 121:724-737.

Pitkin, R., Reynolds, W. A., Williams, G. A., and Hargis, G. K. (1979). Calcium metabolism in normal pregnancy: A longitudinal study. *Am. J. Obstet. Gynecol.*, 133:781-790.

Pitkin, R. M., Reynolds, W. A., Williams, G. A., Kawahara, W., Bauman, A. F., and Hargis, G. K. (1980). Maternal and foetal parathyroid hormone responsiveness in pregnant primates. *J. Clin. Endocrinol. Metab.*, 51:1044-1047.

Podbesek, R. D., Stevenson, R., Zanelli, G. D., Edouard, C., Meunier, P. J., Reeve, J., and Parsons, J. A. (1981). Treatment with human parathyroid hormone fragment (hPTH 1-34) stimulates bone formation and intestinal calcium absorption in the greyhound: Comparison with data from the osteoporosis trial. In *Hormonal Control of Calcium Metabolism*, D. V. Cohn, R. V. Talmage, and J. L. Matthews (eds.). Excerpta Medica, Amsterdam-Oxford-Princeton, pp. 118-123.

Posillico, J. T., Anderson, N., Jr., and Tyrey, T. (1981). Adrenal secretory response of the parathyroid glands (PTG) in vitro to varying calcium concentrations as detected by cytochemical bioassay of PTH. In *Hormonal Control of Calcium Metabolism,* D. V. Cohn, R. V. Talmage, and J. L. Matthews (eds.). Excerpta Medica, Amsterdam-Oxford-Princeton, p. 356.

Posillico, J. T., Tyrey, L., Wells, S. A., Jr., and Anderson, N., Jr. (1981). Correlation between the stimulation of parathyroid hormone secretion and low calcium evoked hyperpolarization in human parathyroid adenoma. *Endocrinology,* 109 (suppl):122. Proceedings of the 63rd meeting of the Endocrine Society.

Raisz, L. G. (1976). Mechanisms of bone resorption. In *Handbook of Physiology,* Vol. 7, Sec. 7, G. D. Aurbach (ed.). pp. 117-136.

Raisz, L. G., Lorenzo, J., Gworek, S., Kream, B., and Rosenblatt, M. (1979). Comparison of the effects of a potent synthetic analog of bovine parathyroid hormone (bPTH) with native bPTH (1-84) and synthetic bPTH (1-34) on bone resorption and collagen synthesis. *Calcif. Tissue Int.,* 29:215-218.

Raisz, L. G., Yanjik, C. H., Bockman, R. S., Bower, B. F. (1979). Comparison of commercially available parathyroid hormone immunoassays in the differential diagnosis of hypercalcaemia due to primary hyperparathyroidism or malignancy. *Ann. Intern. Med.,* 91:739-740.

Rasmussen, H., and Craig, L. C. (1962). Parathyroid hormone: The parathyroid polypeptides. *Recent. Prog. Horm. Res.,* 18:269-295.

Reeve, J., Meunier, P. J., Parsons, J. A., Bernat, M., Bijvoet, D. L. M., Courpon, P., Edouard, C., Klenerman, L., Neer, R. M., Renier, J. C., Slovik, D., Vismans, F. J. F. E., Potts, J. T., Jr. (1980). Anabolic effect of human parathyroid hormone fragment on trabecular bone in involutional osteoporosis: A multicentre trial. *Br. Med. J.,* 280:1340-1344.

Rosenblatt, M. (1981). Parathyroid hormone: Chemistry and structure-activity relations. In *Pathobiology Annual,* Vol. II, H. O. Ioachim (ed.). Raven, New York, pp. 53-86.

Rosenblatt, M., Callahan, E. N., Mahaffey, J. E., Pont, A., and Potts, J. T., Jr. (1977). Parathyroid hormone inhibitors. Design, synthesis and biologic evaluation of hormone analogues. *J. Biol. Chem.,* 252:5847-5851.

Rosenblatt, M., and Potts, J. T., Jr. (1977). Design and synthesis of parathyroid hormone analogues of enhanced biological activity. *Endocrinol. Res. Commun.,* 4:115-133.

Rosenblatt, M., and Potts, J. T., Jr. (1981). Analogues of an in vitro parathyroid hormone inhibition: Modifications at the amino terminus. *Calcif. Tissue Int.,* 33:153-157.

Sachs, C., Bourdeau, A. M., and Dechaux, M. (1981). Chronobiological variations of ionized calcium. In *Hormonal Control of Calcium Metabolism,* D. V. Cohn, R. V. Talmage, and J. L. Matthews (eds.). Excerpta Medica, Amsterdam-Oxford-Princeton, p. 428.

Scothorne, R. J. (1964). Functional capacity of foetal parathyroid glands, with reference to their clinical use as homografts. *Ann. N.Y. Acad. Sci.*, 120:669-676.

Segre, G. V., Harris, S. T., Tully, G., and Neer, R. (1981). Aminoterminal radio-immunoassay for human parathyroid hormones. *Endocrinology*, 109 (suppl): 228. Proceedings of the 63rd meeting of the Endocrine Society.

Segre, G. V., Niall, H. D., Habener, J. F., and Potts, J. T., Jr. (1974). Metabolism of parathyroid hormone: Physiologic and clinical significance. *Am. J. Med.*, 56:774-784.

Segre, G. V., Rosenblatt, M., Reiner, B. L., Mahaffey, J. E., and Potts, J. T., Jr. (1979). Characterization of parathyroid hormone receptors in canine renal cortical plasma membranes using a radioiodinated sulfur-free hormone analog. *J. Biol. Chem.*, 254:6980-6986.

Sherwood, L. M., Hanley, D. A., Takatsuki, K., Birnbaumer, M. E., Schneider, A. B., and Wells, S. A., Jr. (1978). Regulation of parathyroid hormone secretion. In *Hormonal Control of Calcium Metabolism,* D. V. Cohn, R. V. Talmage, and J. L. Matthews (eds.). Excerpta Medica, Amsterdam-Oxford-Princeton, pp. 301-307.

Sherwood, L. M., Mayer, G. P., Ramberg, C. F., Jr., Kronfeld, D. S., Aurbach, G. D., and Potts, J. T., Jr. (1968). Regulation of parathyroid hormone secretion: Proportional control by calcium, lack of effect of phosphate. *Endocrinology,* 83:1043-1051.

Silverman, R., and Yalow, R. S. (1973). Heterogeneity of parathyroid hormone; clinical and physiologic implications. *J. Clin. Invest.,* 52:1958-1971.

Sinha, T.K., Miller, S., Fleming, J., Khaira, R., Edmonson, J., Johnston, C. C., Jr. and Bell, N. H. (1975). Demonstration of a diurnal variation in serum parathyroid hormone in primary and secondary hyperparathyroidism. *J. Clin. Endocrinol. Metab.,* 41:1009-1013.

Slatopolsky, E., Martin, K., and Hruska, K. (1980). Parathyroid hormone metabolism and its potential as a uremic toxin. *Am. J. Phsyiol.,* 230:F1-F12.

Steiner, D. F. (1976). Peptide hormone precursors: Biosynthesis, processing and significance. In *Peptide Hormones,* J. A. Parsons (ed.). MacMillan, London, pp. 49-65.

Stewart, A. F., Horst, R., Deftos, L. J., Cadman, E. C., Lang, R., and Broadus, A. E. (1980). Biochemical evaluation of patients with cancer-associated hypercalcaemia. *N. Engl. J. Med.,* 303:1377-1383.

Stögmann, W., and Fischer, J. A. (1975). Pseudohypoparathyroidism: Disappearance of the resistance to parathyroid extract during treatment with vitamin D. *Am. J. Med.,* 59:140-144.

Suh, S. M., Fraser, D., and Kooh, S. W. (1970). Pseudohypoparathyroidism: Responsiveness to parathyroid extract induced by vitamin D_2 therapy. *J. Clin. Endocrinol. Metab.,* 30:609-614.

Tashjian, A. H., Ontjes, D. A., and Munson, P. L. (1964). Alkylation and oxidation of methionine in bovine parathyroid hormone; effects on hormonal activity and antigenicity. *Biochemistry,* 3:1175-1182.

Tregear, G. W., van Rietschoten, J., Green, E., Keutmann, H., Niall, H. D., Reit, B., Parsons, J. A., and Potts, J. T., Jr. (1973). Bovine parathyroid hormone: Minimum chain length of synthetic peptide required for biological activity. *Endocrinology,* 93:1349-1353.

Tregear, G. W., van Rietschoten, J., Greene, E., Keutmann, H., Niall, H. D., Parsons, J. A., and Potts, J. T., Jr. (1974). Principles and recent applications in the solid-phase synthesis of peptide hormones. In *Endocrinology 1973: Proceedings of the Fourth International Symposium,* S. Taylor (ed.). Heinemann, London, pp. 1-15.

Trowell, O. A. (1959). The culture of mature organs in a synthetic medium. *Exp. Cell. Res.,* 16:118-147.

Yalow, R. S. (1978). Significance of the heterogeneity of parathyroid hormone. In *Endocrinology of Calcium Metabolism,* D. H. Copp and R. V. Talmage (eds.). Excerpta Medica, Amsterdam-Oxford, pp. 308-312.

Yonaga, T. (1978). Action of parathyroid hormones with special reference to its anabolic effect on different kinds of tissues in rats (I). *Bull. Tokyo Med. Dent. Univ.,* 25:237-248.

Zanelli, J. M., and Gaines-Das, R. E. (1980). International collaborative study of N.I.B.S.C. research standard for human parathyroid hormone for immunoassay. *J. Endocrinol.,* 86:291-304.

Zanelli, J. M., and Gaines-Das, R. (1983). The first International Reference Preparation of human parathyroid hormone for immunoassay: Characterization and calibration by international collaborative study. *J. Clin. Endocrinol. Metab.,* 57. (in press).

Zanelli, J. M., and Parsons, J. A. (1980). Bioassay of parathyroid. In *Handbuch der inneren Medizin VI/1A. Knocken. Gelenke. Muskeln.* Springer-Verlag, Berlin, pp. 599-621.

Zanelli, J. M., O'Hare, M. J., Nice, E. C., and Corran, P. H. (1980). Purification and assay of bovine parathyroid hormone by reversed-phase high-performance liquid chromatography. *J. Chromatogr.* 233:59-67.

Zanelli, J. M., Rafferty, B., Stevenson, R. W., and Parsons, J. A. (1980). An homologous and sensitive radioimmunoassay for the synthetic amino terminal (1-34) fragment of human parathyroid hormone: Application to the clearance of this peptide administered in vivo. *J. Immunoassay,* 1:289-308.

12

A Cytochemical Bioassay for Angiotensin II

Julia C. Jones, J. Alaghband-Zadeh, and Graham A. MacGregor / Charing Cross Hospital Medical School, London, England

INTRODUCTION

Despite the fact that the discovery of renin by Tigerstedt and Bergman was as early as 1898, the function of the renin-angiotensin-aldosterone system remains an area of great controversy. With the recent development of radioimmunoassays for the measurement of various components of the system and the development of specific drugs that block or partially inhibit the system, there is increasing evidence that the renin-angiotensin aldosterone system is an important regulator of the peripheral vascular tone and thereby of blood pressure. The system also plays an important role in the regulation of renal blood flow and sodium balance, the latter particularly through its regulatory role on aldosterone secretion.

The accurate measurement of the effector hormone, angiotensin II, is an important factor in the interpretation of many of these studies. However, at present the radioimmunoassay of plasma angiotensin II lacks sufficient sensitivity to measure at the lower end of the normal range and particularly below the normal range. This is important particularly when angiotensin converting-enzyme inhibitors, that block the formation of angiotensin II, have been used, because the levels will then be very low. In addition, there are also problems of cross-reactivity of the angiotensin II antibody with the very high levels of angiotensin I (AI), thereby giving falsely high levels of angiotensin II. In view of these problems, as well as the need to demonstrate biological activity, we

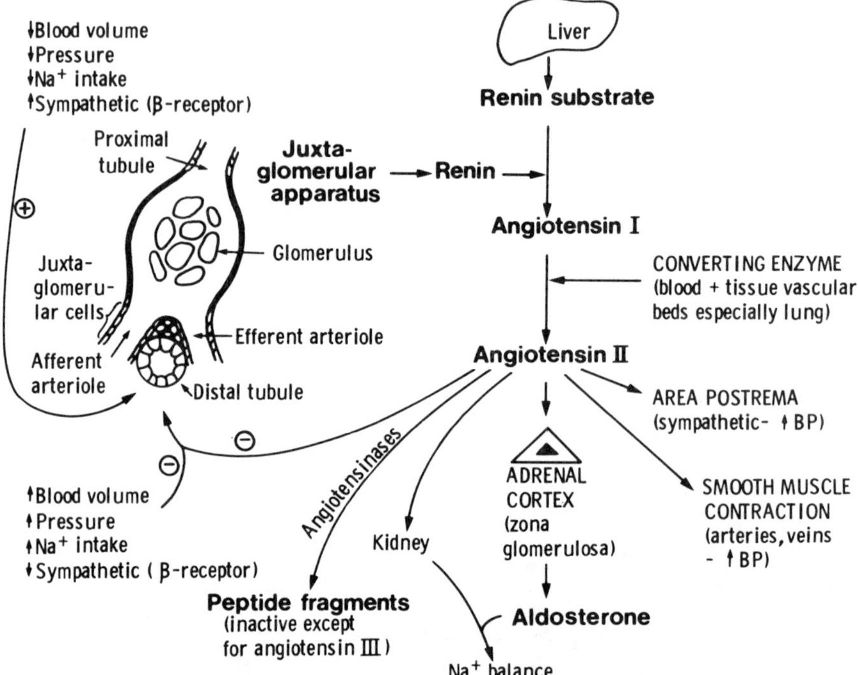

Figure 1   A diagrammatic representation of the renin-angiotensin system.

have developed a highly sensitive cytochemical bioassay for the measurement of human angiotensin II in biological fluids. Before describing the development of this assay and the measurements obtained with it so far, a brief account of the renin system, the actions of angiotensin II, and its function as far as it is known in the human, are briefly described.

Components of the Renin-Angiotensin System

Aspects of the biochemistry and mechanisms of action of the renin-angiotensin system have been reviewed by Page and Bumpus (1974), Oparil and Haber (1974), Peart (1976), Peach (1977), Reid et al. (1978), Moore (1979), Skeggs et al. (1980), and Peach (1981).

The enzyme, renin (MW around 40,000) is synthesized in the juxtaglomerular cells of the kidney. It is known to be released into the bloodstream, where it cleaves its substrate, a large α_2 globulin synthesized by the liver, to release a decapeptide, angiotensin I. Angiotensin I is then converted to the octapeptide,

angiotensin II, by the converting enzyme, which is present within the bloodstream but is also present in many tissues, particularly the lungs (Figure 1).
Angiotensin II is the effector hormone of the renin system, angiotensin I having
no direct physiological action. Angiotensin II is destroyed by angiotensinases,
which are present in the bloodstream and in tissues. Angiotensin III is the only
peptide fragment produced that is known to have any action (Blair-West et al.,
1971; Goodfriend and Peach, 1975; Carey et al., 1978). Aside from the blood,
it is now known that all the necessary components of the renin system are present in several tissues, such as the kidney and within the arterial wall. The exact
function of these tissue systems is not clearly established.

Actions of Angiotensin II

Angiotensin II is known to cause contraction of all types of smooth muscle,
including the smooth muscle cells of the arterioles and of the veins. Weight
for weight, angiotensin II is the most powerful vasoconstrictor of arterioles
known. When infused in small amounts that only raise the circulating angiotensin II levels, as measured by radioimmunoassay, to within the physiological
range, there is a rise in blood pressure (Chinn and Dusterdieck, 1972). However,
part of this increase in blood pressure has been shown, in the greyhound dog,
to be mediated by the area prostrema of the hind part of the brain (Joy and
Lowe, 1970). Angiotensin II, itself, has been shown to facilitate sympathetic
neurotransmission (Clough et al., 1980). Angiotensin II also stimulates the
production and release of aldosterone (Laragh et al., 1960; Biron et al., 1961).
Through aldosterone and also by a direct effect of angiotensin II on the kidney
(Laragh et al., 1963; Langford, 1964), the renin system is an important mechanism controlling sodium excretion.

Control of the Circulating Level of Angiotensin II

The level of circulating angiotensin II is dependent on the concentration of
renin and renin substrate. Under normal circumstances, angiotensin converting
enzyme is present in great excess. The clearance of angiotensin II, under most
circumstances, plays little role in determining the level except where there is
severe liver disease.

The amount of renin released from the juxtaglomerular cells in the kidney
is governed by a complex interplay of factors, including a probably pressure-
sensitive receptor in the afferent arteriole, as well as the sympathetic nervous
system, mediated through a β receptor (Reid et al., 1978). Angiotensin II also
exerts a negative feedback on the release of renin. More controversially, it is
thought that the sodium flux across the macula densa in the distal part of the
tubule may also determine renin release. Therefore, an increase in sympathetic

tone or stimulation with β agonists will cause an increase in renin release. Conversely, drugs that block the sympathetic nervous system, particularly those that block β receptors, will cause a decrease in renin release. Alterations in sodium balance cause profound changes in renin release: a reduction in sodium intake, or a loss in blood volume or extracellular water, leads to a marked rise in renin release, whereas an increase in sodium intake or an increase in blood volume or extracellular water leads to a decrease in renin release. It is not clear at present through which mechanism these changes are mediated.

Renin substrate is synthesized in the liver and therefore depends on adequate liver function. Although it was not regarded as a rate-limiting factor in the prodution of angiotensin II, recent evidence suggests that it may be more important than was thought previously. A reduction in substrate levels occurs when there is a reduction in plasma proteins such as may occur in cirrhosis (Gould et al., 1966) or in the nephrotic syndrome. An increase in renin substrate is known to occur during pregnancy (Skinner et al., 1972) and in women taking oral contraceptives (Skinner et al., 1969). Factors governing the level of substrate in the normal human are less well studied.

Role of the Renin System in the Normal Human

With the development of radioimmunoassay and specific antibodies for various components of the renin system, measurements confirmed that a reduction in sodium intake caused an increase in renin release, angiotensin I, plasma renin activity, and angiotensin II, as well as a parallel increase in urinary and plasma aldosterone (Brown et al., 1972; 1973; 1977). With a reduction in sodium intake there is a loss of sodium from the body and a reduction in blood volume. In spite of this, in the normal human there is no fall in blood pressure. It was speculated that this might be due to a compensatory rise in angiotensin II. With the development of competitive inhibitors of angiotensin II, such as saralasin, it was possible to show by infusing saralasin that angiotensin II does play an important part in maintaining blood pressure in the normal human when sodium intake has been reduced (Brown et al., 1976) or where sodium loss has been induced with diuretics (Posternak et al., 1977). At the same time, there was a fall in aldosterone during the infusion of saralasin. However, all these competitive inhibitors of angiotensin II have agonistic activity and stimulate angiotensin II receptor sites when the level of angiotensin II is normal or high. More recently, inhibitors of the angiotensin converting enzyme have been developed, and Cushman et al. (1977) successfully synthesized a powerful inhibitor of angiotensin converting enzyme that was absorbed orally, namely, Captopril (Squibb). Captopril has been shown to cause a fall in blood pressure in the normal human on a normal sodium intake, i.e., around 150 mmol/day (MacGregor et al., 1980; 1981). The studies with Captopril were also done in subjects on a high and low sodium

intake. As might be expected, the fall in blood pressure was related to the salt
intake. However, the interpretation of these studies is not clear-cut, as inhibition
of converting enzyme may also partially block the breakdown of bradykinin,
a potential vasodilator. However, recent experiments suggest that the major action of the converting enzyme inhibitors is through the fall in angiotensin II that
occurs. But with the presently available radioimmunoassay, it is not certain
whether the production of angiotensin II is completely inhibited or whether
there are still small but undetectable amounts of angiotensin II after converting
enzyme blockade. The present radioimmunoassay is not sensitive enough to
detect levels of angiotensin II below approximately 5 pmol/liter and, in spite
of highly specific antibodies, there could be some cross-reaction with the very
high angiotensin I levels that occur after inhibition of the converting enzyme
(Morton et al., 1979). In normal subjects studied on their normal diet, with
Captopril, there was also a fall in plasma aldosterone as well as an increase in
sodium excretion.

High Blood Pressure

In 1934, Goldblatt et al. demonstrated the effect of placing a single clip on
the renal artery of one kidney of a dog, while leaving the other kidney intact.
Many of the dogs developed persistent hypertension, and it was felt that this
was due to increased renin release by the ischemic kidney. For many years, it
was felt that the renin system might be an important cause of not only renal-
vascular hypertension, but also of essential hypertension. Even in patients
with essential hypertension, but with normal levels of renin or angiotensin II,
Case et al. (1978) found that treatment with converting enzyme inhibitor
caused a fall in blood pressure. On the other hand, the finding that converting
enzyme inhibitors also lowered blood pressure in subjects with normal blood
pressure to the same extent suggested that, in these hypertensive patients, the
renin system could not be the cause of high blood pressure, but was a mechanism that was maintaining blood pressure in the same way as in the normal
subjects (MacGregor et al., 1979; 1981). Some hypertensive patients have high
levels of renin and angiotensin, the most common cause being malignant hypertension where damage to the small afferent arterioles leads to ischemia of the
glomeruli and thereby an increase in renin release, causing the blood pressure
to rise still further. Other causes of high renin and high blood pressure are reno-
vascular hypertension, where there is narrowing of the main artery or of a
smaller artery within the kidney or, more rarely, a tumor that produces renin.
Many patients with essential hypertension have a suppressed renin system with
low values of plasma renin activity and angiotensin II (Dunn and Tannen, 1974).
In these hypertensive patients with low renin levels, the exact cause of which

remains speculative, the radioimmunoassay for angiotensin II may not be sensitive enough to differentiate differences between patients.

BIOASSAY OF ANGIOTENSIN II

The physiological actions of angiotensin II that have been most widely used for bioassay procedures are: the ability to cause a rise in blood pressure when injected intravenously into animals; the vasoconstrictor activity; the ability to cause smooth muscle contraction; and steroid production by isolated adrenal cell preparations.

In Vivo Bioassays

The response to a single intravenous injection of angiotensin II is the same in all animal species studied. After a lag period of 20-30 sec, blood pressure rises sharply to reach a maximum within 1-2 min, then falls to the original level within 3-5 min. Animal species that have been used for the assay of angiotensin include the cat, dog, and rat (Shipley and Tilden, 1947; Skeggs et al., 1951).

The rat pressor assay system is the method most commonly used, as the test animal is easily obtained and is sensitive to smaller amounts of angiotensin. Dekanski (1954) found the dibenamine-treated rat to be 20 times more sensitive than the cat and 80 times more sensitive than the dog. In addition, reproducible results could be obtained over long periods of time, up to 8 hr.

The assay is carried out in nephrectomized rats anesthetized with pentobarbital or urethane (Paladini et al., 1959; Boucher et al., 1964), or in intact rats treated with a ganglion blocker, such as pentolinium tartrate (Peart, 1955; Morris and Robinson, 1964) or dibenamine (Dekanski, 1954), or treated with a combination of α and β blockers (Needleman et al., 1972). The blood pressure is recorded directly from the carotid artery with a mercury manometer, attached to a kymograph, and all doses of angiotensin II are given into the jugular vein. Unknown and standard preparations are compared by means of a dose-response curve or by a four-point assay design (two standards and two unknowns). The rise in blood pressure is directly proportional to the logarithm of the angiotensin dose. The sensitivity of the rat preparation varies, smaller rats having been found to be more sensitive than larger rats. Morris and Robinson (1964) took their minimum response to be 1 ng of angiotensin, which corresponded to a blood pressure rise of 10 mmHg, though it was possible to obtain a response to as little as 0.2 ng angiotensin (equivalent to a blood pressure rise of 4 mmHg).

Due to the low circulating blood levels of angiotensin in experimental animals and humans, it was first necessary to extract chemically, partially purify, and concentrate angiotensin from blood before assay. Accordingly, several extraction

Table 1 Arterial Angiotensin II Levels by the Rat Pressor Assay in
Normotensive Subjects

Comment	Angiotensin II		Reference
	Mean ± SD	Range	
Undetectable in 10 of 24 subjects	0.014	0.00-0.05 GU/liter[a]	Kahn et al. (1952)
Undetectable in 14 of 20 subjects	62.5	0-350 pg/ml	Boucher et al. (1964)
Undetectable in normotensives	—	—	Morris and Robinson (1964)
Undetectable in 19 of 28 subjects	70 ± 20	0-550 pg/ml	Genest et al. (1964)
—	23 ± 3	0-60 pg/ml	Mulrow (1964)
Undetectable in 8 of 30 subjects	95 ± 12	0-190 pg/ml	Massani et al. (1966)
Undetectable in 2 of 15 subjects	130 ± 87	0-260 pg/ml	Nielson and Moller (1967)

[a]Goldblatt units per liter.

procedures were developed. These involved first extracting angiotensin from an
alcoholic filtrate of blood, or from plasma, with organic solvents such as n-
butanol, and then adsorbing the angiotensin onto alumina (Skeggs et al., 1951;
1952) or cation exchange resins such as Dowex 50-X2 (Paladini et al., 1959;
Scornik and Paladini, 1961) or BioRad AG50W-X2 (Massani et al., 1966; Nielson
and Moller, 1967). Mulrow (1964) used a two-column chromatographic procedure with IRC50 and Dowex 50. The angiotensin was then eluted and concentrated for bioassay, or submitted to an additional paper chromatographic step
(Boucher et al., 1964) before assay. The method of Morris and Robinson (1964)
was simpler, depending on the extraction of angiotensin into methanol and
precipitation with ether.

There are several disadvantages associated with in vivo bioassays. The assay
system is unable to differentiate between angiotensin I and II, as angiotensin I
is rapidly converted to angiotensin II in vivo. Apart from requiring large volumes
of blood (50-250 ml), the extraction procedures are lengthy and recoveries often
low or variable. The sample handling capacity is small, and levels of circulating
angiotensin in normotensive subjects are frequently too low to be measured
(Table 1).

In Vitro Bioassays—Isolated Tissue Preparations

In vitro bioassay procedures have made use of preparations of smooth muscle or isolated organs. The vasoconstrictor activity of angiotensin II has been tested on isolated organs, such as the rabbit's ear or the Lowen-Trendelenberg toad hind limb preparation (Houssay and Taquini, 1938; Fasciolo et al., 1938). Various isolated smooth muscle preparations have been used to test for angiotensin activity, such as guinea pig ileum; rabbit or toad intestine; and rat, rabbit, or guinea pig uterus. Tissues that have been used for the quantitative assay of angiotensin include the rat uterus (Schwarz et al., 1955; Needleman et al., 1972), rabbit aortic strip (Helmer, 1957), and the rat colon (Regoli and Vane, 1964).

The tissue is maintained in vitro, suspended in a bath or chamber and bathed in an appropriate medium, or, for superfusion techniques, in a jacket with the solution flowing over the tissue at a constant rate. The tissue is attached by means of stainless steel hooks to an isotonic lever, usually under some degree of tension and connected to a kymograph to record the changes in contractility. Test and standard preparations are introduced into the chamber or the reservoir containing the superfusing solution, and the change in contractile activity recorded over the baseline level.

Rat Uterus Preparation

Uteri from rats either in estrus (Schwarz et al., 1955) or pretreated with stilbestrol (Paiva and Paiva, 1960) react to angiotensin II with a dose-dependent contraction. The uterus is able to differentiate between angiotensin I and II. Needleman et al. (1972) found angiotensin I to have only 2% of the oxytocic activity of angiotensin II, indicating that rat uterus has minimal converting enzyme activity. The height of contraction is the response usually measured, but Schwarz et al. (1955) measured the response time. The reaction to angiotensin II is delayed, and they measured the time interval between injection and maximal contraction. The response time was inversely proportional to the dose.

Rabbit Aortic Strip Assay

Helmer (1957) used spirally cut rabbit aortic strips to assay angiotensin II. A contraction began after 2-4 min with angiotensin II, and a maximum response was reached after 10-20 min. Relaxation was slow, and 60 min was often required to complete the cycle. The aortic strips were sensitive to 3.5 ng angiotensin and in sensitive preparations as little as 700 pg could be detected. This method was used to detect pressor substances in renal vein blood of hypertensives (Helmer, 1961) and for the quantitative measurement of renin in plasma (Helmer and Judson, 1963) using the measurement of angiotensin II after incubation of dialyzed plasma under standardized conditions. (Plasma was adjusted to pH 5.5 and incubated at 37°C for 1 hr.)

Rat Colon Preparation

Regoli and Vane (1964) described a specific assay for angiotensin II using rat ascending colon. Norepinephrine or propranolol added to the solution bathing the tissue diminished interference by catecholamines and bradykinin. The tissue was also relatively insensitive to other substances affecting smooth muscle contraction, such as 5-hydroxytryptamine (serotonin), histamine, and acetylcholine. With this assay system, concentrations as low as 10^{-10} g/ml (100 pg/ml) could be detected. Using a superfusion technique, blood from an intact animal was dripped over the tissue in an organ bath, so that changes in circulating angiotensin II levels could be measured (Hodge et al., 1966a,b).

In Vitro Bioassays—Isolated-Cell Preparations

Adrenal cell preparations have been described by Halkerston and Feinstein (1968), Kloppenberg et al. (1968), and Swallow and Sayers (1969), who used trypsin and/or bacterial collagenase to dissociate the tissue into its component cells. These procedures used the entire gland to obtain the cells, and therefore separate effects on glomerulosa and fasiculata-reticularis cells could not readily be distinguished.

Haning et al. (1970) used crude collagenase with deoxyribonuclease to prepare isolated adrenal cells from the separated capsular and decapsulated portions of rat adrenals. The dispersed cells obtained from the capsules consisted mainly of zona glomerulosa cells, and the maximal contamination with fasiculata-reticularis cells was estimated to be 15%. The in vitro effects of various factors, namely, angiotensin II, ACTH, serotonin, and K^+, on the output and conversion of corticosterone to aldosterone was studied. The capsular cell preparation was found to be relatively insensitive to angiotensin II. However, other workers (Douglas et al., 1978) found capsular cells to contain specific receptors for angiotensin II and to have a high sensitivity for angiotensin II, responding to concentrations of 3×10^{-11} M with a significant increase in aldosterone production. With an improved purification technique employing gravitational sedimentation, Tait et al. (1974) were able to obtain more sensitive zona glomerulosa cells with routinely less than 5% fasiculata contamination. This cell preparation was shown to respond to 2.5×10^{-10} M angiotensin II at a K^+ concentration of 3.6 mM (Tait et al., 1980). The preparation of Mendelsohn and Kachel (1980) was even more sensitive, responding to angiotensin II concentrations of 10^{-12} M with a significant increase in corticosterone and aldosterone production.

Although isolated zona glomerulosa cell preparations are sensitive to low concentrations of angiotensin II, it is important to consider the possibility of other stimulating factors, the effects of the composition of the medium, and possible contamination by other steroid-producing cells. Capsular-cell preparations usually are more sensitive to ACTH than to angiotensin II, and also respond to serotonin and changes in K^+ concentration of the incubation medium. Mendelsohn and Kachel (1981) found that these cells respond to small amounts of serum in a dose-dependent fashion. Furthermore, if the cells were first maximally stimulated with either angiotensin II or high K^+ or serotonin, addition of serum caused a further increase in aldosterone output. The aldosterone-stimulating effect of serum was found to be dialyzable, and appeared to be due to both a serotonin-like substance and to some other independent effect.

RADIOIMMUNOASSAY

The development of radioimmunoassays for angiotensin II became possible when highly purified synthetic peptides became available for use as antigens, and, after iodination, to be used as radioactive tracers. Various assay procedures have been developed, which differ in the following ways:

1. Radioimmunoassay step
    a. production of sensitive and specific anti-angiotensin II antisera
    b. labeling
    c. separation of free and antibody-bound angiotensin II
2. Presence or absence of an extraction and concentration step for angiotensin II in plasma and the nature of the extraction procedure
3. Method of inhibiting converting enzyme and angiotensinase activity in blood or plasma

Radioimmunoassay Procedures

Antisera
Due to the small size of the angiotensin II molecule it was originally considered unlikely that it alone would be antigenic. Angiotensin II was therefore coupled to larger molecules, such as poly-L-lysine (Vallotton et al., 1967; Sundsfjord, 1970), rabbit or bovine serum albumin (Goodfriend et al., 1964, 1968; Gocke et al., 1968; Düsterdieck and McElwee, 1971; Nicholls and Espiner, 1976; Lijnen et al., 1978), or porcine gammaglobulin (Catt et al., 1967; Cain et al., 1972). The octapeptide is usually coupled to these macromolecules by a carbodiimide condensation reaction (Goodfriend et al., 1964) to produce an effective immunogen. Boyd and Peart (1968) were able to produce a high-titre antibody against free angiotensin by immunizing rabbits with angiotensin II amide adsorbed physically onto microparticles of carbon.

   The antibodies raised using these hapten-carrier complexes have a specificity directed largely toward the C-terminal (3-8) hexapeptide of angiotensin II. Therefore, these antisera usually cross-react with the heptapeptide and hexapeptide metabolites of angiotensin II and sometimes with the pentapeptide metabolite. Radioimmunoassays of angiotensin II thus measure not only angiotensin II but also these immunoreactive metabolites which often have either the same or sometimes even greater affinity for the antibody as angiotensin II (AII) (Emmanuel et al., 1973). Most antisera show less than 5% cross-reactivity toward angiotensin I and even less toward plasma renin-substrate. Most antisera recognize the various analogs of angiotensin II, that is, ile[5]-AII, val[5]-AII, and asn[1], val[5]-AII equally, so that angiotensin amide (CIBA-Hypertensin) can be used as the standard to measure the human plasma form, ile[5]-AII.

Label

The presence of a tyrosine residue in the angiotensin molecule facilitates iodination with ^{125}I or [131]I. Labeled peptides of high specific activity for radioimmunoassay have been obtained using the chloramine-T technique of Hunter and Greenwood (1962). A modification of this procedure (Nielsen et al., 1971), using lower concentrations of chloramine-T, sodium metabisulfite, and radioactive iodine, yielded a monoiodinated peptide that was stable for several months.

Separation Step

Separation of antibody-bound angiotensin from free angiotensin has been achieved by various methods, namely, dextran-coated charcoal (Boyd et al., 1967; Catt et al., 1967; Gocke et al., 1968; Page et al., 1969; Cain et al., 1972; Kosunen, 1976; Nicholls and Espiner, 1976; Lijnen et al., 1978); plasma-coated charcoal (Düsterdieck and McElwee, 1971); antibody-coated tubes (Goodfriend et al., 1968); Sephadex G-25 gel filtration or gel centrifugation (Sundsfjord, 1970; Kappelgaard et al., 1976); and by the double antibody technique (Emmanuel et al., 1973).

Extraction

Angiotensin II assays are sensitive enough to measure circulating levels in most normal subjects. However, most assays still require a preliminary plasma extraction and concentration step before the radioimmunoassay of the hormone. Extraction of angiotensin II from plasma has been achieved using the cation exchange resins, Dowex 50-X2 or Sulfethyl Sephadex, or by Fuller's earth.

A simpler assay has been described which employs an ethanol precipitation of plasma, before assay of angiotensin II in the supernatant (Nicholls and Espiner, 1976). Several assays have been described which do not use a preliminary extraction step. The assays of Emmanuel et al. (1973) and Lijnen et al. (1978) gave similar values to those obtained using extracted plasma. Earlier assays on unextracted plasma gave slightly higher values for circulating angiotensin II in normal subjects (Gocke et al., 1968; Goodfriend et al., 1968). Goodfriend et al.

(1968) found it necessary to compare each unknown plasma with standards prepared in the same plasma rendered angiotensin-free by extraction with AG50W-X4 resin. Gocke et al. (1968) included controls for nonspecific binding for each plasma sample.

Inhibition of Converting Enzyme and Angiotensinase

Inhibition is achieved by ethanolic precipitation of blood, or by the addition of various inhibitors to blood before the plasma is separated and extracted, or to the plasma before it is assayed. EDTA and 2,3-dimercaptopropanol (BAL) inhibit converting enzyme; EDTA also inhibits some but not all angiotensinase activity. Angiotensinase activity can be blocked with o-phenanthroline and diisopropylfluorophosphate (DFP). Chilling the blood or plasma also reduces enzyme activity.

Normal Values

The first radioimmunoassay described by Vallotton et al. (1967), gave high values in normal subjects (range 120-350 pg/ml) compared with some bioassay data. These high values were attributed to loss of angiotensin standard onto the glass assay tubes and a nonspecific difference between buffer and plasma. A modified procedure (Page et al., 1969) gave similar values in normal subjects (21 ± 20 pg/ml) to those obtained by other workers. The range of venous angiotensin II levels found in normotensive subjects have been reported to be 8-56 pg/ml (Boyd et al., 1967), 18-110 pg/ml (Gocke et al., 1968), 5-126 pg/ml (Goodfriend et al., 1968), 19 ± 12 pg/ml (Catt et al., 1971), 16 ± 8 pg/ml (Düsterdieck and McElwee, 1971), 23 ± 11 pg/ml (Wilmhurst et al., 1971), 25 ± 14 pg/ml (Cain et al., 1972), 26 ± 1.6 pg/ml (Emmanuel et al., 1973), 31.1 ± 10.7 pg/ml (Kosunen, 1976), and 25.1 ± 13.5 pg/ml (Lijnen et al., 1978). Slight variations in the values may relate to the sodium intake or posture of the subjects, or to variations in the extraction techniques. Radioimmunoassay levels of angiotensin II generally correlate well with plasma renin activity.

The sensitivity of radioimmunoassay is much improved over bioassay procedures, most assays being able to detect about 5 pg/ml or sometimes as little as 2 pg/ml (Düsterdieck and McElwee, 1971), depending on the volume of plasma used. Recovery of angiotensin II added to blood or plasma is quantitative, varying from 80 to 110%, and large numbers of samples can be handled more easily.

Some workers have compared venous and arterial angiotensin II levels, and found that they are similar (Best et al., 1971), or that the venous levels are slightly lower (Catt et al., 1971; Cain et al., 1972; Semple et al., 1976). On further investigation, using paper chromatography of blood extracts, angiotensin

II was found to be the major immunoreactive peptide (85%) in arterial blood but to comprise only 28% of the immunoreactive material in venous blood (Cain et al., 1969); the major immunoreactive peptide in venous blood was found to be the hexapeptide. That these workers were able to obtain similar values on radioimmunoassay of arterial and venous blood was attributed to the cross-reactivity of the antiserum with angiotensin II metabolites. Conversely, Semple et al. (1976) found, by chromatographic separation of plasma extracts, that angiotensin II was the predominant peptide in both normal venous and arterial plasma (55-100%). Thus, these workers concluded that venous angiotensin levels accurately reflected arterial levels.

CYTOCHEMICAL BIOASSAY OF ANGIOTENSIN II

Need for a Sensitive Bioassay

At present there are no sensitive bioassay procedures capable of measuring normal circulating angiotensin II levels without the extraction of large volumes of blood. The radioimmunoassays available for angiotensin II are able to measure normal circulating levels using smaller volumes of blood, but most still incorporate a preliminary extraction and concentration step. Although these radioimmunoassays are much more sensitive than bioassay procedures, they still lack the ability to distinguish between low normal and subnormal levels accurately.

There is also the possibility of interference due to cross-reaction of the anti-angiotensin II antiserum to metabolites of angiotensin II. In particular situations, such as when converting enzyme inhibitors have been used, the build-up of angiotensin I may also affect the result obtained.

The cytochemical technique has been applied to the development of a more sensitive assay for angiotensin II based on its biological rather than immunological activity. The method is based on the redox end point utilized by Chayen et al. (1972) for the assay of ACTH. After treatment with standard angiotensin II (MRC 70/302) or dilutions of plasma, the reducing potency in the cells of the zona glomerulosa, in sections of adrenal, is measured by microdensitometry.

Procedure for Cytochemical Section Assay

Male Wistar rats (WAG strain) of about 200 g body weight were placed on a low-sodium diet ($<$ 5 mmol per 100 g diet) for about 3 weeks. The rats were killed by cervical dislocation, the adrenal glands removed, stripped of surrounding fat, and each adrenal cut into two segments. Each adrenal segment was placed on defatted lens tissue on a stainless steel grid, in a vitreosil dish containing Trowell's T8 medium, pH 7.6, supplemented with 10^{-3} M sodium ascorbate. The dishes were placed in airtight Perspex culture chambers and incubated for 5 hr at

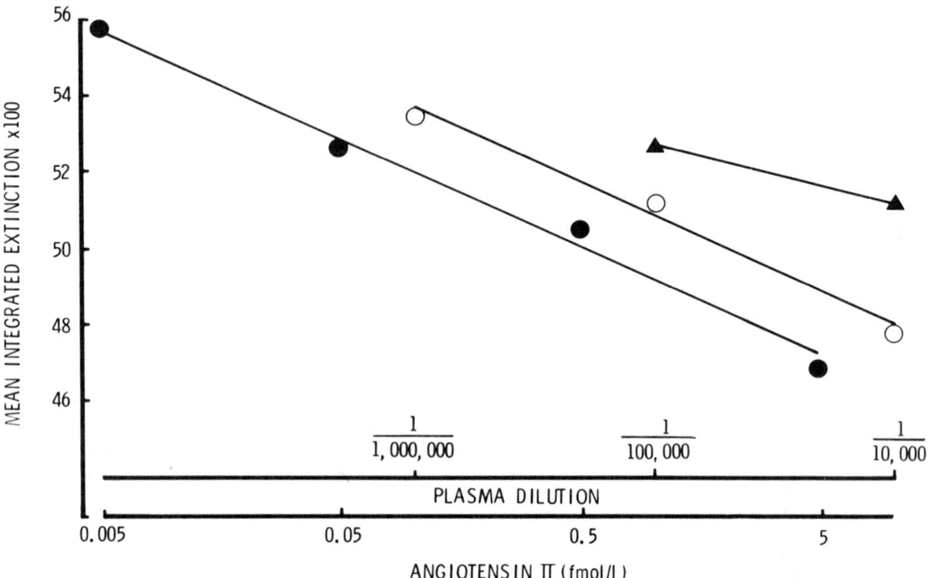

Figure 2   A calibration graph showing the response (mean integrated extinction × 100), measured in the zona glomerulosa, in sections which had been exposed to graded concentrations of ile^5-angiotensin II (MRC 70/302 (●————●; dilutions of a plasma, o————o; and the same plasma to which excess of an angiotensin II antiserum had been added, ▲————▲.

37°C in an atmosphere of 95% oxygen and 5% carbon dioxide. Segments were then chilled separately by immersion in n-hexane maintained at –70°C, by a mixture of solid carbon dioxide and industrial spirit, for 30-60 sec. The chilled adrenal tissue was placed in cold glass tubes and stored at –70°C. Sections of 20 μm thickness were cut in a Bright's cryostat, with the ambient temperature at –30°C, and the knife chilled further by packing around its haft with solid CO_2. The sections were then air-dried at room temperature, before exposure to standards or samples. Ile-5-angiotensin II standard (MRC 70/302) was prepared in plasma from anephric patients. Unknown plasma samples and the standard were diluted in the reaction medium. The reaction medium consisted of Trowell's T8 medium, pH 7.6, containing 10^{-3} M sodium ascorbate and 5% polypeptide (Sigma P5115) as a tissue stabilizer. Duplicate sections were exposed to different dilutions of standard or sample for 1 min, then transferred to a staining solution consisting of a mixture of 1.35% ferric chloride hexahydrate (3 vol) and 0.1% potassium ferricyanide (1 vol). After 4 min the sections were transferred to fresh staining solution for a further 4 min. The sections were then washed in tap water, air-dried, cleared in xylene and mounted in Styrolite. The amount of ferro-ferricyanide (Prussian blue) was measured by its absorption at 680 nm, using a Vickers M85 scanning and integrating microdensitometer. Measurements were made in 10 different areas of the zona glomerulosa in each duplicate section. The relative absorption values were converted to mean integrated extinction values, multiplied by 100 for convenience.

Results and Validation

Standard Response
Figure 2 shows the response obtained in the zona glomerulosa when sections of adrenal were exposed to 10-fold dilutions of angiotensin II (MRC 70/302). A decrease in reducing potency can be seen with increasing concentrations of the standard. A negative linear log-dose relationship was obtained over the range 0.05-5.0 fg/ml. In some experiments a linear response was obtained from 0.005 to 5.0 fg/ml.

Stability of Angiotensin II in Plasma
The level of angiotensin in blood reflects the balance between the production by renin and converting enzyme, and the breakdown by angiotensinases. With normal blood-handling procedures this balance is upset because angiotensin II is rapidly generated in blood or plasma on standing, and is also broken down by plasma angiotensinases. This necessitates the addition of a suitable inhibitor of these activities to blood. For the radioimmunoassay of angiotensin II a mixture of EDTA and o-phenanthroline is commonly employed as an inhibitor. This was considered to be unsuitable for a bioassay procedure, as EDTA is a chelating agent. Trasylol was found to inhibit the generation of angiotensin II without affecting its bioactivity. Consequently, all blood samples were collected into a tube containing heparin and 10,000 kIU Trasylol per 10 ml blood. The blood was spun immediately in the cold and the separated plasma rapidly frozen and stored at -70°C.

Precision
Reproducibility. The reproducibility of the procedure was examined by measuring the reaction product in a number of sections (usually six) which had been exposed to three dilutions of the standard angiotensin II or to two dilutions of a plasma sample. In all cases the standard deviations were less than ±7% of the mean integrated extinction value.

Interassay precision was assessed by repeated measurement of separate aliquots of the same plasma sample taken from a normal subject. One aliquot was measured in each of 15 assays. The mean angiotensin II level was found to be

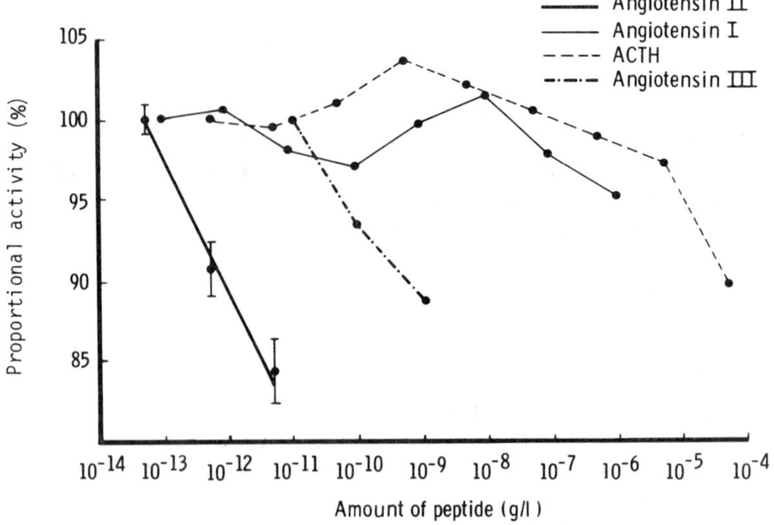

Figure 3   The response (proportional activity) in sections exposed to graded concentrations of ile^5-angiotensin II (MRC 70/302) ●————●; ile^5-angiotensin III (Sigma) o— · — · —o; ile^5-angiotensin I (MRC 71/328) ●————●; and adreno-corticotropin (ACTH, MRC 74/555) o- - - - o.

13.5 pg/ml, with a standard deviation of ±4.54 pg/ml. The mean slope of the standard response in these assays was 4.37 mean integrated extinction units × 100. The mean index of precision (λ) for the standard response was 0.065.

Recovery.   20, 40, and 80 pg/ml of angiotensin II standard were added to plasma samples. When these samples were assayed together with the untreated plasma, the recovery of angiotensin II added to plasma varied between 60 and 88%, with a mean value of 76%.

Sensitivity
The sensitivity of the assay is defined as the smallest concentration of the hormone that can be measured above the effect of the reaction medium alone. For this assay the sensitivity is about 0.5 fg/ml. Using a 1:100 dilution of plasma, as little as 50 fg/ml may be detected.

Specificity
The specificity of the assay was assessed in two ways: first, by testing various substances for cross-reactivity in the system, and, second, by the addition of excess anti-angiotensin II antiserum to plasma and measuring the residual activity.

Cross-reactivity.   The specificity of the assay was tested with structurally related peptides, namely, angiotensin I and angiotensin III, and with ACTH, which acts on the adrenal cortex and is said to stimulate aldosterone release. A response was seen with angiotensin III at concentrations between 0.01 and 1.0 pg/ml. Much higher concentrations of ACTH (0.5 and 5.0 ng/ml) were needed to have an effect on this system (Figure 3). Angiotensin I had no significant effect. A hundred times more angiotensin III and a million times more ACTH was required to demonstrate the same biological effect as was seen with angiotensin II.

Antiserum.   In one experiment a plasma sample, assayed at three dilutions, gave a mean value of 24.3 pg/ml. When this plasma was diluted 1:10,000 and 1:100,000 and excess of an anti-angiotensin II serum was added to it, the mean value was 3.6 pg/ml (Figure 2). Addition of the antibody had reduced the angiotensin II bioactivity of this plasma by 85%. In another experiment, angiotensin II bioactivity was reduced by 88.5% when antiserum was added to plasma.

Plasma Levels of Angiotensin II

Normal Subjects.   The variation in plasma angiotensin II levels was evaluated in five normotensive subjects on three different sodium diets. On a normal sodium intake of 120 mmol/day, the mean angiotensin II level determined by cytochemical bioassay was 12.9 pmol/liter. On a low sodium intake (10 mmol/day) the mean angiotensin II level rose to 45.7 pmol/liter, whereas on a high sodium intake (350 mmol/day) the mean angiotensin II level fell to 5.3 pmol/liter. The angiotensin II levels in these subjects showed a good correlation with plasma renin activity.

Anephric Subjects.   Anephric subjects have been found to have very low levels of plasma renin activity. However, discrepant results have been obtained when angiotensin II has been measured by radioimmunoassay. Some workers have found anephric subjects to have undetectable levels, whereas others have found measureable levels of AII in these subjects. In three anephric patients we have found very small amounts of angiotensin II (0.2, 0.37, and 0.28 pmol/ liter) by the cytochemical technique.

Application of the Assay

This is the first highly sensitive bioassay to be described for angiotensin II. Its sensitivity has several advantages for physiological and clinical studies. Only small volumes of plasma are required, as plasma has to be diluted at least 100 times to fall within the range of the standard dose-response graph. The need for only small volumes of blood will allow serial sampling from subjects. Changes in angiotensin II levels throughout the day, and the half-time of disappearance of a single injection of angiotensin II, can be studied in normal subjects. The extreme sensitivity of the assay will permit measurement of very low levels of angiotensin II beyond the range of current radioimmunoassays. For example, low angiotensin II levels have been found in anephric patients. Other situations in which low levels of angiotensin II may be found include patients treated with the converting enzyme inhibitor, Captopril, and low-renin hypertensives.

REFERENCES

Best, J. B., Bett, J. H. N., Coghlan, J. P., Cran, E. J., and Scoggins, B. A. (1971). Circulating angiotensin-II and aldosterone levels during dietary sodium restriction. *Lancet*, ii:1353-1354.

Biron, P., Koiw, E., Nowaczynski, W., Brouillet, J., and Genest, J. (1961). The effects of intravenous infusions of valine-5 angiotensin II and other pressor agents on urinary electrolytes and corticosteroids, including aldosterone. *J. Clin. Invest.*, 40:338-347.

Blair-West, J. R., Coghlan, J. P., Denton, D. A., Funder, J. W., Scoggins, B. A., and Wright, R. D. (1971). The effect of the heptapeptide (2-8) and hexapeptide (3-8) fragments of angiotensin II on aldosterone secretion. *J. Clin. Endocrinol. Metab.*, 32:575-578.

Boucher, R., Veyrat, R., de Champlain, J., and Genest, J. (1964). New procedures for measurement of human plasma angiotensin and renin activity levels. *Can. Med. Assoc. J.*, 90:194-201.

Boyd, G. W., Landon, J., and Peart, W. S. (1967). Radioimmunoassay for determining plasma levels of angiotensin II in man. *Lancet*, ii:1002-1005.

Boyd, G. W., and Peart, W. S. (1968). The production of high-titre antibody against free angiotensin II. *Lancet*, ii:129-133.

Brown, J. J., Brown, W. C. B., Fraser, R., Lever, A. F., Morton, J. J., Robertson, J. I. S., Rosei, E. A., and Trust, P. M. (1976). The effects of saralasin, an angiotensin II antagonist, on blood pressure and the renin-angiotensin-aldosterone system in normal and hypertensive subjects. *Aust. N.Z. J. Med.*, 6(suppl.)3:48-52.

Brown, J. J., Fraser, R., Lever, A. F., Love, D.R., Morton, J. J., and Robertson, J. I. S. (1972). Raised plasma angiotensin II and aldosterone during dietary sodium restriction in man. *Lancet*, ii:1106-1107.

Brown, J. J., Fraser, R., Lever, A. F., Morton, J. J., Oelkers, W., Robertson, J. I. S., and Young, J. (1973). Further observations on the relationship between plasma angiotensin II and aldosterone during sodium deprivation. In *Mechanisms of Hypertension*, Vol. 302, M. P. Sambhi (ed.). *Excerpta Medica*, 148-154.

Brown, J. J., Fraser, R., Lever, A. F., Morton, J. J., and Robertson, J. I. S.

(1977). Renin-angiotensin system in the regulation of aldosterone in humans. In *Hypertension.* J. Genest, E. Koiw, and O. Kuchel (eds.). McGraw-Hill, New York, pp. 874-888.

Cain, M. D., Catt, K. J., and Coghlan, J. P. (1969). Effect of circulating fragments of angiotensin II on radioimmunoassay in arterial and venous blood. *J. Clin. Endocrinol. Metab.,* 29:1639-1643.

Cain, M. D., Coghlan, J.P., and Catt, K. J. (1972). Measurement of angiotensin II in blood by radioimmunoassay. *Clin. Chim. Acta,* 39:21-34.

Carey, R. M., Vaughan, E. D., and Peach, M. J. (1978). Activity of (des-aspartyl)-angiotensin II and angiotensin II in man. *J. Clin. Invest.,* 61:20-31.

Case, D. B., Atlas, S. A., Laragh, J. H., Sealey, J. E., Sullivan, P. A., and Mc-Kinstry, D. N. (1978). Clinical experience with blockade of the renin-angiotensin-aldosterone system by an oral converting enzyme inhibitor (SQ14,225, captopril) in hypertensive patients. *Prog. Cardiovasc. Dis.,* 21:195-206.

Catt, K. J., Cain, M. C., and Coghlan, J. P. (1967). Measurement of angiotensin II in blood. *Lancet,* ii:1005-1007.

Catt, K. J., Cran, E., Zimmet, P. E., Best, J. B., Cain, M. D., and Coghlan, J. P. (1971). Angiotensin II blood levels in human hypertension. *Lancet,* i:459-464.

Chayen, J., Loveridge, N., and Daly, J. R. (1972). A sensitive bioassay for adrenocorticotrophic hormone in human plasma. *Clin. Endocrinol.* (Oxf) 1:219-233.

Chinn, R. H., and Düsterdieck, G. (1972). The response of blood pressure to infusion of angiotensin II: Relation to plasma concentrations of renin and angiotensin. *Clin. Sci.,* 42:489-504.

Clough, D. P., Hatton, R., Conway, J., and Adigun, S. A. (1980). Angiotensin activates sympathetic reflexes in the anaesthetized cat. *Clin. Sci.,* 59:287s-289s.

Cushman, D. Q., Cheung, H. S., Sabo, E. F., and Ondetti, M. A. (1977). Design of potent competitive inhibitors of angiotensin-converting enzyme. *Biochemistry,* 16:5484-5491.

Dekanski, J. (1954). The quantitative assay of angiotonin. *Br. J. Pharmacol.,* 9:187-191.

Douglas, J., Aguilera, G., Kondo, T., and Catt, K. (1978). Angiotensin II receptors and aldosterone production in rat adrenal glomerulosa cells. Endocrinology, 102:685-696.

Dunn, M. J., and Tannen, R. L. (1974). Low renin hypertension. *Kidney Int.,* 5:317-325.

Düsterdieck, G., and McElwee, G. (1971). Estimation of angiotensin II concentration in human plasma by radioimmunoassay: Some applications to physiological and clinical states. *Eur. J. Clin. Invest.,* 2:32-38.

Emmanuel, R. L., Cain, J. P., and Williams G. H. (1973). Double antibody radio-immunoassay of renin activity and angiotensin II in human peripheral plasma. *J. Lab. Clin. Med.*, 81:632-640.

Fasciolo, J. C., Houssay, B. A., and Taquini, A. C. (1938). The blood-pressure raising secretion of the ischaemic kidney. *J. Physiol.* (Lond.), 94:281-293.

Genest, J., Boucher, R., de Champlain, J., Veyrat, R., Chretien, M., Biron, P., Tremblay, G., Roy, P., and Cartier, P. (1964). Studies of the renin-angiotensin system in hypertensive patients. *Can. Med. Assoc. J.*, 90:263-268.

Gocke, D. J., Sherwood, L. M., Oppenhoff, I., Gerten, J., and Laragh, J. H. (1968). Measurement of plasma angiotensin II and correlation with renin activity. *J. Clin. Endocrinol. Metab.*, 28:1675-1678.

Goldblatt, H., Lynch, J., Hanzal, R. F., and Summerville, W. (1934). Studies on experimental hypertension. I. The production of persistent elevation of systolic blood pressure by means of renal ischaemia. *J. Exp. Med.*, 59:347-379.

Goodfriend, T. L., Ball, D.L., and Farley, D. B. (1968). Radioimmunoassay of angiotensin. *J. Lab. Clin. Med.*, 72:648-662.

Goodfriend, T. L., and Peach, M. J. (1975). Angiotensin III (des-aspartic acid)-angiotensin II, evidence and speculation for its role as an important agonist in the renin-angiotensin system. *Circ. Res.*, 36(suppl. I):38-48.

Goodfriend, T. L., Levine, L., and Fasman, G. D. (1964). Antibodies to brady-kinin and angiotensin: Use of carbodiimides in immunology. *Science*, 144: 1344-1346.

Gould, A. B., Skeggs, L. T., and Kahn, J. R. (1966). Measurement of renin and substrate concentrations in human serum. *Lab. Invest.*, 15:1802-1813.

Halkerston, I. D. K., and Feinstein, M. (1968). Preparation of ACTH responsive isolated cells from rat adrenals. *Fed. Proc. Abstracts*, 27:626.

Hanning, R., Tait, S. A. S., and Tait, J. F. (1970). In vitro effects of ACTH, angiotensins, serotonin and potassium on steroid output and conversion of corticosterone to aldosterone by isolated adrenal cells. *Endocrinology*, 87: 1147-1167.

Helmer, O. M. (1957). Differentiation between two forms of angiotonin by means of spirally cut strips of rabbit aorta. *Am. J. Physiol.*, 188:571-577.

Helmer, O. M. (1961). Pressor substances in renal vein blood of hypertensives. *Med. Clin. North Am.*, 45:309-320.

Helmer, O. M., and Judson, W. E. (1963). The quantitative determination of renin in the plasma of patients with arterial hypertension. *Circulation*, 27: 1050-1060.

Hodge, R. L., Lowe, R. D., and Vane, J. R. (1966a). Increased angiotensin for-mation in response to carotid occlusion in the dog. *Nature* (Lond.), 211:491-493.

Hodge, R. L., Lowe, R. D., and Vane, J. R. (1966b). The effects of alteration of blood-volume on the concentration of circulating angiotensin in anaesthetised dogs. *J. Physiol.* (Lond.), 185:613-626.

Houssay, B. A., and Taquini, A. C. (1938). Acción vasocontrictora de la sangre venosa del riñón isquemiado. *Rev. Soc. Argent. Biol.*, 14:5-14.

Hunter, W. M., and Greenwood, F. C. (1962). Preparation of iodine-131 labelled human growth hormone of high specific activity. *Nature* (Lond.), 194:495-496.

Joy, M. D., and Lowe, R. D. (1970). Evidence that the area postrema mediates the central cardiovascular response to angiotensin II. *Nature* (Lond.), 228:1303-1304.

Kahn, J. R., Skeggs, L. T., Shumway, N. P., and Wisenbaugh, P. E. (1952). The assay of hypertensin from the arterial blood of normotensive and hypertensive human beings. *J. Exp. Med.*, 95:523-529.

Kappelgaard, A. M., Nielsen, M. D., and Giese, J. (1976). Measurement of angiotensin II in human plasma: Technical modifications and practical experience. *Clin. Chim. Acta*, 67:299-306.

Kloppenberg, P. W. C., Island, D. P., Liddle, G. W., Michelakis, A. M., and Nicholson, W. E. (1968). A method of preparing adrenal cell suspensions and its applicability to the in vitro study of adrenal metabolism. *Endocrinology*, 82:1053-1058.

Kosunen, K. J. (1976). A simple method for measurement of angiotensin II in plasma. *Scand. J. Clin. Invest.*, 36:467-472.

Langford, H. G. (1964). Tubular action of angiotensin. *Can. Med. Assoc. J.*, 90:332-333.

Laragh, J. H., Angers, M., Kelly, W. G., and Lieberman, S. (1960). The effect of epinephrine, norepinephrine, angiotensin II and others on the secretory rate of aldosterone in man. *JAMA*, 174:234-240.

Laragh, J. H., Cannon, P. J., Bentzel, C. J., Sicinski, A. M., and Meltzer, J. I. (1963). Angiotensin II, norepinephrine and renal transport of electrolytes and water in normal man and in cirrhosis with ascites. *J. Clin. Invest.*, 42:1179-1192.

Lijnen, P. J., Amery, A. K., Fagard, R. H., and Katz, F. H. (1978). Radioimmunoassay of angiotensin II in unextracted plasma. *Clin. Chim. Acta*, 88:403-412.

MacGregor, G. A., Markandu, N. D., and Roulston, J. E. (1979). Does the renin-angiotensin system maintain blood pressure in both hypertensive and normotensive subjects? A comparison of propranolol, saralsasin and captopril. *Clin. Sci.*, 57:145s-148s.

MacGregor, G. A., Markandu, N. D., Roulston, J. E., Jones, J. C., and Morton, J. J. (1980). The renin-angiotensin-aldosterone system in the maintenance of blood pressure, aldosterone secretion and sodium balance in normotensive subjects. *Clin. Sci.*, 59:95s-99s.

MacGregor, G. A., Markandu, N. D., Roulston, J. E., Jones, J. C., and Morton, J. J. (1981). Maintenance of blood pressure by the renin-angiotensin system in normal man. *Nature*, 291:329-331.

Massani, Z. M., Finkielman, S., Worcel, M., Agrest, A., and Paladini, A. C. (1966). Angiotensin blood levels in hypertensive and nonhypertensive diseases. *Clin. Sci.*, 30:473-483.

Mendelsohn, F. A. O., and Kachel, C. D. (1981). Stimulation by serum of aldos- and III on aldosterone production by isolated rat adrenal zona glomerulosa cells: Importance of metabolism and conversion of peptides in vitro. *Endocrinology*, 106:1760-1768.

Mendelsohn, F. A. O., and Kachel, C. D. (1981). Stimulation by serum of aldosterone production from rat adrenal glomerulosa cells in vitro: Relationships to K^+, serotonin and angiotensin II. *Acta Endocrinol.* (Kbh)., 97:231-242.

Moore, A. F. (1979). A review of the renin-angiotensin system: Historical and methodological perspectives. *Methods Find. Exp. Clin. Pharmacol.*, 1:23-44.

Morris, R. E., and Robinson, P. R. (1964). A method for determination of angiotensin II in human blood. *Johns Hopkins Med. J.*, 114:127-145.

Morton, J. J., Casals-Stenzel, J., Lever, A. F., Millar, J. A., Riegger, A. J. G., and Tree, M. (1979). Inhibitors of the renin-angiotensin system in experimental hypertension with a note on the measurement of angiotensin I, II, and III during infusion of converting enzyme inhibitor. *Br. J. Clin. Pharmacol.*, 7 (suppl. 2):233-241.

Mulrow, P. J. (1964). The role of the renin-angiotensin system in the hypertension associated with renal vascular disease. *Can. Med. Assoc. J.*, 90:277-280.

Needleman, P., Johnson, E. M., Vine, W., Flanigan, E., and Marshall, G. R. (1972). Pharmacology of antagonists of angiotensin I and II. *Circ. Res.*, 31: 862-867.

Nicholls, M. G., and Espiner, E. A. (1976). A sensitive, rapid radioimmunoassay for angiotensin II. *N.Z. Med. J.*, 83:399-403.

Nielsen, M. D., Jorgensen, M., and Giese, J. (1971). I-125 labelling of angiotensin I and II. *Acta Endocrinol. (Kbh)*, 67:104-116.

Nielson, I., and Moller, I. (1967). Simultaneous determination of renin activity and angiotensin concentration levels in human plasma. *Acta Med. Scand.*, 182:263-269.

Oparil, S., and Haber, E. (1974). The renin-angiotensin system. *N. Engl. J. Med.*, 291:389-401.

Page, I. H., and Bumpus, F. M. (1974). *Angiotensin: Handbook of Experimental Pharmacology, Vol. XXXVII.* Springer-Verlag, New York.

Page, L. B., Haber, E., Kimura, A. Y., and Purnode, A. (1969). Studies with the radioimmunoassay for angiotensin II, and its application to measurement of renin activity. *J. Clin. Endocrinol. Metab.*, 29:200-206.

Paiva, A. C. M., and Paiva, T. B. (1960). The importance of the N-terminal end of angiotensin II for its pressor and oxytocic activities. *Biochem. Pharmacol.*, 5:187-191.

Paladini, A. C., Braun-Menendez, E., Del Frade, I. S., and Massani, Z. M. (1959). The estimation of angiotensin in blood. *J. Lab. Clin. Med.*, 53:264-271.

Peach, M. J. (1977). Renin-angiotensin system: Biochemistry and mechanisms of action. *Physiol. Rev.*, 57:313-370.

Peach, M. J. (1981). Molecular actions of angiotensin. *Biochem. Pharmacol.*, *30: 2745-2751.*

Peart, W. S. (1955). A new method of large-scale preparation of hypertensin, with a note on its assay. *Biochem. J.*, 59:300-302.

Peart, W. S. (1976). The renin-angiotensin system. In *Peptide Hormones*, J. A. Parsons (ed.). Macmillan Press, London, pp. 179-196.

Posternak, L., Brunner, H. R., Gavras, H., and Brunner, D. B. (1977). Angiotensin II blockade in normal man: Interaction of renin and sodium in maintaining blood pressure. *Kidney Int.*, 11:197-203.

Regoli, D., and Vane, J. R. (1964). A sensitive method for the assay of angiotensin. *Br. J. Pharmacol.*, 23:351-359.

Reid, I. A., Morris, B. J., and Ganong, W. F. (1978). The renin-angiotensin system. *Ann. Rev. Physiol.*, 40:377-410.

Schwarz, H., Masson, G. M. C., and Page, I. H. (1955). A method of assay of the oxytocic activity of angiotonin preparations on the rat's uterus. *J. Pharmacol. Exp. Ther.*, 114:418-429.

Scornik, O. A., and Paladini, A. C. (1961). Angiotensin blood levels in dogs with experimental renal hypertension. *Am. J. Physiol.*, 201:526-530.

Semple, P. F., Boyd, A. S., Dawes, P. M., and Morton, J. J. (1976). Angiotensin II and its heptapeptide (2-8), hexapeptide (3-8) and pentapeptide (4-8) metabolites in arterial and venous blood in man. *Circ. Res.*, 39:671-678.

Shipley, R. E., and Tilden, J. H. (1947). A pithed rat preparation suitable for assaying pressor substances. *Proc. Soc. Exp. Biol. Med.*, 64:453-455.

Skeggs, L. T., Dorer, F. E., Levine, M., Lentz, K. E., and Kahn, J. R. (1980). The biochemistry of the renin-angiotensin system. *Adv. Exp. Med. Biol.*, 130: 1-27.

Skeggs, L. T., Kahn, J. R., and Shumway, N. P. (1951). The isolation of hypertensin from the circulating blood of normal dogs with experimental renal hypertension by dialysis in an artificial kidney. *Circulation*, 3:384-389.

Skeggs, L. T., Kahn, J. R., and Shumway, N. P. (1952). The isolation and assay of hypertensin from blood. *J. Exp. Med.*, 95:241-246.

Skinner, S. L., Lumbers, E. R., and Symonds, E. M. (1969). Alterations by oral contraceptives of normal menstrual changes in plasma renin activity concentration and substrate. *Clin. Sci.*, 36:67-76.

Skinner, S. L., Lumbers, E. R., and Symonds, E. M. (1972). Analysis of changes in the renin-angiotensin system during pregnancy. *Clin. Sci.*, 42:479-488.

Sundsfjord, J. A. (1970). Radioimmunoassay of angiotensin II in plasma. *Acta Endocrinol. (Kbh)*, 64:181-192.

Swallow, R. L., and Sayers, G. (1969). A technic for the preparation of isolated rat adrenal cells. *Proc. Soc. Exp. Biol. Med.*, 131:1-4.

Tait, J. F., Tait, S. A. S., Bell, J. B. G., Hyatt, P. J., and Williams, B. C. (1980). Further studies on the stimulation of rat adrenal capsular cells: Four types of response. *J. Endocrinol.*, 87:11-27.

Tait, J. F., Tait, S. A. S., Gould, R. P., and Mee, M. S. R. (1974). The properties of adrenal zona glomerulosa cells after purification by gravitational sedimentation. *Proc. R. Soc. Lond.*, 185:375-407.

Tigerstedt, R., and Bergman, P. G. (1898). Niere und Kreislauf. *Scand. Arch. Physiol.*, 8:223-271.

Vallotton, M. B., Page, L. B., and Haber, E. (1967). Radioimmunoassay of angiotensin in human plasma. *Nature (Lond.)*, 215:714-715.

Wilmshurst, E. G., Zeunert, H. A., and Gordon, R. D. (1971). Radioimmunoassay of angiotensin II in peripheral venous plasma: Effects of diet and posture and application to measurement of renal venous renin ratio. *Aust. N.Z. J. Med.*, 3:228-232.

13

Hypothalamic Regulating Hormones

J. A. H. Wass and G. M. Besser / St. Bartholomew's Hospital, London, England

INTRODUCTION

The science of neuroendocrinology began in the late 1940s and early 1950s with the studies of Geoffrey Harris, who stressed the importance of the observation that almost all blood reached the adenohypophysis via a portal system in the hypophyseal stalk that connects it to the median eminence area of the hypothalamus. Later work demonstrated neurosecretory cells within the central nervous system, particularly in the hypothalamus, and these were shown to respond to classic means of synaptic activation. Despite this early work, it was not until 1969 that two groups of investigators, headed by Andrew Schally and Roger Guillemin, elucidated the structure and synthesized thyrotropin-releasing hormone (TRH) and, later, luteinizing hormone and follicle-stimulating hormone releasing hormone (gonadotropin-releasing hormone; GnRH).

This earlier work has had a profound and wide impact. New hormones and neurotransmitters have been discovered in the brain, new diseases have been described, and a more profound understanding of older ones obtained. Neuroendocrinology, besides being of immense physiological interest, has become of immediate relevance to the clinician and now embraces a wide range of other scientific disciplines.

In the light of these new and exciting findings, this brief chapter will deal with the nature and biological function of the known hypothalamic hormones.

To date, little success has been achieved with measuring them either cytochemically or by radioimmunoassay, and this has resulted in a paucity of information regarding their physiological function and regulation.

Historical Introduction

Popa and Fielding in 1930 were the first to describe the capillary system connecting the hypothalamus with the hypophysis (Popa and Fielding, 1930). However, they incorrectly believed that blood flow occurred upward from the sinusoids of the adenohypophysis to the floor of the infundibular recess of the third ventricle. However, during a study in monkeys, Wislocki and King (1936) concluded that the blood flowed down the hypophyseal portal vessels in the direction of the adenohypophysis. The concept of a neurovascular link between the hypothalamus and the anterior lobe of the hypophysis was first described by Green and Harris (1947), who demonstrated that most of the blood reaching the adenohypophysis first traversed the portal capillaries lying in the substance of the median eminence. They further suggested that humoral substances, later called releasing factors, were liberated from the nerve endings of the hypothalamic tracts into the primary plexus of capillaries of the hypophyseal portal vessels and were carried by these to the adenohypophysis, where they regulated its secretion. Harris also postulated that nervous stimuli from higher brain centers might cause the liberation of these substances into capillary sinusoids of the median eminence. They later directly observed, in anesthetized rats, that the direction of blood flow in these vessels was from the median eminence to the adenohypophysis, and this gave further weight to their belief that the central nervous system regulated the activity of the adenohypophysis by means of a humoral relay through the hypophyseal portal vessels. Bargmann and Scharrer (1951) provided the first evidence that neurones of the hypothalamus could synthesize and liberate hormones. They proposed the presence of a neurosecretory pathway in the posterior pituitary based on the concept that the peptide hormones, oxytocin and vasopressin, were synthesized in hypothalamic nuclei; packaged into neurosecretory vesicles; and transported along axons to nerve terminals in the neurohypophysis, where they were released into the systemic circulation in response to appropriate physiological stimuli.

THYROTROPIN-RELEASING HORMONE: ISOLATION AND STRUCTURE

The existence of a hypothalamic factor regulating thyroid-stimulating hormone (TSH) secretion from the anterior pituitary was first demonstrated by Greer (1951). This worker found that bilateral symmetrical lesions between the suprachiasmatic and caudal ventromedial nuclei of the hypothalamus decreased the (pyro) Glu - His - Pro - NH$_2$

    1   2   3

Figure 1   Structure of thyrotropin-releasing hormone (TRH).

size of goiters induced by propylthiouracil in rats. He suggested that this was due to an interference in the secretion of a hypothalamic hormone. Ganong et al. (1955) induced lesions in the anterior end of the median eminence in dogs and found a selective inhibition of the thyroid, without alteration of other anterior pituitary tropic hormone release. Harris and Woods (1958) stimulated the hypothalamus and confirmed that the anterior end of the median eminence was involved. In 1966, TRH was isolated from porcine hypothalami and was shown to contain three amino acids, histidine, proline, and glutamic acid, in equimolar ratios (Schally et al., 1966). However, it was not until 1969 that the structure of porcine TRH was determined (Folkers et al., 1969; Schally et al., 1969). The work in Guillemin's laboratory of Burgus et al. (1969; 1979) on ovine TRH parallelled that of Schally's group. It is probable that both bovine and human TRH have the same structure (see Figure 1) (Schally and Bowers, 1971).

Studies on the Actions of TRH

Bowers et al. (1968), using nearly pure TRH from porcine hypothalami, showed for the first time that there was a rise in circulating TSH in humans, as measured both by radioimmunoassay and bioassay; levels reached a maximum between 10 and 30 min and thereafter decreased. The results of this study further demonstrated that TRH is not species specific, since porcine TRH was active in mouse and rat TRH bioassays as well as in human studies. Thyroxine (T$_4$) and triiodothyronine (T$_3$) block the stimulatory effect of TRH and TSH release, thus suggesting a feedback action of thyroid hormones directly on the pituitary gland.

Tashjian et al. (1971) demonstrated that synthetic TRH caused the release of prolactin from rat pituitary tumor cells in vitro. Later this was shown to occur in humans (Jacobs et al., 1971). This prolactin response is partially inhibited by dopamine agonists. It is not clear whether TRH is a physiological prolactin releasing hormone. Somatostatin, however, inhibits TSH secretion, and studies using somatostatin antibodies suggest that this may be a physiologically important inhibitor from the hypothalamus of pituitary TSH secretion (Hall et al., 1973).

The presence of TRH in nerve terminals suggests that it may also act as a neurotransmitter (Hökfelt et al., 1975). It is also present in the islets of Langerhans of the pancreas (Martino et al., 1978), but its function in the gut is not yet elucidated (Jackson, 1982).

Radioimmunoassay for TRH

The availability of synthetic TRH made possible the establishment of a radioimmunoassay. Bassiri and Utiger (1972) conjugated TRH to bovine serum albumin in order to make it immunogenic. By means of the antisera derived from these studies, examination of the distribution of TRH in rat brain was possible. Brownstein et al. (1974) dissected discrete nuclei and found the greatest concentration in the median eminence. Outside the hypothalamus, which contained 31.2% of the total brain content of TRH, most of the large amounts of TRH have been found in preoptic areas (Winokur and Utiger, 1974). These results suggest that TRH, in addition to its hypophysiotropic function, may possibly serve as a central neurotransmitter or neuromodulator.

Mechanism of Action

TRH acts directly on the pituitary tissue since, in vitro, it releases TSH from the pituitaries of sheep, rats, and goats. In pituitary tissue cultures, TRH stimulates the synthesis as well as the release of TSH (Mittler et al., 1969). Hypothalamic fragments, incubated with the necessary amino acids, have been shown to synthesize TRH (Mitnick and Reichlin, 1971), but the exact cellular site of TRH formation is at present unknown. ^{125}I-labeled TRH accumulates in the pituitary thus again indicating that TRH acts directly on the pituitary. It is rapidly inactivated in rat and human plasma; the half-life of TRH in the blood of a rat is about 4 min (Redding and Schally, 1971).

TRH is specifically bound by plasma membrane receptors of bovine anterior pituitary glands, and this appears to be the first event in TRH action (Wilber and Seibel, 1973). Adenylate cyclase activity is stimulated by the addition of TRH, and derivatives of cyclic AMP can also stimulate TSH release in vitro; thus, cyclic AMP may be associated with the action of TRH on the pituitary cell (Labrie et al., 1972).

Uses of TRH

Following the first clinical studies (Hall et al., 1970), synthetic TRH has been shown to be useful in the evaluation of the pituitary reserve of both thyrotropin and prolactin secretion. It is most useful in differentiating hypothalamic and pituitary hypothyroidism from primary hypothyroidism and in the diagnosis of thyrotoxicosis, euthyroid Graves' disease, and thyroid autonomy (Ormston et al., 1971; Hall et al., 1972). The TRH test is fully reviewed by Burger and Patel (1977).

LUTEINIZING HORMONE RELEASING HORMONE: GONADOTROPIN-RELEASING HORMONE

Harris (1957) showed that localized electrical stimulation of the rat hypothalamus in the region of the median eminence and preoptic suprachiasmatic regions caused ovulation. He deduced that this was due to an effect of nerve fibers in the hypothalamus releasing a hormone which passed to the anterior pituitary, thereby causing ovulation. Previous physiological and pharmacological experiments by Sawyer et al. (1947) had demonstrated for the first time the importance of the adrenergic system in the neural control of release of LH, since if dibenamine, an adrenergic blocking agent, was given intravenously within 1 min of copulation, it prevented ovulation.

McCann et al. (1960) were the first to demonstrate an LH releasing factor in rat hypothalamic extracts. They injected crude acid extracts of the stalk median eminence and then showed LH release by bioassay. The substance responsible for LH releasing activity was called LH releasing factor. It was concluded that this material differed from histamine, serotonin, substance P, epinephrine, and vasopressin. Campbell, Feuer, and Harris (1964) demonstrated that crude extracts of the tissue of the median eminence from rabbit, cattle, or monkey could cause ovulation in the estrus rabbit, thus suggesting that this releasing factor was the same substance in different mammals.

By using a sensitive bioassay method devised by them in ovariectomized estrogen-progesterone-blocked rats, Igarashi and McCann (1964) were the first to demonstrate, by direct evidence, the existence of a follicle stimulating hormone releasing factor in hypothalamic stalk median eminence extract.

The isolation and synthesis of luteinizing hormone releasing hormone was conducted independently in the laboratories of Schally and Guillemin. Matsuo et al. (1971) first determined the correct amino acid sequence of porcine LH and FSH releasing hormone; it was shown to be a decapeptide (Figure 2). This was confirmed by Baba et al. (1971) using conventional degradation methods for sequential analysis. In this second study C-terminal pentapeptide amide was synthesized using solid-phase methods. The structure of ovine hypothalamic LHRH was found in Guillemins' laboratory to be identical (Burgus et al., 1972). Subsequent work has shown that the gonadotropin-releasing hormone of all mammalian species is identical.

Schally et al. (1971) proposed that only one hypothalamic hormone was responsible for the release of both LH and FSH, and there is much evidence to support this. Chemical and enzymatic inactivation of the decapeptide was always (pyro) Glu - His - Trp - Ser - Tyr - Gly - Leu - Arg - Pro - Gly - NH$_2$

   1   2   3   4   5   6   7   8   9   10

Figure 2   Structure of gonadotropin-releasing hormone (GnRH).

accompanied by a loss of both LH and FSH-RH activity, and no separation between LH and FSH activity could be obtained. To date, no separate FSH-RH
has been isolated, and it is clear that LHRH releases both LH and FSH from the
anterior pituitary and is better referred to as GnRH (the gonadotropin-releasing
hormone). All the physiological control of LH and FSH can be accounted for by
interaction of one GnRH with the differential feedback effects of different gonadal steroids and inhibin altering independently the LH or FSH responses (Mortimer, 1977).

Actions of GnRH

This decapeptide releases specifically both LH and FSH, in normal subjects:
there is no release of other anterior pituitary hormones. The release is dose
related between doses of 50 and 500 μg; when this is given in a single intravenous bolus, less FSH than LH is secreted (Besser et al., 1972). After an intravenous bolus, peak circulating hormone levels occur between 10 and 30 min, usually
simultaneously. If rapid sampling is carried out, e.g., at 10 min intervals, gonadotropin secretion can be shown to be pulsatile and asynchronous (Mortimer et al.,
1973). In female subjects it has been quite clearly shown that serial injections of
the same dose of GnRH throughout the normal menstrual cycle result in a dis-
tinctive pattern of gonadotropin secretion. There is a relatively small response
in the follicular and luteal phases, but the greatest response is at midcycle, coin-
cident with the rise in plasma estradiol (Nillius and Wide, 1972; Yen et al.,
1972). It is evident that estrogens may modify hypothalamic function by a positive feedback in women at midcycle and a negative feedback in men. Thus, men
treated with estrogen for 3 days show suppressed basal and GnRH-stimulated LH
and FSH secretion.

Not only does GnRH stimulate the release of LH and FSH, but also there is
evidence that the synthesis of these two hormones is stimulated by GnRH
(Redding et al., 1972). Thus, the addition of nanogram amounts of LHRH to the
incubation medium daily for 5 days augmented the total content of LH and FSH
in the tissue and medium of rat anterior pituitaries. Similarly, LH and FSH secretion may be maintained in GnRH-deficient patients for more than a year with
long-term GnRH therapy (Mortimer, 1977).

Radioimmunoassay for GnRH

The availability of synthetic GnRH led to the development of several radioimmunoassays (Arimura et al., 1973; Jeffcoate et al., 1973). This has allowed the
demonstration of the very short half-life in blood. It has proved very difficult to
detect endogenous levels in the peripheral circulation, but its secretion appears
pulsatile and reaches a peak of about 2 pg/ml, detectable in the circulation in
women just before the preovulatory surge (Mortimer et al., 1976). Using im-
munostaining techniques it has been shown that GnRH is found predominantly
in the median eminence and the tuberoinfundibular tract of the hypothalamus
(Hökfelt et al., 1978).

Mechanisms of Effect

The mechanisms by which GnRH induces LH and FSH release are not completely known. It does appear to be associated with changes in cyclic AMP, and
this can stimulate LH and FSH release in vitro (Borgeat et al., 1972). Furthermore, GnRH incubated with rat and bovine anterior pituitary tissue, caused an
increase in intracellular concentration of cyclic AMP, by stimulating adenyl
cyclase in the pituitary cells.

Degradation and Metabolism

GnRH is rapidly degraded in blood by enzymic cleavage between residues 6 and
7. It is excreted by the kidneys (Redding et al., 1973). Numerous analogs of
GnRH have been developed which involve substitution at position 6 and result
in a longer duration of action and more powerful stimulation of both LH and
FSH secretion.

It appears important that GnRH is released in a pulsatile manner, since prolonged stimulation with GnRH causes the release of the readily releasable gonadotropins from the gonadotropes and then a decrease in gonadotropin secretion.
Continued availability of LH and FSH for release appears to depend on pulsatile GnRH secretion (Belchetz et al., 1978).

Uses of GnRH

The major use of this hypothalamic peptide has been in investigation of patients
with pituitary and hypothalamic disease. Although the pattern of response of
the gonadotropins to intravenous injection of GnRH is not so helpful as that
with TRH in diagnosis of hypothalamic disease, it is possible to assess the re-
serves of readily releasable LH and FSH if an intravenous bolus is given and
samples are taken before, and 20 and 60 min after, the injection. Treatment with
GnRH has been given by regular subcutaneous injection, and successfully in-

duced puberty, potency, spermatogenesis, and ovulation (Mortimer, 1977b).

However, great care needs to be exercised when using GnRH therapy. A paradoxical result of the development of long-acting analogs has been their use as contraceptive agents, as prolonged and excessive stimulation of the gonadotropes with these analogs results in a decrease in synthesis and release of gonadotropins (Nillius et al., 1978).

SOMATOSTATIN

Reichlin (1960) first showed the hypothalamic control of growth. He demonstrated impaired growth in rats in whom lesions were placed in the ventral hypothalamus. Even though these rats became hypothyroid and hypogonadal, they still did not grow despite appropriate replacement therapy. Roth et al. (1963) showed that, following section of the pituitary stalk, there was no rise in growth hormone after insulin-induced hypoglycemia. Later, Deubin and Meites (1964) demonstrated that there was a growth hormone releasing factor, although the chemical nature of this still is uncertain.* It is clear, however, that growth hormone is under the predominantly stimulatory influence of the hypothalamus because growth hormone deficiency results from pituitary stalk section.

The first evidence for a hypothalamic factor inhibiting growth hormone secretion from the anterior pituitary came from the work of Krulich et al. (1968); while looking for a factor releasing growth hormone, they discovered a fraction of rat and sheep hypothalami that inhibited growth hormone release. They postulated a dual mechanism of growth hormone control. In 1973, Brazeau and Guillemin isolated this growth hormone release inhibiting factor, somatostatin, from 500,000 sheep hypothalami (Brazeau et al., 1973). It was sequenced and found to contain 14 amino acid residues (a tetradecapeptide). It has a cyclic structure, joined by two intramolecular disulfide bonds between the two cysteine residues (Figure 3). The linear sequence was synthesized and purified and its biological activity was found to be the same as the cyclic form in vitro.

This isolation and synthesis has allowed antibodies to be developed, and using these, immunocytochemical methods have demonstrated the widespread distribution of somatostatin throughout the body.

Distribution

The highest concentration of somatostatin in the rat brain is found in the median eminence and arcuate nucleus, but it is also present in other hypothalamic nuclei, as well as throughout the extrahypothalamic brain, especially the cerebral cortex and preoptic area and in the spinal cord (Brownstein et al., 1975).

*See note added in proof, p. 346.

$$H_2 - N - Ala - Gly - Cys - Lys - Asn - Phe - Phe - Trp - Lys - Thr - Phe - Thr - Ser - Cys - COOH$$

$$\quad\quad\quad\; 1 \quad\; 2 \quad\; 3 \quad\; 4 \quad\; 5 \quad\; 6 \quad\; 7 \quad\; 8 \quad\; 9 \quad\; 10 \quad 11 \quad 12 \quad 13 \quad 14$$

Figure 3   Structure of somatostatin.

It is also present in the human hypothalamus and nervous system in a distribution analogous to that found in the rat (Patel and Reichlin, 1978; Cooper et al., 1982). In fractionated homogenates, somatostatin is localized in the synaptosome fraction, thus suggesting its possible role as a neurotransmitter (Epelbaum et al., 1977). Somatostatin is present in ganglion cells of the spinal ganglia as well as in the intestinal wall. Many laboratories have also found high concentrations of somatostatin in the stomach, pancreas and intestines (Arimura et al., 1975; Polak et al., 1975). Large amounts are detectable in the gastric antrum, duodenum, and pancreatic islets, and smaller amounts may be found in the remainder of the gut. Somatostatin has been shown to be localized to the D cells, which are present in the islets and in the pylorus and the fundus of the stomach (Dubois, 1975; Polak et al., 1975).

Function

Some of the first experiments done in the human showed that exogenously administered somatostatin suppressed a wide variety of hormones, including growth hormone, TSH, insulin, glucagon, gastrin, and other polypeptide gastrointestinal hormones; for a review see Wass (1982).

In view of this widespread distribution, it is clear that somatostatin has a number of possible roles. In the hypothalamus, it seems that its properties of inhibiting both basal and stimulated release of growth hormone and TSH make it a probable physiological endocrine regulator of both growth hormone and TSH secretion by the anterior pituitary. There is now good evidence for this (Arimura and Schally, 1976; Berelowitz et al., 1978). Second, it appears likely that it is a neurotransmitter or neuromodulator in the extrahypothalamic brain and spinal cord. Thus, it is released from rat brain tissues in vitro in response to membrane depolarization by calcium-dependent mechanisms (Iversen et al., 1978). Axonal transport has been demonstrated in the rat vagus (Gilbert et al., 1980). The localization of somatostatin in the D cells of the pancreatic islets, bearing a close relationship to those cells secreting glucagon and insulin, suggests that it plays a local paracrine regulatory role in the control of both insulin and glucagon secretion. Last, in the gastrointestinal tract, it may have a paracrine action on a number of different hormone-secreting cells, as well as an

endocrine action on the rest of the gastrointestinal tract. At present no definite
evidence for such a role exists.

Assays

Because of its broad spectrum of biological action, a number of bioassays for the
peptide have been developed which depend on suppression of growth hormone,
insulin, or glucagon secretion (Vale et al., 1975). They are useful for measuring
biological potency, but their use for measuring somatostatin content of tissue
extracts and blood is limited by the presence of other hormones and biologically
active peptides that would interfere with the biological responses. Furthermore,
they are time consuming and expensive. The isolation and synthesis of synthetic
somatostatin has enabled antisera for radioimmunoassay to be raised. Radioimmunoassays for somatostatin, using antisera raised in rabbit or sheep, have
now been developed in many laboratories. Studies in tissues have usually utilized
an unextracted radioimmunoassay, but there are a large number of discrepancies
between the estimates of somatostatin content of tissue extracts; these discrepancies may be due to a number of factors, including different antibody specificities, the presence of somatostatin-binding protein in tissues, different methods
of storage and extraction of tissues, and different purity of synthetic somatostatin preparations (Penman et al., 1979).

The radioimmunoassay of somatosatin in blood presents several problems.
First, endogenous somatostatin is rapidly degraded in the presence of plasma and
this is also true for the iodinated somatostatin tracer (Brazeau et al., 1978; Kronheim et al., 1976). The second problem relates to the fact that somatostatin
binds to a high molecular weight protein present in the plasma (Arimura et al.,
1978). Third, somatostatin circulates in very low concentrations in peripheral
blood. The use of an extraction step overcomes these problems. Thus, extraction
separates somatostatin from larger molecular weight binding proteins and the
degrading enzymes; further, a concentration step can be included, thus enabling
measurement of low circulating levels of the peptide (Penman et al., 1979).
Normally, in the human, circulating levels vary between 10 and 80 pg/ml.

CORTICOTROPIN-RELEASING FACTOR

The first study demonstrating that the adenohypophyseal secretion of adrenocorticotropic hormone (ACTH) is under hypothalamic control was that of De
Groot and Harris (1950). They observed that, in rabbits, electrical stimulation
of the posterior median eminence led to increased ACTH release, indirectly determined by the production of lymphopenia. They further demonstrated that
the placement of lesions in the posterior median eminence and mammillary body
and portal vasculature abolished the lymphopenic response to stress.

The fact that a chemical substance capable of releasing an adenohypophyseal hormone could be extracted from the hypothalamus was demonstrated by Guillemin and Rosenberg (1955) and by Saffran and Schally (1955). They showed that crude hypothalamic extracts induced the release of ACTH from cultured adenohypophyseal tissue.

It seems clear that corticotropin-releasing factor (CRF) is a peptide, since the CRF activity of hypothalamic extracts is readily destroyed by proteolytic enzymes. For some time it was thought that vasopressin was chemically identical to CRF (Martini and Morpurgo, 1955; McCann, 1957). The administration of vasopressin will release ACTH, although it is less effective than hypothalamic extracts containing CRF (Sirett and Purves, 1972). Recent studies using immunological methods have revealed that large amounts of vasopressin can be secreted into hypophyseal portal blood by axon terminals (Zimmerman, 1976). Until recently it has proved difficult to isolate CRF, and this has been attributed to chemical instability (Schally et al., 1973).

More recent work by Gillies and Lowry (1979) has shown that the corticotropin-releasing activity of rat pituitary stalk-median eminence (SME) is abolished by incubation with specific arginine vasopressin antiserum. Chromatographic separation of SME extracts yielded two fractions with CRF activity. On the basis of these findings Gillies et al. (1979) suggested that the corticotropin-releasing activity of the hypothalamus is due to vasopressin and that it requires a synergistic factor in order to exhibit its full biological activity. This hypothesis has, however, been disputed (Buckingham, 1979).

Most recently, Vale et al. (1981) have isolated, characterized, and synthesized a 41 amino acid peptide with CRF activity. It has been shown to be highly active in stimulating the secretion of ACTH and β-endorphin in vitro from anterior pituitary cell cultures; the potency of this peptide is of the same order of magnitude as that described for other hypophysiotropic peptides (Spiess et al., 1981). This peptide has now been shown to be effective in stimulating ACTH release in vivo in the human (Grossman et al., 1982).

PROLACTIN-INHIBITING FACTOR

Prolactin differs from the other anterior pituitary hormones and is predominantly under the inhibitory influence of the hypothalamus. Everett (1954) first demonstrated that removal of the hypophysis from its normal relationship with the hypothalamus actually results in increased hypophyseal secretion. In 1968, van Maanen and Smelik suggested that catecholamines, particulary dopamine, may act directly on the anterior pituitary to inhibit prolactin secretion. Lu and Meites (1971) demonstrated that the administration of L-dopa and monoamine oxidase inhibitors impede prolactin release in rats, while methyl dopa, which inhibits dopa decarboxylase, increases prolactin release. MacLeod and Lehmeyer (1974) further demonstrated that prolactin release was inhibited by the incubation of hypophyses in culture with dopamine. This was prevented by pretreating the animals with dopamine antagonists. Thus there is good evidence for dopamine receptors on prolactin-secreting cells, and it seems that dopamine can act directly on the anterior pituitary, in very low concentrations, to inhibit prolactin secretion. Furthermore, catecholamine and dopamine terminals have been shown to abut directly on the portal capillary bed (Fuxe et al., 1970). The fact that Schally et al. (1976), in trying to isolate PIF, found that the fraction with the greatest PIF activity was devoid of peptides and contained catecholamines adds further weight to the argument that the major PIF is dopamine itself, but other PIFs, for example, γ-aminobutyric acid (GABA) may exist (Thorner, 1977).

MEASUREMENT OF HYPOTHALAMIC REGULATORY HORMONES IN BIOLOGICAL SYSTEMS

Two types of assay are used to measure the low levels of hypothalamic regulatory hormones that may be expected to occur in biological fluids. The first uses radioimmunoassay. However, it must be recognized that there are several problems in the radioimmunoassay of these small peptides. Because these molecules are small, antisera have to be raised using antigen modified by binding to larger peptide molecules, e.g., thyroglobulin. Iodination may also be a problem. Conventionally, iodination by the lactoperoxidase or the chloramine-T techniques is to a tyrosine or histidine residue; there are difficulties if none are present. Thus, while it is possible to iodinate the histidine residue of TRH, special analogs containing tyrosine have to be made for somatostatin. Once iodinated, careful purification of the ligand is important so that a compound of high specific activity, which is carrier free and preferably monoiodinated, is used. Hypothalamic peptides are present in low concentrations and therefore require highly sensitive assays; this may require an extraction technique with a concentration step. Extraction is performed using organic solvents or adsorption onto glass followed by elution, but it is important to remember that recovery of the peptide should be at least 50%; as always, a rigorous check for parallelism in an assay after extraction is essential. The second important problem relates to enzymatic degradation of these small peptides, and this may complicate assays unless precautions are taken. The tissue in the nervous system destroys the hypothalamic peptides, although the rate may differ with different peptides. The same problem pertains to the gut as well as the plasma. Thus, unless precautions are taken, the hormone may disappear: it may be degraded to yield immunoactive but biologically inactive fragments which cause discrepancies between the immunological and biological potencies. Last, the enzymes may degrade the iodinated peptide. Although theoretically these problems may be circumvented by using enzyme inhibitors, this has not proved ideal; a good example of this is the plasma somatostatin assay (Penman et al., 1979). A chromatographic validation step is essential to show identity in size of the measured and standard peptide. Recently it has become important to further validate the system using high-pressure liquid chromatography.

The second type of assay is some form of in vitro bioassay. In general, such in vitro bioassays of the hypothalamic hormones have used preparations of pituitary cells. Perifusion systems of dispersed pituitary cells (Lowry, 1974) have obvious advantages over static systems in that they resemble more closely the physiological situation and allow sequential investigation of drugs or hormones. Although biological assays of these hormones have not been adequately developed, biological assay remains essential for the validation of the biological relevance of any immunoassay system because they measure biologically active molecules (Besser et al., 1971).

It should be clear from this discussion that there are many unresolved problems in the measurement of the hypothalamic regulatory hormones, and until these are solved, the physiology of these hormones cannot be fully assessed. Their widespread distribution makes the determination of different levels in blood difficult to assess and their physiological role hard to establish. Special techniques need to be used to investigate the neurotransmitter and modulator role of these peptides, and as yet it has proved impossible to investigate fully their paracrine role. To date there are few adequately validated assays of the hypothalamic hormones, but it is certainly important to continue the study of their measurement in order to elucidate more fully their physiological roles.

EDITORS' NOTE: THE CYTOCHEMICAL
BIOASSAYS OF RELEASING HORMONES

Cytochemical bioassays have been described for the assay of TRH (Gilbert et al., 1975) and of CRF (Buckingham and Hodges, 1977). Up to the present, they have been used for research investigations; no clinical applications have yet been reported.

They depend on the maintenance culture, for up to 5 hr, of segments of the anterior pituitary gland of either the guinea pig (Gilbert et al., 1975) or rats (Buckingham and Hodges, 1977). These are then exposed to various concentrations of synthetic TRH, or of substances with CRF-like activity, or of plasma. The segments release the appropriate pituitary hormone into the culture medium. The hormone, released into the culture medium, is then assayed by the relevant cytochemical bioassay.

The procedures appear to be sensitive and precise. Concentrations as low as 10 fg/ml of TRH have been assayed. With extracts of rat hypothalamus, the index of precision of 10 dose-response graphs was 0.046 ± 0.002 (SEM).

The development of section bioassays for the pituitary hormones that are released has greatly improved the feasibility of these assays of releasing hormones. Thus it is likely that the section assay of luteinizing hormone (Buckingham et al., 1979a,b; Buckingham and Hodges, 1981) will open the way for a cytochemical bioassay of GnRH.

NOTE ADDED IN PROOF

Growth Hormone-Releasing Factor (Somatocrinin)

It is known that peptide hormones can be ectopically produced, and a small number of cases of acromegaly have been reported to be associated with tumors which have growth hormone-releasing activity (Thorner et al., 1982).

Recently, in November 1982, a 44 amino acid peptide was isolated by Guillemin and colleagues (Guillemin et al., 1982), from an acromegalic patient with a palpable pancreatic tumor which was secreting such a peptide. The structure was ascertained to be H-Tyr-Ala-Asp-Ala-Ile-Phe-Thr-Asn-Ser-Tyr-Arg-Lys-Val-Leu-Gly-Gln-Leu-Ser-Ala-Arg-Lys-Leu-Leu-Gln-Asp-Ile-Met-Ser-Arg-Gln-Gln-Gly-Glu-Ser-Asn-Gln-Glu-Arg-Gly-Ala-Arg-Ala-Arg-Leu-NH_2. A report of its synthesis was also made (Guillemin et al., 1982). Human pancreatic growth hormone releasing factor (hpGRF) is specific in the stimulation of growth hormone secretion. Other fractions with growth hormone-stimulating properties were also isolated from this tumor; one with 37 and one with 40 amino acids. These had less growth hormone-stimulating properties. Shortly after this report, Rivier et al. reported a growth hormone-releasing factor with the first 40 amino acids of that reported by Guillemin's group (Rivier et al., 1982). This was purified from another pancreatic islet tumor described by Thorner et al. (1982) in an acromegalic patient with a tumor in the tail of her pancreas, whose acromegaly disappeared when this was removed. More recently, the structure of rat hypothalamic GRF has been reported (Brazeau et al., 1982). This is also a secretagogue of growth hormone in cultured rat pituitary cells which is specific and inhibited by somatostatin-28 and -14.

At present, no bioassay systems are available for this peptide.

In humans, hpGRF 1-40 has been reported to selectively stimulate growth hormone secretion, and 1 μg/kg causes a rise in growth hormone to between 3-40 ng/ml (Thorner et al., 1983). Most recently, it has been reported by Grossman et al. (1983) to cause growth hormone secretion in children with GH deficiency nonresponsive to insulin-induced hypoglycemia. It is clear that children with isolated growth hormone deficiency may have GRF deficiency and may respond to injections of this compound, which is easier to synthesize than human growth hormone itself. It therefore has great therapeutic potential.

Brazeau, P., Ling, N., Bohlen, P., Esch, F., Ying, S-Y., and Guillemin, R. (1982). Growth hormone releasing factor, somatocrinin, releases pituitary growth hormone in vitro. *Proc. Natl. Acad. Sci.*, 79:7909-7913.

Grossman, A., Savage, M. O., Wass, J. A. H., Lytras, N., Sueiras-Diaz, J., Coy, D. H., and Besser, G. M. (1983). Growth hormone-releasing factor in growth hormone deficiency: The demonstration of a hypothalamic defect in growth hormone release. *Lancet,* in press.

Guillemin, R., Brazeau, P., Bohlen, P., Esch, F., Ling, N., and Wehrenberg, W. (1982). Growth hormone-releasing factor from a human pancreatic tumour that caused acromegaly. *Science,* 218:585-587.

Rivier, J., Spiess, J., Thorner, M., and Vale, W. (1982). Characterization of a growth hormone-releasing factor from a human pancreatic islet tumour. *Nature,* 300:276-278.
Thorner, M. O., Perryman, R. L., Cronin, M. J., Rogol, A. D., Draznin, M., Johanson, A., Vale, W., Horvath, E., and Kovacs, K. (1982). Somatotroph hyperplasia: Successful treatment of acromegaly by removal of a pancreatic islet tumor secreting a growth hormone-releasing factor. *J. Clin. Invest.,* 70:965-977.
Thorner, M. O., Spiess, J., Vance, M. L., Rogol, A. D., Kaiser, D. L., Webster, J. D., Rivier, J., Borges, J. L., Bloom, S. R., Cronin, M. J., Evans, W. S., MacLeod, R. M., and Vale, W. (1983). Human pancreatic growth-hormone-releasing factor selectively stimulates growth-hormone secretion in man. *Lancet,* January 1/8:24-28.

REFERENCES

Arimura, A., Lundqvist, G., Rothman, J., Chang, R., Fernandez-Durango, R., Elde, R., Coy, D. H., Meyers, C., and Schally, A. V. (1978). Radioimmunoassay of somatostatin. *Metabolism,* 27(Suppl. 1):1139-1144.

Arimura, A., Sato, H., Dupont, A., Nishi, N., and Schally, A. V. (1975). Somatostatin: Abundance of immunoreactive hormone in rat stomach and the pancreas. *Science,* 189:1007-1009.

Arimura, A., Sato, H., Kumasaka, P., Worobec, R. B., Debeljuk, L., Dunn, J. and Schally, A. V. (1973). Production of antiserum to LH-releasing hormone (LH-RH) associated with gonadal atrophy in rabbits: Development of radioimmunoassays for LHRH. *Endocrinology,* 93:1092-1103.

Arimura, A., and Schally, A. V. (1976). Increase in basal and thyrotropin releasing hormone (TRH) stimulated secretion of thyrotropin (TSH) by passive immunisation with antiserum to somatostatin. *Endocrinology,* 98:1069-1072.

Baba, Y., Matsuo, H., and Schally, A. V. (1971). Structure of the porcine LH and FSH releasing hormone. 2. Confirmation of the proposed structure by conventional sequential analyses. *Biochem. Biophys. Res. Commun.,* 44:459-463.

Bargmann, W., and Scharrer, E. (1951). The site of origin of the hormones of the posterior pituitary. *Am. Sci.,* 39:255-259.

Bassiri, R., and Utiger, R. D. (1972). The preparation and specificity of antibody to thyrotropin releasing hormone. *Endocrinology,* 90:722-727.

Belchetz, P. E., Plant, T. M., Nakai, Y., Keogh, E. J., and Knobil, E. (1978). Hypophysial responses to continuous and intermittent delivery of hypothalamic gonadotropin-releasing hormone. *Science,* 202:631-633.

Berelowitz, M., Pimstone, B., Shapiro, B., Kronheim, S., and De Wit, D. (1978). Tissue growth hormone release inhibiting hormone like immunoreactivity in experimental hypothyroidism and hypopituitarism. *Clin. Endocrinol. (Oxf),* 9:185-191.

Besser, G. M., McNeilly, A. S., Anderson, D. C., Marshall, J. C., Harsoulis, P., Hall, R., Ormston, B. J., Alexander, L., and Collins, W. P. (1972). Hormonal responses to synthetic luteinising hormone and follicle stimulating hormone-releasing hormone in man. *Br. Med. J.*, 3:267-271.

Besser, G. M., Orth, D. N., Nicholson, W. E., Bygny, R. L., Abe, K., and Woodham, J. P. (1971). Dissociation of the disappearance of bioactive and radioimmunoactive ACTH from plasma in man. *J. Clin. Endocrinol. Metab.*, 32: 595-603.

Borgeat, P., Chavancy, G., Dupont, A., Labrie, F., Arimura, A., and Schally, A. V. (1972). Stimulation of adenosine 3:5-Cyclic monophosphate accumulation in anterior pituitary gland in vitro by synthetic luteinising hormone releasing hormone. *Proc. Nat. Acad. Sci., USA*, 69:2677-2681.

Bowers, C. Y., Schally, A. V., Hawley, D. W., Gual, C., and Parlow, A. (1968). Effect of thyrotropin-releasing factor in man. *J. Clin. Endocrinol. Metab.*, 28:978-982.

Brazeau, P., Epelbaum, J., Tannenbaum, G. S., Rorstad, P., and Martin, J. B. (1978). Somatostatin: Isolation, characterisation and blood determination. *Metabolism*, 27(Suppl. 1):1133-1137.

Brazeau, P., Vale, W., Burgus, R., Ling, N., Butcher, M., Rivier, J., and Guillemin, R. (1973). Hypothalamic polypeptide that inhibits the secretion of immunoreactive pituitary growth hormone. *Science*, 179:77-79.

Brownstein, M., Arimura, A., Sato, H., Schally, A. V., and Kizer, J. S. (1975). The regional distribution of somatostatin in the rat brain. *Endocrinology*, 96: 1456-1461.

Brownstein, M. J., Palkovits, M., Saavedra, J. M., Bassiri, R. M., and Utiger, R. D. (1974). Thyrotrophin-releasing hormone in specific nuclei of rat brain. *Science*, 185:267-269.

Buckingham, J. C. (1980). Corticotrophin releasing factor. *Pharmacol. Rev.*, 31:253-275.

Buckingham, J. C., and Hodges, J. R. (1981). A cytochemical bioassay method for the determination of luteinizing hormone in biological fluids and tissues. *Br. J. Pharmacol.*, 73:111-118.

Buckingham, J. C., Chayen, J., Hodges, J. R., Robertson, W. R., and Weisz, J. (1979a). A cytochemical section assay method for the determination of luteinizing hormone. *J. Endocrinol.*, 81:160P.

Buckingham, J. C., Chayen, J., Hodges, J. R., Robertson, W. R., and Weisz, J. (1979b). A cytochemical section bioassay method for the determination of luteinizing hormone. *Acta Endocrinol. (Kbh)* 225(Suppl.):131.

Buckingham, J. C., and Hodges, J. R. (1977). The use of corticotrophin production by adenohypophyseal tissue in vitro for the detection and estimation of potential corticotrophin releasing factors. *J. Endocrinol.*, 72:187-193.

Burger, H. G., and Patel, Y. C. (1977). TSH and TRH: Their physiological regulation and the clinical applications of TRH. In *Clinical Neuroendocrinology*, L. Martini and G. M. Besser (eds.). Academic, New York, pp. 67-131.

Burgus, R., Butcher, M., Amos, M., Ling, N., Monahan, M., Rivier, J., Fellows, R., Blackwell, R., Vale, W., and Guillemin, R. (1972). Primary structure of the ovine hypothalamic luteinising hormone releasing factor (LRF). *Proc. Nat. Acad. Sci. USA*, 69:278-282.

Burgus, R., Dunn, T. F., Desiderio, D., and Guillemin, R. (1969). Structure moleculaire du facteur hypothalamique hypophysiotrope TRF d'origine ovine: Mise en evidence par spectrometrie de masse de la sequence. *C. R. Acad. Sci. [D] (Paris)*, 269:1870-1893.

Campbell, H. J., Feuer, G., and Harris, G. W. (1964). The effects of intra pituitary infusion of median eminence and other brain extracts on anterior pituitary gonadotrophic secretion. *J. Physiol. (Lond.)*, 170:474-486.

Cooper, V., Fernstrom, M. H., Rorstad, O. P., Leeman, S. E., and Martin, J. B. (1982). The regional distribution of somatostatin, substance P and neurotensin in the human brain. *Brain Res.*, In press.

De Groot, J., and Harris, G. W. (1950). Hypothalamic control of the anterior pituitary gland and blood lymphocytes. *J. Physiol. (Lond.)*, 111:335-346.

Deuben, R. R., and Meites, J. (1964). Stimulation of pituitary growth hormone release by a hypothalamic extract in vitro. *Endocrinology*, 74:408-414.

Dubois, M. P. (1975). Immunoreactive somatostatin is present in discrete cells of the endocrine pancreas. *Proc. Nat. Acad. Sci. USA*, 72:1340-1343.

Epelbaum, J., Brazeau, P., Tsang, D., Brawer, J., and Martin, J. B. (1977). Subcellular distribution of radioimmunoassayable somatostatin in rat brain. *Brain Res.*, 126:309-323.

Everett, J. W. (1954). Luteotrophic function of autographs of the rat hypophysis. *Endocrinology*, 54:685-690.

Folkers, K., Enzmann, F., Boler, J., Bowers, C. Y., and Schally, A. V. (1969). Discovery of modification of the synthetic tripeptide sequence of the thyrotrophin releasing hormone having activity. *Biochem. Biophys. Res. Commun.*, 37:123-126.

Fuxe, K., Hökfelt, T., and Ungerstedt, U. (1970). Morphological and functional aspects of central monoamines. *Int. Rev. Neurobiol.*, 13:93-126.

Ganong, W. F., Frederickson, D. S., and Hume, D. M. (1955). Effect of hypothalamic lesions on thyroid function in the dog. *Endocrinology*, 57:355-362.

Gilbert, D. M., Bitensky, L., Besser, G. M., and Chayen, J. (1975). Measurable effects of low concentrations (100 fg/ml) of thyrotrophin releasing hormone. *J. Endocrinol.*, 64:25P.

Gilbert, R. F. T., Emson, P. C., Fahrenkrug, J., Lee, C. M., Penman, E., and Wass, J. A. H. (1980). Axonal transport of neuropeptides in the cervical vagus nerve of the rat. *J. Neurochem.*, 34:108-113.

Gilles, G., Estivariz, F. E., and Lowry, P. J. (1979). Investigations on the nature of corticotrophin releasing factor using the perfused isolated rat anterior pituitary cell column. *J. Endocrinol.*, 80:56P.

Gillies, G., and Lowry, P. J. (1979). Corticotrophin releasing factor may be modulated vasopressin. *Nature,* 278:463-464.

Green, J. D., and Harris, G. W. (1947). The neurovascular link between the neurohypophysis and adenohypophysis. *J. Endocrinol.,* 5:136-145.

Greer, M. A. (1951). Evidence of hypothalamic control of the pituitary release of thyrotrophin. *Proc. Soc. Exp. Biol. Med.,* 77:603-608.

Grossman, A., Kruseman, A. C. N., Perry, L., Tomlin, S., Schally, A. V., Coy, D. H., Rees, L. H., Comaru-Schally, A. M., and Besser, G. M. (1982). A new hypothalamic hormone, CRF, specifically stimulates the release of ACTH and cortisol in man. *Lancet,* i:921-922.

Guillemin, R., and Rosenberg, B. (1955). Humoral hypothalamic control of anterior pituitary: A study with combined tissue cultures. *Endocrinology,* 57: 599-607.

Hall, R., Amos, R., Garry, R., and Buxton, R. L. (1970). Thyroid-stimulating hormone responses to synthetic thyrotrophin releasing hormone in man. *Br. Med. J.,* 2:274-277.

Hall, R., Besser, G. M., Schally, A. V., Coy, D. H., Evered, D., Goldie, D. J., Kastin, A. J., McNeilly, A. S., Mortimer, C. H., Phenekos, C., Tunbridge, W. M. G., and Weightman, D. (1973). Action of growth-hormone-release inhibitory hormone in healthy men and in acromegaly. *Lancet,* ii:581-584.

Hall, R., Ormston, B. J., Besser, G. M., Cryer, R. J., and McKendrick, M. (1972). The thyrotrophin releasing hormone test in diseases of the pituitary and hypothalamus. *Lancet,* i:759-763.

Harris, G. W. (1937). The induction of ovulation in the rabbit, by electrical stimulation of the hypothalamo-hypophyseal mechanism. *Proc. R. Soc. Lond.,* 122:374-393.

Harris, G. W., and Woods, J. W. (1958). The effect of electrical stimulation of the hypothalamus or pituitary gland on thyroid activity. *J. Physiol. (Lond.),* 143:246-274.

Hökfelt, T., Efendic, S., Hellerström, C., Johansson, O., Luft, R., and Arimura, A. (1975). Cellular localisation of somatostatin in endocrine-like cells and neurones of the rat with special references to the A_1-cells fo the pancreatic islets and to the hypothalamus. *Acta Endocrinol.,* 80(Suppl. 200):5-41.

Hökfelt, T., Johansson, O., Ljundahl, A., Lundberg, M., Schultzberg, M., Fuxe, K., Goldstein, M., Steinbusch, H., Verhofstad, A., and Elde. R. (1978). Neurotransmitters and neuropeptides: Distribution patterns and cellular localization as revealed by immunocytochemistry. In *Central Regulation of the Endocrine System,* K. Fuxe, T. Hökfelt, and R. Luft (eds.). Plenum, New York, pp. 31-48.

Igarashi, M., and McCann, S. M. (1964). A hypothalamic follicle stimulating hormone releasing factor. *Endocrinology,* 74:446-452.

Iversen, L. L., Iversen, S. D., Bloom, F., Douglas, C., Brown, M., and Vale, W. (1978). Calcium dependent release of somatostatin in neurotensin from rat brain in vitro. *Nature,* 273:161-163.

Jackson, I. M. D. (1982). Thyrotrophin-releasing hormone. *N. Engl. J. Med.,* 306:145-155.

Jacobs, L. S., Snyder, P. J., Wilber, J. F., Utiger, R., and Daughaday, W. (1971). Increased serum prolactin after administration of synthetic thyrotropin releasing hormone (TRH) in man. *J. Clin. Endocrinol. Metab.,* 33:996-998.

Jeffcoate, S. L., Fraser, H. M., Gunn, A., and Holland, D. T. (1973). Radioimmunoassay of luteinising hormone releasing factor. *J. Endocrinol.,* 57:189-190.

Kronheim, S., Berelowitz, M., and Pimstone, B. L. (1976). A radioimmunoassay for growth hormone release inhibiting hormone; method and quantitative tissue distribution. *Clin. Endocrinol., (Oxf.),* 5:619-630.

Krulich, L., Dhariwal, A. P. S., and McCann, S. M. (1968). Stimulatory and inhibitory effects of purified hypothalamic extracts on growth hormone release from rat pituitary in vitro. *Endocrinology,* 83:783-790.

Labrie, F., Barden, N., Poirier, G., and DeLean, A. (1972). Binding of thyrotropin-releasing hormone to plasma membranes of bovine anterior pituitary gland. *Proc. Nat. Acad. Sci. USA,* 69:283-287.

Lowry, P. J. (1974). A sensitive method for the detection of corticotrophin releasing factor using a perfused pituitary cell-column. *J. Endocrinol.,* 62:163-164.

Lu, K. H., and Meites, J. (1971). Inhibition by L-dopa and monoamine oxidase inhibitors of pituitary prolactin release; stimulation by methyldopa and D amphetamine. *Proc. Soc. Exp. Biol. Med.,* 137:480-483.

MacLeod, R. M., and Lehmeyer, J. E. (1974). Studies on the mechanism of the dopamine mediated inhibition of prolactin secretion. *Endocrinology,* 94:1077-1085.

Martini, L., and Morpurgo, C. (1955). Neurohumoral control of the release of adrenocorticotrophic hormone. *Nature (Lond.),* 175:1127-1128.

Martino, E., Lernmark, A., Seo, H., Steiner, D. F., and Refetoff, S. (1978). High concentration of thyrotrophin-releasing hormone in pancreatic islets. *Proc. Nat. Acad. Sci. USA,* 75:4265-4267.

Matsuo, H., Baba, Y., Nair, R. M. G., Arimura, A., and Schally, A. V. (1971). Structure of the porcine LH and FSH releasing hormone. 1. The proposed amino acid sequence. *Biochem. Biophys. Res. Commun.,* 43:1334-1339.

McCann, S. M. (1957). The corticotrophin releasing activity of extracts of the posterior lobe of the pituitary in vivo. *Endocrinology,* 60:664-676.

McCann, S. M., Taleisnik, S., and Friedman, H. M. (1960). LH releasing activity in hypothalamic extracts. *Proc. Exp. Biol. Med.,* 104:432-434.

Mitnick, M. A., and Reichlin, S. (1971). Thyrotropin releasing hormone: Biosynthesis by rat hypothalamic fragments in vitro. *Science*, 172:1241-1243.

Mittler, J. C., Redding, T. W., and Schally, A. V. (1969). Stimulation of thyrotropin (TSH) secretion by TSH-releasing factor (TRF) in organ cultures of anterior pituitary. *Proc. Soc. Exp. Biol. Med.*, 130:406-409.

Mortimer, C. H. (1977a). Gonadotropin-releasing hormone. In *Clinical Neuroendocrinology*, L. H. Martini and G. M. Besser (eds.). Academic, New York, pp. 213-236.

Mortimer, C. H. (1977b). Clinical applications of the gonadotrophin releasing hormone. *Clin. Endocrinol. Metab.*, 6:167-179.

Mortimer, C. H., Besser, G. M., Goldie, D. J., Hook, J., and McNeilly, A. S. (1973). Asynchronous changes in circulating LH and FSH after the gonadotrophin releasing hormone. *Nature (New Biol.)*, 246:22-23.

Mortimer, C. H., McNeilly, A. S., Rees, L. H., Lowry, P. J., Gilmore, D., and Dobbie, H. G. (1976). Radioimmunoassay and chromatographic similarity of circulating endogenous gonadotrophin releasing hormone and hypothalamic extracts in man. *J. Clin. Endocrinol. Metab.*, 43:882-888.

Nillius, S. J., Berquist, C., and Wide, L. (1978). Inhibition of ovulation in women by chronic treatment with a stimulatory LRH-analogue—a new approach to birth control? *Contraception*, 17:537-545.

Nillius, S. J., and Wide, L. (1972). Variation in LH and FSH response to LH-releasing hormone during the menstrual cycle. *J. Obstet. Gynaecol. Br. Commonw.*, 79:865-873.

Ormston, B. J., Garry, R., Cryer, R. J., Besser, G. M., and Hall, R. (1971). Thyrotrophin-releasing hormone as a thyroid function test. *Lancet*, ii:10-14.

Patel, Y. C., and Reichlin, S. (1978). Somatostatin in hypothalamus, extrahypothalamic brain and peripheral tissues of the rat. *Endocrinology*, 102:523-530.

Penman, E., Wass, J. A. H., Lund, A., Lowry, P. J., Stuart, J., Dawson, A., Besser, G. M., and Rees, L. H. (1979). Development and validation of a specific radioimmunoassay for somatostatin in human plasma. *Ann. Clin. Biochem.*, 16:15-25.

Polak, J. M., Pearse, A. G. E., Grimelius, L., Bloom, S. R., and Arimura, A. (1975). Growth hormone release inhibiting hormone in gastrointestinal and pancreatic D-cells. *Lancet*, i:1220-1222.

Popa, G., and Fielding, U. (1930). A portal circulation from the pituitary to the hypothalamic region. *J. Anat.*, 65:88-91.

Redding, T. W., Kastin, A. J., Gonzalez-Barcena, D., Coy, D. H., Coy, E. J., Schalch, D. S., and Schally, A. V. (1973). The half life, metabolism and excretion of tritiated luteinizing hormone-releasing hormone (LH-RH) in man. *J. Clin. Endocrinol. Metabl.*, 37:626-631.

Redding, T. W., and Schally, A. V. (1971). The distribution of radioactivity following the administration of labelled thyrotropin-releasing hormone (TRH) in rats and mice. *Endocrinology,* 89:1075-1081.

Redding, T. W., Schally, A. V., Arimura, A., and Matsuo, H. (1972). Stimulation of release and synthesis of luteinising hormone (LH) and follicle stimulating hormone (FSH) in tissue cultures of rat pituitaries in response to natural and synthetic LH and FSH releasing hormone. *Endocrinology,* 90:764-770.

Reichlin, S. (1960). Growth and the hypothalamus. *Endocrinology,* 67:760-773.

Roth, J., Glick, S. M., Yalow, R. S., and Berson, S. A. (1963). Secretion of human growth hormone: Physiologic and experimental modification. *Metabolism,* 12:577-579.

Saffran, M., and Schally, A. V. (1955). Release of corticotrophin by anterior pituitary tissue in vitro. *Can. J. Biochem. Physiol.,* 33:408-415.

Sawyer, C. H., Markee, J. E., and Hollinshead, W. H. (1947). Inhibition of ovulation in the rabbit by the adrenergic blocking agent dibenamine. *Endocrinology,* 41:395-402.

Schally, A. V., Arimura, A., and Kastin, A. J. (1973). Hypothalamic regulatory hormones. *Science,* 179:341-350.

Schally, A. V., Arimura, A., Kastin, A. J., Matsuo, H., Baba, Y., Redding, T. W., Nair, R. M. G., Debeljuk, L., and White, W. F. (1971). Gonadotrophin releasing hormone: One polypeptide regulates secretion of luteinising and follicle stimulating hormone. *Science,* 173:1036-1038.

Schally, A. V., and Bowers, C. Y. (1971). *Proceedings of the Mid-West Conference of Thyroid Endocrinology,* VI, 1970, 25-63.

Schally, A. V., Bowers, C. Y., Redding, T. W., and Barrett, J. F. (1966). Isolation of thyrotrophin releasing factor (TRF) from porcine hypothalamus. *Biochem. Biophys. Res. Commun.,* 25:165-169.

Schally, A. V., Dupont, A., Arimura, A., Takahara, J., Redding, T. W., Clemens, J., and Shaar, C. (1976). Purification of a catecholamine-rich fraction with prolactin release-inhibiting factor (PIF) activity from porcine hypothalami. *Acta Endocrinol. (Kbh),* 82:1-14.

Schally, A. V., and Kastin, A. J. (1966). Purification of a bovine hypothalamic factor which elevates pituitary MSH levels in rats. *Endocrinology,* 79:768-772.

Schally, A. V., Redding, T. W., Bowers, C. Y., and Barrett, J. F. (1969). Isolation and properties of porcine thyrotropin releasing hormone. *J. Biol. Chem.,* 244:4077-4088.

Sirett, N. E., and Purves, H. D. (1972). Assay of corticotrophin-releasing factor (CRF) in ACTH primed "grafted" rats. *Neuroendocrinology,* 10:83-93.

Spiess, J., Rivier, J., Rivier, C., and Vale, W. (1981). Primary structure of corticotropin releasing factor from ovine hypothalamus. *Proc. Nat. Acad. Sci. USA,* 78:6517-6521.

Tashjian, A. H., Barowsky, N. J., and Jensen, D. K. (1971). Thyrotrophin releasing hormone: Direct evidence of the stimulation of prolactin production by pituitary cells in culture. *Biochem. Biophys. Res. Commun.*, 43:516-523.

Thorner, M. O. (1977). Prolactin: Clinical physiology and the significance and management of hyperprolactinaemia. In *Clinical Neuroendocrinology*, L. Martini and G. M. Besser, (eds.), Academic, New York, pp. 320-361.

Vale, W., Brazeau, P., Rivier, C., Brown, M., Boss, B., Rivier, J., Burgus, R., Ling, N., and Guillemin, R. (1975). Somatostatin. *Recent Progr. Horm. Res.*, 31:365-397.

Vale, W., Spiess, J., Rivier, C., and Rivier, J. (1981). Characterisation of a 41 residue ovine hypothalamic peptide that stimulates secretion of corticotropin and beta-endorphin. *Science*, 213:1394-1397.

van Maanen, J. H., and Smelik, P. G. (1968). Induction of pseudopregnancy in rats following local depletion of monoamines in the median eminence of the hypothalamus. *Neuroendocrinology*, 3:177-186.

Wass, J. A. H. (1982). Somatostatin and its physiology in man in health and disease. In *Clinical Neuroendocrinology*, Vol. 2, L. Martini and G. M. Besser (eds.). Academic, New York. pp. 359-395.

Wilber, J. F., and Seibel, M. J. (1973). Thyrotropin-releasing hormone interactions with an anterior pituitary membrane receptor. *Endocrinology*, 92:888-893.

Winokur, A., and Utiger, R. D. (1974). Thyrotropin-releasing hormone: Regional distribution in rat brain. *Science*, 185:265-267.

Wislocki, G. B., and King, L. S. (1936). The permeability of the hypophysis and the hypothalamus to vital dyes, with a study of the hypophyseal vascular supply. *Am. J. Anat.*, 58:421-472.

Yen, S. S. C., VandenBerg, G., Rebar, R., and Ehara, Y. (1972). Variation of pituitary responsiveness to synthetic LRF during different phases of the menstrual cycle. *J. Clin. Endocrinol. Metab.*, 35:931-934.

Zimmerman, E. A. (1976). Localisation of hypothalamic hormones by immunocytochemical techniques. *Front. Neuroendocrinol.*, 4:25-62.

14

Natriuretic Hormone

H. E. de Wardener, Stephen Fenton, and J. Alaghband-Zadeh / Charing Cross
Hospital Medical School, London, England

The evidence for the existence of a circulating substance which controls urinary
sodium excretion other than aldosterone is based on a multiplicity of "whole-
animal" experiments. The first was performed in dogs receiving large amounts of
vasopressin and salt-retaining steroids in which it was demonstrated that an infusion of saline caused a rise in urinary sodium excretion even when the glomerular
filtration rate and renal blood flow were sharply reduced by inflating a balloon
in the thoracic aorta (de Wardener et al., 1961). The contribution of a circulating natriuretic substance to this rise in urinary sodium excretion was revealed by
experiments which demonstrated that the rise in urinary sodium excretion could
not be explained entirely by the dilutional effects of the saline on the blood.
Cross-circulation experiments were performed between two dogs, both of which
were given large amounts of aldosterone and vasopressin, but only one dog was
given saline. The dog which received saline had a substantially greater natriuresis
than the other, although the dilutional changes in the blood of the two dogs, and
the change in arterial pressure, renal plasma flow (PAH), and inulin clearance
were the same. Since the dog with the volume expansion had the greater rise in
urinary sodium excretion, it was proposed that part of the natriuresis was produced by a change in the concentration of some circulating substance other than
aldosterone.

Nevertheless, the hypothesis at first did not receive wide acceptance. Rather
it was emphasized that notwithstanding the experiments just described, the
natriuresis was more probably due to the dilutional effects of the saline. This
alternative view was reinforced by the discovery that physical factors, such as

peritubular hydrostatic pressure, packed cell volume, and possibly peritubular oncotic pressure, could influence sodium excretion.

To avoid these objections, experiments were then performed in which the fluid volume of an animal is rapidly expanded without diminishing its plasma protein concentration or packed cell volume. In these experiments the presence of a circulating substance which controls urinary sodium excretion is simultaneously assayed either by a denervated kidney in situ, or an isolated kidney in vitro, each perfused at a controllable pressure. The ideal way to perform these experiments is to expand an animal's blood volume with blood with which it is in equilibrium (Bahlmann et al., 1967). The blood of an animal is continuously exchanged with the contents of a reservoir which is initially primed with either dog's blood, or a solution of bovine albumin in Ringer's solution. After 1-2 hr of such an exchange the blood in the reservoir is in equilibrium with the blood in the animal. The animal's blood volume is then expanded by about 30% in 15-20 min by lowering the fluid level in the reservoir. Variations of this experiment have been performed by several groups, who have found a 50-100% rise in urinary sodium excretion by the "assay" kidney (de Wardener, 1977).

Several other groups have performed almost the same experiment, but instead of using equilibrated blood, the blood volume has been expanded with homologous blood from another animal. The rise in urinary sodium excretion is similar to that in the equilibrated blood experiments. Two groups of workers who have cross-circulated a donor rat "expanded" in this way with an "unexpanded" recipient rat have only been able to obtain a natriuresis in the recipient rat if the blood volume expansion in the donor rat is sustained, i.e., the loss of sodium and water in the urine is replaced by reinfusing the urine intravenously (Sonnenberg et al., 1972; Knock and de Wardener, 1980).

MECHANISM OF ACTION

In some of the experiments described above, the rise in urinary sodium excretion was not accompanied by a rise in renal blood flow or glomerular filtration rate, which suggested that the rise in sodium excretion was probably due to a diminution in tubular reabsorption (Kaloyanides and Azer, 1971). This impression has been strengthened by finding that plasma from a volume-expanded animal inhibits sodium transport in vitro. Lichardus et al. (1968), using an extract of plasma, and Nutbourne et al. (1970), using whole blood, observed a reduction of the short-circuit current in frog skin by plasma or blood obtained from a volume-expanded animal. And Clarkson et al. (1970), using fragments of dog and rabbit renal tubules, demonstrated that plasma from a blood-volume-expanded dog inhibits net sodium and potassium transport in vitro. These experiments showed that the blood of a blood-volume-expanded animal has the ability to reduce sodium transport across cell membranes, but it did not reveal the nature of the inhibition. It now appears probable that it is due, at least in part, to a change in the concentration of a circulating (Na^+-K^+)-ATPase inhibitor. Gonick et al. (1977) found that a low molecular weight fraction obtained from serum has a significantly greater ability to inhibit renal (Na^+-K^+)-ATPase when the serum comes from a rat that has received a saline infusion. The presence of a circulating (Na^+-K^+)-ATPase inhibitor in the human was first demonstrated by Poston et al. (1980). Their observations were made in normal subjects given 9 mg 9α-fludrocortisone per day. Blood was obtained after the subjects had returned into sodium balance, even though they continued to take 9α-fludrocortisone, so that they were therefore in a state of "mineralocorticoid escape." The ouabain-sensitive component of the sodium efflux rate constant [that due to (Na^+-K^+)-ATPase] of the leukocytes of these subjects, while on 9α-fludrocortisone, was significantly lower than it had been before the administration of the mineralocorticoid. Furthermore, when leukocytes from a control subject who had not received 9α-fludrocortisone were incubated in the serum of a person who had been given 9α-fludrocortisone, the ouabain-sensitive sodium efflux of the leukocytes obtained from the control subject fell to about the same concentration as that of the leukocytes of the mineralocorticoid-escaped person from whom the serum had been obtained. These changes in the leukocytes obtained from a control subject could only be due to the presence, in the plasma of the person taking 9α-fludrocortisone, of a substance which inhibits (Na^+-K^+)-ATPase. Flier et al. (1979) have demonstrated that the blood of a toad (*Bufo marinus*) which lives in salt water also contains a (Na^+-K^+)-ATPase inhibitor. They were unable to demonstrate a similar substance in the blood of a frog (*Rana pipiens*) which lives in fresh water. More recently, Gruber et al. (1980) have obtained an extract from dog blood which inhibits brain (Na^+-K^+)-ATPase, and have found that the inhibition is greater when the extract is obtained from the blood of a dog which has been given a large infusion of saline.

As inhibition of (Na^+-K^+)-ATPase (e.g., with ouabain) is associated with a stimulation of glucose-6-phosphate dehydrogenase (G6PD) activity, Alaghband-Zadeh, Clarkson, et al. (1981) have studied the capacity of plasma from the normal human to stimulate G6PD activity. As predicted, they have found that plasma from the normal human stimulates G6PD and that this ability is also related to sodium intake.

Flier et al. (1979) and Gruber et al. (1980) also found that the material they were studying contained a substance that had properties, such as adherence to ouabain-binding sites or an affinity for antibodies raised against a cardiac glycoside, which suggested that it was structurally related to ouabain. And these two groups of workers have therefore concluded that the plasma's capacity to inhibit

(Na^+-K^+)-ATPase is due to a substance which is structurally similar to and occupies the same receptor sites as, cardiac glycosides. Whether or not this is true, there is a general agreement that the animal substance is far more potent than the plant glycoside (see below).

NATURE OF THE CIRCULATING (Na^+-K^+)-ATPase INHIBITOR

It is possible that several substances are responsible for the changes in sodium transport described above. There appears to be one, the high molecular weight natriuretic substance, which has a molecular weight of around 50,000. And there appear to be at least two low molecular weight substances. One of these can only be obtained after some disruptive maneuver (acidification, tryptic digestion, or treatment with hypertonic sodium chloride) has been applied to the high molecular weight substance. The other low molecular weight natriuretic substance separates behind the salts on G-25 (F4) Sephadex separation of freeze-dried urine or plasma, without any preliminary procedure. This is the substance upon which most work has been performed, and most of the investigations which have been performed to study the nature of this substance have been carried out on material obtained from urine. It has still to be shown, however, that the substance in the urine is the same as that in the plasma.

After a bolus injection, the natriuresis, which this low molecular weight natriuretic substance (F4) causes, is rapid in onset and of short duration. The substance inhibits (Na^+-K^+)-ATPase (de Wardener et al., 1981) and has a molecular weight below 500; it is very polar, in that it is most soluble in water and insoluble in chloroform and ether; it is relatively resistant to heating at high and low pH; and it is unaffected by proteolytic enzymes, except for prolidase which destroys its activity completely; it is also destroyed by nitrous acid (Clarkson et al., 1979).

Some workers have claimed that the low molecular weight natriuretic material which separates behind the salts on Sephadex G-25 separation is a peptide. There is little evidence to support this conclusion except for the effect of prolidase. It is possible, however, that as the prolidase enzyme preparation is an impure preparation obtained from pig kidney, the inactivation it causes is brought about by a nonproteolytic enzyme. This conclusion is reinforced by recent observations that the most purified substance may not contain any amino acids after hydrolysis.

SITE OF PRODUCTION

There is considerable evidence from in vivo experiments that the natriuresis in an assay kidney perfused with blood from an animal, the blood volume of which has been expanded with equilibrated blood, is not abolished by bilateral nephrectomy, adrenalectomy, or hypophysectomy. Nevertheless, Godon and Dechenne (1978) claim that a high molecular weight natriuretic substance does originate from the kidney. There is also some evidence that the liver and the left auricle may also secrete natriuretic substances. The main focus of attention, however, has been on the brain as the source at least of one natriuretic substance. Theoretically it is not unreasonable to assume that sodium balance should be controlled from a neural area closely related to that which regulates water balance, since both are eventually related to the regulation of body volume. And comparative physiology reveals several instances of neurosecretory substances which influence sodium excretion (Lee and de Wardener, 1974). The most dramatic experiment in favor of such a hypothesis is that of Kaloyanides et al. (1977). They were unable to obtain a natriuresis from an isolated kidney perfused by a blood-volume-expanded donor dog which had been decapitated by a constricting vice, although the dog's blood pressure, renal blood flow, and glomerular filtration rate were well maintained. Many other experimental studies suggest that the brain is involved in urinary sodium excretion (Clarkson et al., 1974; de Wardener, 1977; Silva-Netto et al., 1980). There is also some fragmentary clinical evidence from patients with head injuries.

Four recent observations on the hypothalamus are of particular interest (see below), in relation to the low molecular weight natriuretic substance obtained from the urine and the plasma, the properties of which are described above. Haupert and Sancho (1979) and Lichstein and Samueloy (1980) have obtained a low molecular weight fraction, from bovine and rat hypothalamus, respectively, which inhibits $(Na^+\text{-}K^+)$-ATPase, and ouabain binding. The substance responsible for these changes resists acid hydrolysis and treatment with trypsin and "other proteolytic enzymes." Similar results have been obtained by Fishman (1979) with extracts of pig brain.

CLINICAL RELEVANCE

The potential relevance of a circulating substance which can influence $(Na^+\text{-}K^+)$-ATPase is immense. The enzyme is situated on the cell membrane of all cells. Its action opposes the diffusion of sodium into, and potassium out of, cells so that it is in part responsible for intracellular and extracellular ionic concentration. Presumably, the importance of a circulating $(Na^+\text{-}K^+)$-ATPase inhibitor on the function of various organs will depend on the number of binding sites, and their avidity, for such a substance. Theoretically such a substance might be of importance in cardiac and arteriolar contraction and brain function, as well as in regulating urinary sodium excretion. The possible involvement of a circulating sodium transport inhibitor in chronic renal failure was put forward many years ago. More recently its possible role in essential hypertension and other diseased states has been suggested.

Chronic Renal Failure

Bricker (1967) has pointed out that as the kidneys are destroyed and the number of nephrons decreases, the kidney's capacity to keep the patient in sodium balance remains remarkably unimpaired even though sodium intake remains unchanged. The kidneys in chronic renal failure therefore continue to excrete into the urine the same amount of sodium, though the quantity filtered at the glomerulus is reduced. In other words, in order that urinary sodium excretion remain unchanged, the fraction of sodium reabsorbed from the filtered sodium (fractional reabsorption) is decreased. This nice adjustment in tubular reabsorption of sodium is not due to a coincidental deterioration of the surviving nephrons because the kidney can adjust to changes in sodium intake. Nor is the fall in tubular reabsorption of sodium due to a hyperfiltration in the remaining nephrons "overloading" the nephron, for if in a uremic animal glomerular filtration rate is further lowered by constricting the aorta, the kidney adapts by a further decrease in tubular reabsorption (Bricker and Licht, 1980).

Bricker suggested that the mechanism of adaptation is a rise in the circulating concentration of a natriuretic substance, and he and others have presented much evidence in support of this hypothesis. They have demonstrated that fractionation of serum and urine from uremic patients on G-25 Sephadex yields a low molecular weight substance which is natriuretic and inhibits sodium transport in the frog skin. At first these observations were received with some caution. As they had been obtained from uremic material, it was always possible that the results were due to some metabolic end product. But more recently, Bohan et al. (1980) have shown that the increased quantities of the low molecular weight material fraction obtained from urine from a chronically uremic dog is not only natriuretic and antinatriferic but that it can also displace ouabain from membrane receptors in renal cortical tissue. These results suggest that the low molecular weight natriuretic material in chronic renal failure may be the same as that obtained from the normal human, and that Bricker's original suggestion to account for the fall in fractional reabsorption of sodium in chronic renal failure is correct.

Role of the Natriuretic Hormone in Essential Hypertension

There are many similarities between genetic strains of hypertension in the rat and essential hypertension in the human. In the hypertensive rat the genetic fault resides in the kidney, for renal cross-transplantation experiments have

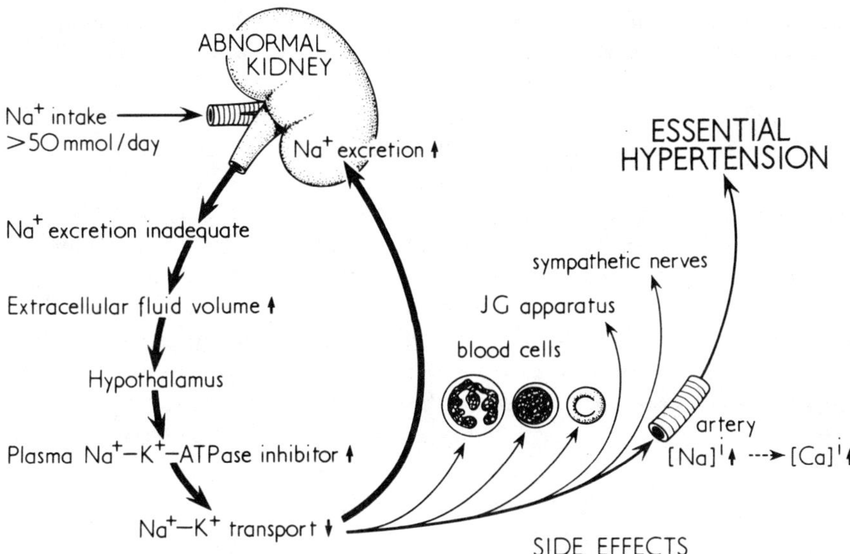

Figure 1 Hypothesis for the possible role of a circulating sodium transport inhibitor in the etiology of essential hypertension. (From de Wardener and MacGregor, 1980. Reprinted from *Kidney International*.)

demonstrated that the hypertension "follows" the kidney. And there is considerable, though mainly circumstantial, evidence that the genetic fault in the kidney is related to a difficulty in sodium excretion which tends to increase the extracellular fluid volume. It is not surprising, therefore, that in these genetic forms of hypertension there are a number of observations, including widespread disturbances of sodium transport in erythrocytes, which suggest that the blood appears to contain an increased concentration of a circulating $(Na^+\text{-}K^+)$-ATPase inhibitor. How such a substance could cause a rise in blood pressure has been discussed by Blaustein (1977), who points out that inhibition of $(Na^+\text{-}K^+)$-ATPase in the smooth muscle of the arteriole could raise the intracellular concentration of calcium and thus increase arteriolar tone.

There are numerous observations in the human that essential hypertension is associated with a generalized impairment of sodium-potassium transport in erythrocytes and leukocytes (Zumkley and Losse, 1980). The more relevant observations demonstrate that this impairment is of the ouabain-sensitive component of the sodium transport, though there is some evidence that other mechanisms may also be involved. de Wardener and MacGregor (1980) have proposed the following hypothesis to link this phenomenon into the etiology of essential hypertension (Figure 1). They have suggested that essential hypertension is due to an inherited variability in the ability of the kidney to eliminate sodium and that this variability becomes increasingly obvious the greater the sodium intake. The difficulty in eliminating sodium at first causes an increase in extracellular fluid and total blood volume, including the central blood volume. This causes an increase in the concentration of a circulating sodium-transport inhibitor which increases urinary sodium excretion until it matches the intake of sodium. But the raised concentration of the circulating sodium-transport inhibitor causes an increase in tone of the smooth muscle in the veins and arteries. The increase in venous tone is the cause of the observed diminished venous compliance with the resultant shift of blood from the periphery to the center (Trippodo and Frolich, 1981). de Wardener and MacGregor propose that it is a persistent increase in intrathoracic blood volume which maintains the

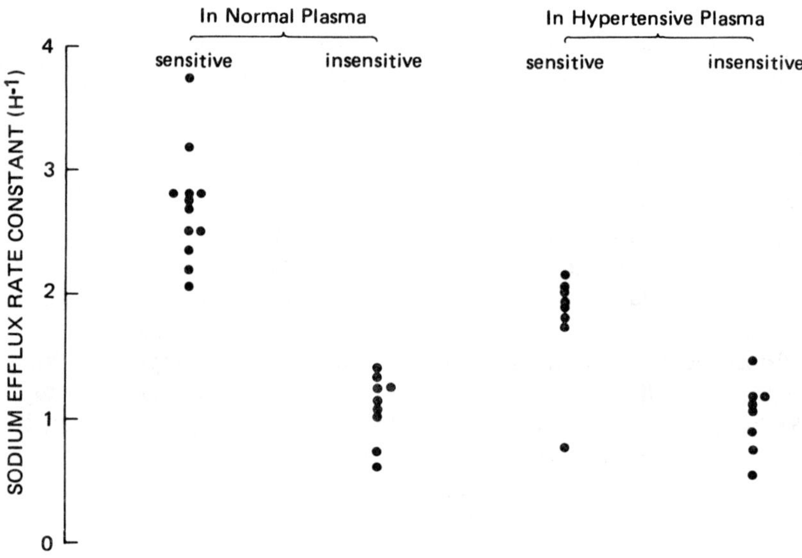

Figure 2   Ouabain-sensitive and insensitive sodium efflux rate constant of white cells from normotensive subjects incubated either in the plasma of other normotensive subjects or in the plasma of hypertensive subjects. (From H. E. de Wardener and G. A. MacGregor, in *Intracellular Electrolytes and Arterial Hypertension*. New York: Thieme-Stratton, 1980.)

stimulus for an increased concentration of the sodium-transport inhibitor, in spite of a normal or even reduced blood volume. The increase in arteriolar tone causes the arterial pressure to rise.

This hypothesis is supported by the results of recent experiments. Poston et al. (1981) have shown that incubation of leukocytes from normotensive subjects in the plasma of patients with essential hypertension reduces the ouabain-sensitive component of the total sodium efflux rate constant of the leukocytes and raises the intracellular sodium concentration to the same concentration as that found in the white cells of patients with essential hypertension (Figure 2). This indicates that the plasma of patients with essential hypertension contains an increased concentration of a $(Na^+\text{-}K^+)$-ATPase inhibitor.

METHODS USED TO DETECT NATRIURETIC HORMONE

As has been described above there have been many methods used to detect "natriuretic activity" in the plasma and in extracts of plasma and urine (de Wardener, 1977). As it was found that plasma or extracts of plasma that were natriuretic also impaired short-circuit current across anuran membranes, the short-circuit current of the frog skin or toad bladder has been used as a test of natriuretic activity. In some accounts it is clear that the distinction between what the two techniques measured has been confused, and it has been assumed that because an extract inhibits short-circuit current, it is synonymous with it having natriuretic activity, and vice versa.

The methods for measuring natriuretic activity are cumbersome and insensitive, as are those for the measurement of changes in short-circuit current. It is for this reason that other techniques continue to be sought. Recently, since there has been increasing evidence that the natriuretic hormone inhibits $(Na^+\text{-}K^+)$-ATPase, new assays have emerged (Lichardus et al., 1980). Some consist of observing the effect of a preparation of $(Na^+\text{-}K^+)$-ATPase obtained from kidney and brain. Others, which are more sensitive, utilize the displacement of radioactive ouabain from the surface of red cells or kidney receptors.

Use of a Cytochemical Measurement of $(Na^+\text{-}K^+)$-ATPase to Measure the Capacity of Biological Fluids or Extracts to Inhibit $(Na^+\text{-}K^+)$-ATPase

The method developed by Chayen et al. (1981) was used to demonstrate the ouabain-sensitive $(Na^+\text{-}K^+)$-ATPase activity in the nephron. After the exposure of kidney segments to dilutions of natriuretic extract or plasma, the kidney segments were chilled rapidly; 10 μm sections were cut and reacted for $(Na^+\text{-}K^+)$-ATPase activity (de Wardener et al., 1981).

Capacity of Biological Fluids to Stimulate G6PD Activity Used as a Marker to Measure the Capacity of Biological Fluids to Inhibit (Na$^+$-K$^+$)-ATPase

Assay Procedure

Female guinea pigs of the Duncan Hartley strain weighing 300-450 g were fed with the Labsure brand RGP pellets (Christopher Hill Group, Poole, Dorset) irradiated diet (1 M rad) for at least 1 week before use. Drinking water was allowed ad libitum supplemented with Abidec vitamin drops (Parke Davis), so that 500 ml of water contained vitamin A, 6000 IU; vitamin D, 600 IU; vitamin B_1, 1.5 μg; vitamin B_6, 0.75 μg; nicotinamide, 7.5 μg; ascorbic acid, 75 μg; and vitamin B_2, 0.6 μg.

The guinea pig was killed by cervical fracture. Both kidneys were removed and cut into segments which were placed in sealed pots [after the method of Trowell's adult organ maintenance culture system (Trowell, 1959)], with a nonproliferative culture medium and under an atmosphere of 95% O_2:5% CO_2 at 37°C for 5 hr (Chayen et al., 1974). The medium was then replaced with fresh medium containing various concentrations of the test material. After exposure times of between 2 and 8 min, the segments were chilled and 16 μm thick sections were prepared from each segment. Details of the chilling and sectioning procedures are given in Chapter 3. The sections were reacted to disclose glucose-6-phosphate dehydrogenase activity.

The sections were incubated at 37°C in an atmosphere of nitrogen in an incubation medium containing glucose-6-phosphate (5 mM), the coenzyme NADP (3 mM), neotetrazolium chloride (5 mM), potassium cyanide (10 mM), the intermediate hydrogen-acceptor phenazine methosulfate (0.67 mM), polyvinyl alcohol (G.18 grade: 12% w/v) dissolved in a 0.05 mM glycylglycine-sodium hydroxide buffer (pH 8.2). The reaction was usually sufficiently strong for adequate measurement after 3-6 min and was stopped by immersing the sections in distilled water. The activity of the G6PD was manifested by the deposition of an intensely colored formazan.

The sections were washed in distilled water and left to dry. They could be stored dry, in the dark, until they were to be measured. They were then mounted in the water-miscible Farrants' medium and left to settle for a few minutes. The amount of formazan in the proximal tubules was measured using a Vickers M85 scanning and integrating microdensitometer at a wavelength of 585 nm, which is the isobestic point of the two formazans of neotetrazolium chloride (Butcher and Altman, 1973); with a $\times$40 objective; a mask of 8 μm diameter or larger (A4, up to the breadth of the cells to be measured), and the smallest size of scanning spot. Measurements were made in 20 proximal tubules from duplicate sections from each segment.

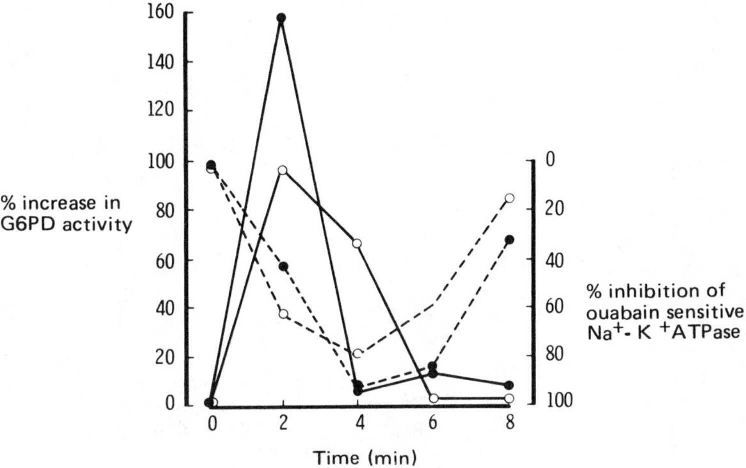

Figure 3   The changes of glucose-6-phosphate dehydrogenase activity are expressed as the percentage of increase over that produced by culture medium alone, in proximal convoluted tubules in segments of guinea pig kidney exposed to purified natriuretic extract from urine: 1 U G6PD-stimulating activity per milliliter (●———●); plasma from a hypertensive subject at 1:50 dilutions (o——— o), for 2-8 min. The changes in ouabain-sensitive sodium potassium adenosine triphosphatase activity are expressed as the percentage of the total ouabain-sensitive activity in the proximal convoluted tubules in serial sections of the same segments exposed to the purified natriuretic extract from urine ●-----●; and the plasma from the hypertensive subject o----- o. (From Fenton et al., 1982).

Stimulation of Renal G6PD Activity and Inhibition of Sodium-Potassium-Dependent ATPase Activity in Serial Sections from the Same Segments

In order to demonstrate that inhibition of (Na^+-K^+)-ATPase activity is associated with a rise in G6PD activity as suggested by Dikstein (1971) and others, a series of experiments was devised to establish this link in vitro and to demonstrate the effect of natriuretic extract and of human plasma on these two enzyme activities. The effect of ouabain was also studied in the same segments of kidney as a control, for it is a potent inhibitor of (Na^+-K^+)-ATPase activity and has been shown to stimulate G6PD activity in vitro.

Individual segments of guinea pig kidney that had been cultured for 5 hr in the manner described were exposed to natriuretic extract (1 and 0.2 U G6PD-

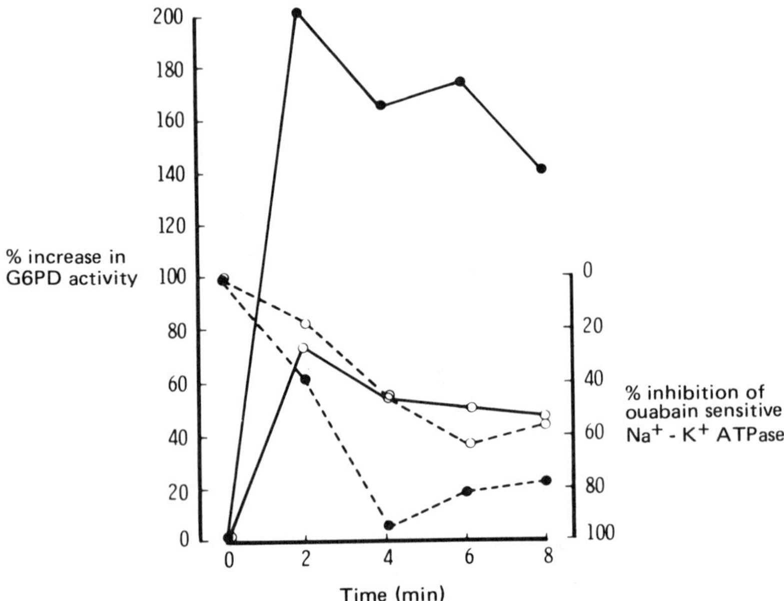

Figure 4 The changes of glucose-6-phosphate dehydrogenase activity expressed as the percentage of increase over that produced by culture medium alone in proximal convoluted tubules in segments of guinea pig kidney exposed to two concentrations of ouabain (4×10^{-6} M o ———— o; 4×10^{-4} M ● ———— ●) for 2-8 min. The changes in ouabain-sensitive sodium-potassium adenosine triphosphatase activity are expressed as the percentage of inhibition of the total ouabain-sensitive activity in the proximal convoluted tubules in serial sections of the same segments exposed to the two concentrations of ouabain (4×10^{-6} M o - - - - - o; 4×10^{-4} M ● - - - - - ●). (From Fenton et al., 1982.)

stimulating activity per milliliter, or human plasma from two individuals (1:50 dilution) or ouabain (4×10^{-4} M and 4×10^{-6} M) for 2, 4, 6, and 8 min. Six sections were obtained from each segment: two were incubated to demonstrate G6PD activity, two were incubated to demonstrate total ATPase activity, and two were incubated in the presence of ouabain (2×10^{-4} M) to demonstrate the ouabain-insensitive (Na+-K+)-ATPase activity; subtraction of the two latter activities reveals the ouabain-sensitive (Na+-K+)-ATPase activity.

A simultaneous rise in G6PD activity and a fall in (Na+-K+)-ATPase activity was evident at 2 min after exposure to all three test materials (Figure 3). After exposure to natriuretic extract or human plasma, G6PD activity fell thereafter toward the control value, whereas the inhibition of (Na+-K+)-ATPase continued to increase and was maximal between 4 and 6 min; at 8 min the (Na^+-K^+)-ATPase activity was approaching the control value. After exposure to ouabain the G6PD activity was maximal at 2 min, but remained high throughout the course of the experiment (Figure 4); similarly, the inhibition of (Na^+-K^+)-ATPase activity continued to increase and was maximal by 4-6 min, but unlike the change in (Na^+-K^+)-ATPase activity demonstrated with the natriuretic extract and human plasma, there was only a small rise in activity by 8 min.

These results demonstrate that stimulation of renal G6PD activity in vitro by natriuretic extract, human plasma, and ouabain is associated with a simultaneous inhibition of renal (Na^+-K^+)-ATPase activity. The similarity of the responses induced by the natriuretic extract and the plasma suggest that they contain a substance common to both. We would suggest that the ability of natriuretic extract and plasma to stimulate renal G6PD activity in vitro may be used as a marker of their capacity to inhibit renal (Na^+-K^+)-dependent ATPase activity in vitro.

Validation of the Cytochemical Technique to Measure the Capacity of Plasma to Stimulate G6PD in Proximal Tubules

Time Response
Glucose-6-phosphate dehydrogenase activity was present in most tubules of the cortex but declined during the 5 hr culture. Exposure of the segments to the

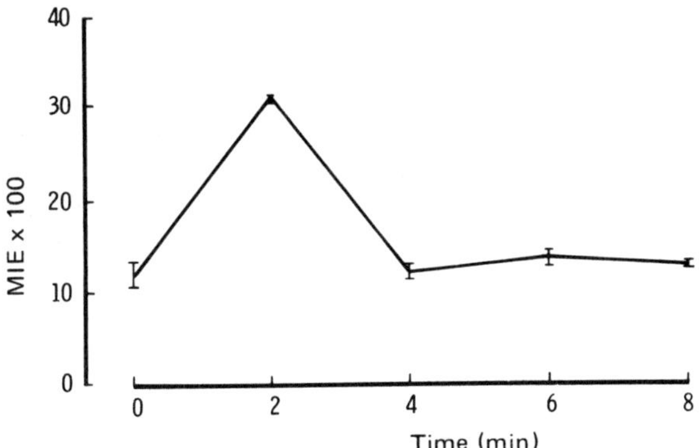

Figure 5 The changes in glucose-6-phosphate dehydrogenase activity (mean integrated extinction $\times$ 100) in proximal convoluted tubules in segments of guinea pig kidney exposed to 100 pg/ml purified natriuretic extract from urine ($\bullet$———$\bullet$) for various times (0, 2, 4, 6, and 8 min). Bars represent SEM (n = 20). (From Fenton et al., 1982.)

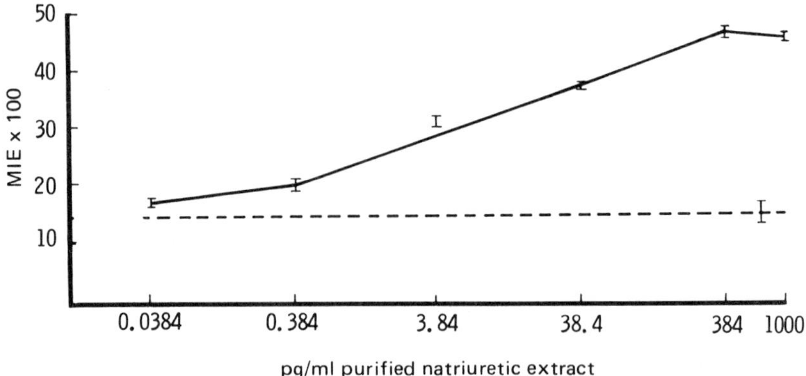

Figure 6 The changes in glucose-6-phosphate dehydrogenase activity (mean integrated extinction × 100) in proximal convoluted tubules in segments of guinea pig kidney exposed for 2 min to concentrations of 0.0384-1000 pg/ml purified natriuretic extract from urine (● ——— ●). The lower broken line (● ----- ●) shows the glucose-6-phosphate dehydrogenase activity in a segment exposed to culture medium alone. Bars represent SEM (n = 20). (From Fenton et al., 1982.)

purified natriuretic extract at a concentration of 100 pg/ml for 2, 4, 6, and 8 min stimulated G6PD activity in most of the tubules of the cortex. The effect was particularly marked in the proximal convoluted tubules, which were positively identified by staining serial sections for alkaline phosphatase activity. In these tubules the rise in G6PD activity was maximal at 2 min (Figure 5).

Dose Response
Segments exposed for 2 min to fresh culture medium alone or with fresh medium containing a series of graded concentrations of the purified natriuretic extract gave a positive linear rise in G6PD activity with the logarithm of the concentration of the extract from 0.384 to 384 pg/ml (Figure 6). Further increase in concentration of the extract did not produce any further increase in G6PD activity. This maximum stimulation of G6PD activity, achieved with 384 pg/ml was assigned an arbitrary value of 1 unit G6PD stimulating activity per milliliter. In subsequent dose responses, the natriuretic extract was used as a standard and expressed in these unit.

Nature of the Response
It was necessary to ascertain that the activity indicated by the formazan deposits observed in the sections exposed to the natriuretic extract were indeed

due to enzyme activity and not merely a result of reducing moieties acting directly on the neotetrazolium salt. Three segments of guinea pig kidney were cultured as previously described: one was exposed for 2 min to fresh culture medium alone; the remaining two segments were exposed separately to different concentrations (38.4 and 384 pg/ml) of the natriuretic extract that had previously been shown to produce an easily measurable formazan deposit. Eight serial sections were taken from each segment: two sections from each were incubated in the presence of substrate, coenzyme, and neotetrazolium; two with substrate and neotetrazolium; two with coenzyme and neotetrazolium; and two with neotetrazolium salt alone. In all three segments, only those sections with substrate, coenzyme, and neotetrazolium salt gave the previously described formazan deposit. The deposit was greater in the sections from the segments exposed to the larger concentration of natriuretic extract and least in the sections from the segment exposed to culture medium alone. This suggests that the observed formazan deposit is a direct result of enzyme activity.

Specificity

Many substances known to act on the nephron were tested at physiological and pharmacological concentrations for their effect on G6PD activity in this system at 2 min. These included angiotensin II, aldosterone, vasopressin, dopamine, ouabain, norepinephrine, epinephrine, parathyroid hormone, calcitonin, 1,25-dihydroxy vitamin D, thyrotropin-releasing hormone, luteinizing hormone releasing hormone, and prolactin. Substances used to separate the purified natriuretic extract obtained from normal urine, and nonnatriuretic fractions from the HPLC separation, were also investigated. Of the substances tested at physiological and pharmacological concentrations, only ouabain showed any effect. The stimulation of G6PD at 2 min with 2×10^{-4} M ouabain (72.9 ng/ml), was similar to that produced by 0.6 U of natriuretic extract (23.04 pg/ml).

Effect of Normal Human Plasma

Blood from an individual on a normal diet was taken into heparinized tubes and immediately separated by centrifugation at room temperature. The plasma was stored at -70°C.

Stimulation of G6PD activity by a suitable dilution of this plasma had the same time course as that of the natriuretic extract. The dose response was tested on sections from segments of guinea pig kidney, maintained for 5 hr and then exposed for 2 min as described above, to fresh medium containing plasma at 1:100, 1:1000, and 1:10,000 dilution. These dilutions gave a positive linear rise in G6PD activity. The slope of this response and that obtained to dilutions of the natriuretic extract from urine showed no divergence from parallelism (Figure 7). The potency extimate ±SEM for this plasma, expressed in the arbitrary units defined for the standard, was 9.6 ± 0.4 U G6PD-stimulating activity per milliliter.

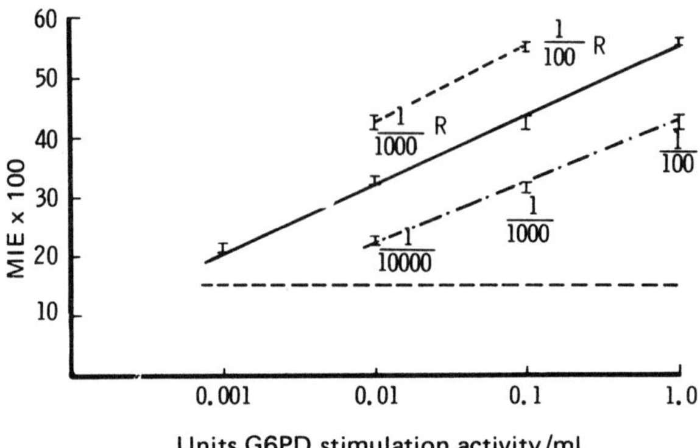

Figure 7   The response produced by dilutions of normal human plasma, 1:100, 1:1000, 1:10,000 (– · – · – ·), in glucose-6-phosphate dehydrogenase activity (mean integrated extinction × 100) in proximal convoluted tubules in segments of guinea pig kidney. It is parallel to that produced by graded concentrations of purified natriuretic extract from urine (● ——— ●). Upper dotted line (● ····· ●) shows the effect of the same plasma after the addition of 100 U/ml glucose-6-phosphate dehydrogenase stimulating activity of purified natriuretic extract from urine. The lower broken line (-----) shows the glucose-6-phosphate dehydrogenase activity in a segment exposed to culture medium alone. Bars represent SEM (n = 20). (After Fenton et al., 1982.)

The index of precision (λ) for this assay was 0.12 with fiducial limits (P = 0.95) of 75-130% [calculated according to Borth (1960)]. Aliquots (120 ul) of plasma from this normal subject on a normal diet were snap-frozen to -70°C and used as a quality control in subsequent assays.

Precision

The values of any two dilutions of plasma have been in agreement to ±10%. On two occasions five serial sections were measured for a single concentration of natriuretic extract, and a single dilution of plasma: the results (MIE × 100 ± SEM) were 10.2 ± 0.3 and 9.4 ± 0.28, respectively. The mean index of precision (λ) ± SEM, for nine consecutive assays was 0.07 ± 0.04. This value for λ is well within the accepted limits for bioassays (Loraine and Bell, 1971). Fiducial limits (P = 0.95) ranged from 74 to 134% and from 90 to 111%.

Sensitivity

The lowest concentration of natriuretic extract to produce a significant rise in G6PD activity over that produced by culture medium alone was 0.001 U/ml.

With normal human plasma the rise in G6PD activity could be detected at dilutions as great as 1:10,000.

Reproducibility
This was estimated by assaying aliquots of a single normal human plasma at two dilutions in nine other consecutive assays. The mean value of this quality control plasma ± SEM was 9.6 ± 0.35 U/ml (n = 9). The index of precision (λ) ± SEM was 0.068 ± 0.003, with fiducial limits (P = 0.95) ranging from 75 to 130% and from 90 to 111%.

Recovery
Two aliquots of a plasma were used. Each was assayed at two dilutions (1:1000 and 1:10,000), but to one aliquot, 90 U/ml of natriuretic extract was added. The recovery was calculated from

$$\frac{\text{Observed potency}}{\text{Expected potency}} \times 100\%$$

The plasma had a potency of 9.3 U G6PD stimulating activity per milliliter, and the strenghtened plasma had an observed potency of 99.6 U/ml, giving a recovery of 103.3%. The index of precision (λ) for this experiment was 0.04; fiducial limits (P = 0.95) were 88-113%.

RESULTS OBTAINED USING CYTOCHEMICAL TECHNIQUES FOR $(Na^+\text{-}K^+)$-ATPase AND G6PD ACTIVITY TO MEASURE THE RELATIVE CONCENTRATION OF NATRIURETIC HORMONE IN PLASMA AND HYPOTHALAMUS

Normal Human

$(Na^+\text{-}K^+)$-ATPase Activity
de Wardener et al. (1981) obtained plasma from five healthy normal subjects on day 5 of a high-sodium diet (normal diet + 200 mmol "Slow Sodium") and on day 5 of a low-sodium diet (10 mmol sodium per day). The 24 hr urinary sodium excretion was 333 ± 28 mmol (mean ± SEM) on day 5 of the high-sodium diet and 14.7 ± 7.3 mmol on day 5 of the low-sodium diet.

Inhibition of total ATPase activity by a purified low molecular weight natriuretic extract and by a sample of normal human plasma was found in all tubules studied, the inhibition in the proximal and distal tubules being greatest after exposure for 6 min and in the thick ascending limb of the loop of Henle (TAL) after exposure for 4 min to the dilutions of the plasma of 1:20-1:500. Plasma samples from each subject on a high- and low-sodium intake were tested for the

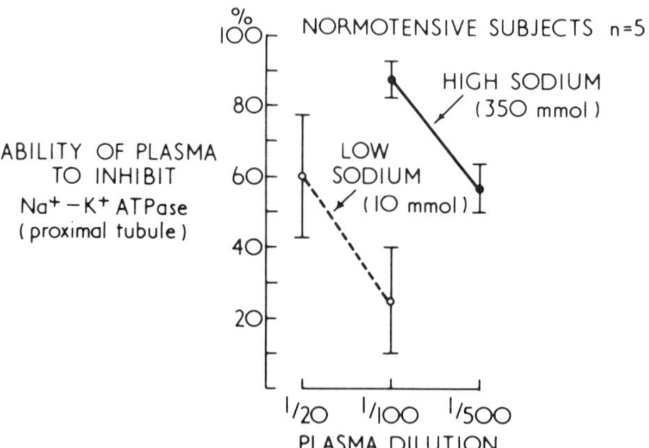

Figure 8   Inhibition of ouabain-sensitive (Na+-K+)-ATPase in proximal tubules of guinea pig kidney by dilutions of plasma from five healthy subjects each on a high- (● ——— ●) and on a low-sodium diet (o -----o). (From de Wardener et al., 1981.)

response at 6 min on segments from one guinea pig; thus, each estimation was a within-animal assay. The results (Figure 8) showed that (1) the plasma obtained from the subjects when on a high-sodium diet gave a stronger inhibition in all three tubules than did the plasmas obtained when the subjects were on a low-salt diet, and (2) that the inhibition induced by the "high-salt plasmas" was dose dependent in that the lower dilution gave greater inhibition; this was generally true also of the "low-salt plasmas" except in the TAL for which the 6 min exposure was not optimal. Some estimate of the relative potencies of the high-salt and low-salt plasmas can be gained from the fact that almost equivalent inhibition was obtained by a dilution of 1:500 of the former and 1:20 of the latter. On this basis the high-salt plasmas could be considered 25 times more potent than the low-salt plasmas.

G6PD Activity

The experiments were performed in 12 normal subjects aged 20-23 years in 5 of whom it had already been demonstrated that the identical plasmas inhibited renal (Na+-K+)-ATPase when the sodium intake was raised (see above). The 12 normal subjects were studied on day 5 of the same high- and low-sodium diets described above. Urinary sodium excretion was 336 ± 15.1 mmol per 24 hr (mean ± SEM) on the high-sodium diet, and 11 ± 3.3 (mean ± SEM) on the low-sodium diet. Plasma was obtained on day 5 of each diet and tested at dilutions

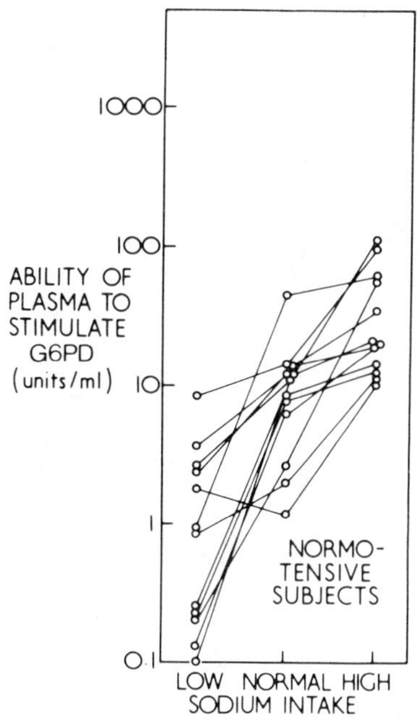

Figure 9 The ability of plasma from normal subjects on a low-, normal and high-sodium intake to stimulate guinea pig renal glucose-6-phosphate dehydrogenase activity in vitro.

of 1:100 and 1:1000 on segments of guinea pig kidney. The ability of the plasma to stimulate renal G6PD in vitro was expressed in arbitrary units of G6PD-stimulating activity (see above). On the high-sodium diet the ability of the plasmas to stimulate renal G6PD was 39.3 ± 10.6 U/ml (mean ± SEM; range 10-113 U/ml), while on the low-sodium diet it was 1.8 ± 0.5 U/ml (mean ± SEM; range 0.06-8.6 U/ml) ($P < 0.005$) (Figure 9).

A further study by MacGregor et al. (1981b) on a group of "normal subjects" aged 40-67 years (mean, 54 years) on a normal sodium intake demonstrated that the ability of the plasma of this older group to stimulate G6PD (34.5 ± 10.4 U/ml) was significantly greater than that of the younger group, who were also on a normal sodium intake (11.5 ± 3.4 U/ml) ($P < 0.05$) and that there was a significant correlation between age and ability of the plasma to stimulate G6PD ($P < 0.005$). The finding that the ability of plasma to stimulate G6PD rises with age is interesting in view of the fall in glomerular filtration rate that also occurs with age. As salt intake does not diminish with age, sodium balance must be

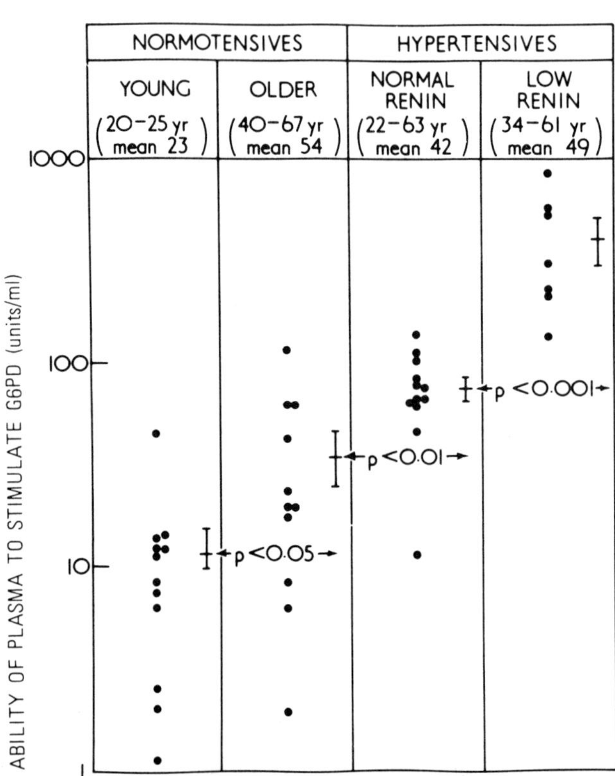

Figure 10   The ability of plasma to stimulate guinea pig renal glucose-6-phosphate dehydrogenase activity in vitro in normotensive subjects and hypertensive patients. (From MacGregor et al., 1981b.)

maintained by an inhibition of tubular reabsorption of sodium. The demonstration that the ability of plasma to stimulate G6PD and therefore to inhibit $(Na^+\text{-}(Na^+\text{-}K^+)$-ATPase increases with age suggests that this is one mechanism by which tubular reabsorption of sodium is inhibited.

Essential Hypertension

MacGregor et al. (1981b) measured the ability of plasma to stimulate G6PD in 23 normotensive subjects aged 20-67 years (mean 38) and 19 patients with essential hypertension aged 22-63 years (mean 45) (Figure 10). The mean

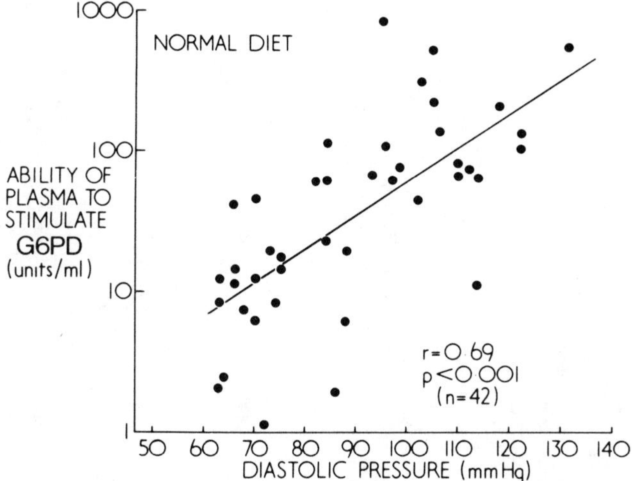

Figure 11 The ability of plasma to stimulate guinea pig renal glucose-6-phosphate dehydrogenase activity in vitro plotted against the diastolic blood pressure in 42 individuals on a normal diet. (From MacGregor et al., 1981a.)

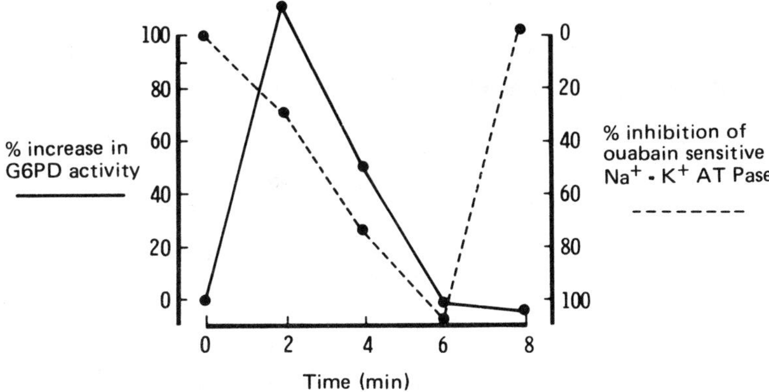

Figure 12 Changes in ouabain-sensitive (Na+-K+)-ATPase activity and G6PD activity induced by hypothalamic extracts after 2-8 min exposure: ●-----● (Na+-K+)-ATPase activity and ●———● G6PD activity.

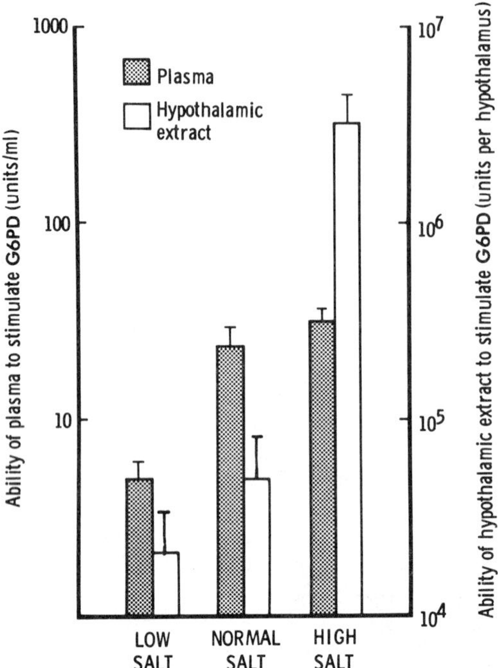

Figure 13 The ability of plasma and hypothalamic extracts from groups of rats on a low-, normal, or high-salt diet to stimulate G6PD activity.

urinary sodium excretion of the two groups was 146 and 162 mmol/day. The mean ability of the plasma to stimulate G6PD in the hypertensive patients was 195 ± 52 U/ml and in the normal subjects it was 22.5 ± 5.8 U/ml (P < 0.001). In the whole group of 42 individuals there was a significant correlation between systolic pressure, mean blood pressure, and diastolic pressure and the ability of plasma to stimulate G6PD activity (P < 0.001) (Figure 11).

The patients with low plasma renin activity, who were also the ones with the highest pressure, had a much greater ability of their plasma to stimulate G6PD than the patients with normal plasma renin activity (400 ± 87 μ/ml and 74.9 ± 9.2 U/ml, respectively; P < 0.001). It is also interesting that in the 19 hypertensive patients there was a significant correlation between the ability of the plasma to stimulate G6PD and the plasma renin activity (P < 0.001). There was no such correlation in the normal subjects. As ouabain directly inhibits renin secretion, it is possible that the fall in plasma renin activity, which affects up to 40% of hypertensive patients, is due in part to the rise in the ability of plasma to inhibit $(Na^+\text{-}K^+)$-ATPase.

Hypothalamus

Alaghband-Zadeh, Fenton, et al. (1981) in preliminary experiments found that acetone extracts of hypothalamus contained large quantities of a substance which, at dilutions of 1:10,000-1:10,000,000, stimulated G6PD and inhibited $(Na^+\text{-}K^+)$-ATPase activity in the same manner as plasma and natriuretic extract (Figure 12). At these dilutions, extracts of other tissues, including kidney, cerebral cortex, pituitary, liver, pancreas, adrenal, and small bowel, did not stimulate G6PD activity.

Three groups of six Wistar rats were fed for 4 weeks on one of three diets containing different quantities of sodium. The mean 24 hr urinary sodium in each group was 162, 1020, and 2396 μmol. The mean G6PD stimulating activity in the extracts of whole hypothalamus was 2.14×10^4 U/ml (n = 5), 5.21×10^4 U/ml (n = 5), and 162×10^4 U/ml (n = 5), respectively, and the mean G6PD stimulatory activity of the plasmas was 4.95 U/ml (n = 6), 26.5 U/ml (n = 4), and 30.1 U/ml (n = 6) (Figure 13). These results suggest that the hypothalamus is the source of the circulating $(Na^+\text{-}K^+)$-ATPase inhibitor, the concentration of which is related to sodium intake.

CONCLUSION

The ability to detect a $(Na^+\text{-}K^+)$-ATPase inhibitor in the plasma and extracts of various organs will accelerate the progress of investigations into the nature and function of the natriuretic hormone in health and disease. Its possible role in chronic renal failure and essential hypertension has been mentioned above. Until the structure of the natriuretic hormone is known it will be necessary to continue using bioassays to detect its presence, and the cytochemical assay is the most sensitive for this purpose.

REFERENCES

Alaghband-Zadeh, J., Clarkson, E. M., Fenton, S., MacGregor, G. A., and de Wardener, H. E. (1981). The effect of sodium intake on the ability of human plasma to stimulate renal glucose-6-phosphate dehydrogenase (G6PD) in vitro. *J. Physiol. (Lond.)*, 315-3:43-44p.

Alaghband-Zadeh, J., Fenton, S., Millett, J., Hancock, K., and de Wardener, H. E. (1981). The effect of sodium intake on the content of a substance in the rat hypothalamus which stimulates glucose-6-phosphate dehydrogenase. (G6PD) in vitro. *Clin. Sci.*, 61:43p.

Bahlmann, J. S., McDonald, J., Ventom, M. G., and de Wardener, H. E. (1967). The effect on urinary sodium excretion of blood volume expansion without changing the composition of the blood. *Clin. Sci.*, 32:403-413.

Blaustein, M. P. (1977). Sodium ions, calcium ions, blood pressure regulation and hypertension: A reassessment of a hypothesis. *Am. J. Physiol.*, 232(3): C165-C173.

Bohan, T., Potter, L., and Bourgoignie, J. J. (1980). Ouabain radioreceptor assay for natriuretic factor. In *Hormonal Regulation of Sodium Excretion,* B. Lichardus, R. W. Schrier, and J. Ponec (eds.). Elsevier/North Holland, Amsterdam, pp. 393-397.

Borth, R. (1960). Simplified mathematics for multiple bioassays. *Acta Endocrinol. (Kbh),* 35:454-468.

Bricker, N. S. (1967). The control of sodium excretion with normal and reduced nephron populations. The pre-eminence of Third Factor. *Am. J. Med.,* 43:313-321.

Bricker, N. S., and Licht, A. (1980). Natriuretic hormone: Biologic effects and progress in identification and isolation. In *Hormonal Regulation of Sodium Excretion,* B. Lichardus, R. W. Schrier, and J. Ponec (eds.). Elsevier/North Holland, Amsterdam, pp. 399-407.

Butcher, R. G., and Altman, F. B. (1973). Studies on the reduction of tetrazolium salts. II. The measurement of the half reduced and fully reduced formazans of neotetrazolium chloride in tissue sections. *Histochemie,* 37: 351-363.

Chayen, J., Bitensky, L., and Daly, J. R. (1974). Minireview: Cytochemical bioassay of hormones. *Life Sci.,* 15:191-201.

Chayen, J., Frost, G. T. B., Dodds, R. A., Bitensky, L., Pitchfork, J., Bayliss, P. H., and Barnett, R. J. (1981). The use of a hidden metal-capture reagent for the measurement of Na^+-K^+-ATPase activity: A new concept in cytochemistry. *Histochemistry,* 71:533-541.

Clarkson, E. M., Koutsaimanis, K. G., Davidman, M., DuBois, M., Penn, W. P., and de Wardener, H. E. (1974). The effect of brain extracts on urinary sodium excretion of the rat and the intracellular sodium concentration of renal tubule fragments. *Clin. Sci. Mol. Med.,* 47:201-213.

Clarkson, E. M., Raw, S. M., and de Wardener, H. E. (1979). Further observations on a low-molecular weight natriuretic substance in the urine of normal man. *Kidney Int.,* 16:710-721.

Clarkson, E. M., Talner, L. B., and de Wardener, H. E. (1970). The effect of plasma from blood volume expanded dogs on sodium, potassium and PAH transport of renal tubule fragments. *Clin. Sci.,* 38:617-627.

de Wardener, H. E. (1977). Natriuretic hormone. *Clin. Sci. Mol. Med.,* 53:1-8.

de Wardener, H. E., and MacGregor, G. A. (1980). Dahl's hypothesis that a saluretic substance may be responsible for a sustained rise in arterial pressure: Its possible role in essential hypertension. *Kidney Int.*, 18:1-9.

de Wardener, H. E., MacGregor, G. A., Clarkson, E. M., Alaghband-Zadeh, J., Bitensky, L., and Chayen, J. (1981). Effect of sodium intake on ability of human plasma to inhibit renal Na^+-K^+-Adenosine triphosphatase in vitro. *Lancet*, i:411-412.

de Wardener, H. E., Mills, I. H., Clapham, W. F., and Hayter, C. J. (1961). Studies on the efferent mechanism of the sodium diruesis which follows the administration of intravenous saline in the dog. *Clin. Sci.*, 21:249-258.

Dikstein, S. (1971). Stimulability, adenosine triphosphatases and their control by cellular redox processes. *Naturwissenschaften*, 58:439-443.

Fenton, S., Clarkson, E., MacGregor, G., Alaghband-Zadeh, J., de Wardener, H. E. (1982). An assay of the capacity of biological fluids to stimulate renal glucose-6-phosphate dehydrogenase activity in vitro as a marker of their ability to inhibit sodium potassium-dependent adenosine triphosphatase activity. *J. Endocr.*, 94:99-110.

Fishman, M. C. (1979). Endogenous digitalis-like activity in mammalian brain. *Proc. Natl. Acad. Sci. USA*, 76:4661-4663.

Flier, J. S., Maratos-Flier, E., Pallotta, J. A., and McIsaac, D. (1979). Endogenous digitalis-like activity in the plasma of the toad *Bufo marinus*. *Nature*, 279:341-343.

Godon, J. P., and Dechenne, C. (1978). In vitro production of a natriuretic material of renal origin. *Renal Physiol.*, i:201-210.

Gonick, H. C., Kramer, H. J., Paul, W., and Lu, E. (1977). Circulating inhibitor of sodium-potassium-activated adenosine triphosphatase after expansion of extracellular fluid volume in rats. *Clin. Sci. Mol. Med.*, 53:329-334.

Gruber, K. A., Whitaker, J. M., and Buckalew, Jr., V. M. (1980). Endogenous digitalis-like substances in plasma of volume expanded dogs. *Nature*, 287:743-745.

Haupert, G. T., and Sancho, J. M. (1979). Na transport inhibitor from bovine hypothalamus. *Proc. Natl. Acad. Sci. USA*, 76:4658-4660.

Kaloyanides, G. J., and Azer, M. (1971). Evidence of a humoral mechanism in volume expansion natriuresis. *J. Clin. Invest.*, 50:1603-1612.

Kaloyanides, G. J., Cohen, L., and DiBona, G. F. (1977). Failure of selected endocrine organ ablation to modify the natriuresis of blood volume expansion in the dog. *Clin. Sci. Mol. Med.*, 52:351-356.

Knock, C. A., and de Wardener, H. E. (1980). Evidence in vivo for a circulating natriuretic substance in rats after expanding the blood volume. *Clin. Sci.*, 59:411-421.

Lee, J., and de Wardener, H. E. (1974). Neurosecretion and sodium excretion. *Kidney Int.*, 6:323-330.

Lichardus, B., Pliska, V., Uhrin, V., and Barth, T. (1968). A cow as a model for investigating natriuretic activity. *Lancet*, i:127-129.

Lichardus, B., Schrier, R. W., and Ponec, J. (1980). *Hormonal Regulation of Sodium Excretion.* Elsevier/North Holland, Amsterdam.

Lichstein, D., and Samueloy, S. (1980). Endogenous "ouabain-like" activity in rat brain. *Biochem. Biophys. Res. Commun.*, 96:1518-1523.

Loraine, J. A., and Bell, E. T. (1971). In *Hormone Assays and Their Clinical Applications*, J. A. Loraine and E. T. Bell (eds.). Livingstone, Edinburgh, p. 14.

MacGregor, G. A., Fenton, S., Alaghband-Zadeh, J., Markandu, N., Roulston, J., and de Wardener, H. E. (1981a). Evidence for a raised concentration of a circulating sodium transport inhibitor in essential hypertension. *Br. Med. J.*, 283:1355-1357.

MacGregor, G. A., Fenton, S., Alaghband-Zadeh, J., Markandu, N. D., Roulston, J. E., de Wardener, H. E. (1981b). An increase in a circulating inhibitor of Na^+-K^+, and dependent ATPase: a possible link between salt intake and the development of essential hypertension. *Clin. Sci.*, 61:17-20s.

Nutbourne, D. M., Howse, J. D., Schrier, R. W., Talner, L. B., Ventom, M. G., Verroust, P. J., and de Wardener, H. E. (1970). The effect of expanding the blood volume of a dog on the short-circuit current across an isolated frog skin incorporated in the dog's circulation. *Clin. Sci.*, 38:629-648.

Poston, L., Sewell, R. B., Wilkinson, S. P., Richardson, P. J., Williams, R., Clarkson, E. M., MacGregor, G. A., and de Wardener, H. E. (1981). Evidence for a circulating sodium transport inhibitor in essential hypertension. *Br. Med. J.*, 282:847-849.

Poston, L., Wilkinson, S. P., Sewell, R., and Williams R. (1980). Inhibitor of leucocyte Na transport during mineralocorticoid "escape." *Clin. Sci.*, 58:9.

Silva-Netto, C. R., de Mello Aires, M., and Malnic, G. (1980). Hypothalamic stimulation and electrolyte excretion: A micropuncture study. *Am. J. Physiol.*, 239:F206-F214.

Sonnenberg, H., Veress, A. T., and Pearce, J. W. (1972). A humoral component of the natriuretic mechanism in sustained blood volume expansion. *J. Clin. Invest.*, 51:2631-2644.

Trippodo, N. C., and Frolich, E. (1981). Similarities of genetic (spontaneous) hypertension. Man and rat. *Circ. Res.*, 48:309-319.

Trowell, O. A. (1959). The culture of mature organs in a synthetic medium. *Exp. Cell Res.*, 16:118-147.

Zumkley, H., and Losse, H. (1980). *Intracellular Electrolytes and Arterial Hypertension.* Georg Thieme Verlag, Stuttgart, New York.

Author Index

*Numbers indicate the page on which an author's work is referred to; numbers in
italic indicate page on which the complete reference is listed.*

Subject Index

about the book . . .

This authoritative volume provides comprehensive coverage of *Cytochemical Bioassays*—
a powerful, in vitro technique one thousand times more sensitive than equivalent radioimmunoassays. With contributions by 29 leading international experts, this single-source
reference provides the information required for thorough understanding and effective
application of this advanced methodology.

For each hormone examined, *Cytochemical Bioassays* offers complete guidlelines to the
assay—from basic endocrinology to procedures and analyzing the results. Research and
clinical endocrinologists, biochemists, clinical chemists, pathologists, molecular and cell
biologists, and geneticists will welcome this important information as a vital aid to their
work.

about the editors . . .

J. CHAYEN is Head of the Division of Cellular Biology at the Kennedy Institute of
Rheumatology, London, England. In addition, he serves as Head of the designated laboratory for cytochemical bioassays and a member of the Expert Advisory Panel on Biological
Standardization of the World Health Organization. Dr. Chayen received the Ph.D. degree
from King's College, London and the D.Sc. degree from the University of London. He
is a member of the Endocrine Society, Society for Endocrinology, Biochemical Society,
Royal Society of Medicine, and Institute of Physics, and a Fellow of the Institute of Biology.

LUCILLE BITENSKY is Head of the Laboratory of Medical Histochemistry and Deputy
Head of the Division of Cellular Biology, Kennedy Institute of Rheumatology, London,
England. Dr. Bitensky has also served as a Visiting Professor at the Royal Free Hospital
School of Medicine, University of London (1972-1981) and McGill University, Montreal,
Canada (1978). She received the Ph.D. and D.Sc. degrees from the University of London
and the MB.B.Ch degree from University of Witwatersrand, South Africa. Dr. Bitensky
is a Fellow of the Royal College of Pathologists and a member of the Royal College of
Physicians of London, as well as the Endocrine Society, Society for Endocrinology,
Biochemical Society, and Royal Society of Medicine.

Printed in the United States of America   ISBN: 0–8247–7001–3